Engineering textiles

The Textile Institute and Woodhead Publishing

The Textile Institute is a unique organisation in textiles, clothing and footwear. Incorporated in England by a Royal Charter granted in 1925, the Institute has individual and corporate members in over 90 countries. The aim of the Institute is to facilitate learning, recognise achievement, reward excellence and disseminate information within the global textiles, clothing and footwear industries.

Historically, The Textile Institute has published books of interest to its members and the textile industry. To maintain this policy, the Institute has entered into partnership with Woodhead Publishing Limited to ensure that Institute members and the textile industry continue to have access to high calibre titles on textile science and technology.

Most Woodhead titles on textiles are now published in collaboration with The Textile Institute. Through this arrangement, the Institute provides an Editorial Board which advises Woodhead on appropriate titles for future publication and suggests possible editors and authors for these books. Each book published under this arrangement carries the Institute's logo.

Woodhead books published in collaboration with The Textile Institute are offered to Textile Institute members at a substantial discount. These books, together with those published by The Textile Institute that are still in print, are offered on the Woodhead web site at: www.woodheadpublishing.com. Textile Institute books still in print are also available directly from the Institute's website at: www.textileinstitutebooks.com.

A list of Woodhead books on textile science and technology, most of which have been published in collaboration with The Textile Institute, can be found at the end of the contents pages.

Woodhead Publishing in Textiles: Number 81

Engineering textiles

Integrating the design and manufacture of textile products

Y. E. El Mogahzy

The Textile Institute

CRC Press
Boca Raton Boston New York Washington, DC

WOODHEAD PUBLISHING LIMITED

Cambridge England

Published by Woodhead Publishing Limited in association with The Textile Institute
Woodhead Publishing Limited, Abington Hall, Granta Park, Great Abington, Cambridge CB21 6AH, England
www.woodheadpublishing.com

Published in North America by CRC Press LLC, 6000 Broken Sound Parkway, NW, Suite 300, Boca Raton, FL 33487, USA

First published 2009, Woodhead Publishing Limited and CRC Press LLC
© Woodhead Publishing Limited, 2009
The author has asserted his moral rights.

British Library Cataloguing in Publication Data
A catalogue record for this book is available from the British Library.

Library of Congress Cataloging in Publication Data
A catalog record for this book is available from the Library of Congress.

Woodhead Publishing ISBN 978-1-84569-048-9 (book)
Woodhead Publishing ISBN 978-1-84569-541-5 (e-book)
CRC Press ISBN 978-1-4200-9372-8
CRC Press order number: WP9372

The publishers' policy is to use permanent paper from mills that operate a sustainable forestry policy, and which has been manufactured from pulp which is processed using acid-free and elemental chlorine-free practices. Furthermore, the publishers ensure that the text paper and cover board used have met acceptable environmental accreditation standards.

Typeset by SNP Best-set Typesetter Ltd., Hong Kong
Printed by TJ International Ltd, Padstow, Cornwall, England

Contents

Author contact details

Yehia Elbudrawy El Mogahzy
Professor of Polymer & Fiber Engineering
Auburn University
P.O. Box 3381
Auburn, Alabama 36831
USA

Tel: +1-334-332-9430
E-mail: elmogye@auburn.edu

Woodhead Publishing in Textiles

To Mona, my wife, the love of my life, and to my children, Taha, Amal, Amany, Mustafa, Farieda, Sharif, Narcy and Ahmed.

Preface

This book is entitled *Engineering textiles: Integrating the design and manufacture of textile products.* The idea for the book stemmed from the enormous utilization of fibrous materials in different applications and the significant transition from exclusively traditional fibrous products to advanced function-focus products that touch every aspect of human life. Unlike many books published in this area, this book is directed at all engineers including material, mechanical, electrical, civil, chemical, polymer and fiber engineers. It covers many aspects of product development, design conceptualization and design analysis that will be useful for engineers in various fields. Indeed, the main objective of this book is to stimulate and create an integrated effort by engineers in different areas to conduct engineering design projects based on multi-disciplinary engineering knowledge. Although the book is primarily written for engineers, it is made so simple that technologists, and business and marketing personnel can easily follow and learn a great deal about product development and cost aspects. Furthermore, this book is also directed towards students learning about development and design of various products, particularly those involving fibrous materials. It is my sincere hope that this book will be useful for both industrial and educational readers in different fields.

Yehia El Mogahzy

Part I

Concepts of fabric engineering, product
development and design

1

Introduction: textile fiber-to-fabric engineering

Abstract: What began as a craft and art industry is now a major industrial empire which has an impact on all aspects of life and provides a wide range of products from fashionable clothing to high tech fibrous systems. This chapter provides a brief overview of the evolution of the textile and fiber industry. In particular, the chapter discusses the turning point in the 1990s, which led to a historical transition from a manufacturing focused, to a development focused industry. This significant transition made a 'fiber-to-fabric engineering' approach, which is the theme of this book, inevitable to meet the new challenges facing the industry. Fiber-to-fabric engineering is the science and art of optimizing the utilization of fibers (natural or synthetic, conventional or high-performance) in various applications (traditional or function-focus and existing or potential) using solid engineering concepts so that their ultimate benefits can be realized by all mankind. Its basic elements are discussed throughout the 15 chapters of this book.

Key words: commodity products, traditional fibrous products, function-focus fibrous products, fiber-to-fabric engineering.

1.1   The textile and fiber industry, past and present

The history of the textile industry reflects the evolution of the industrial world from the time of the ancient Egyptians until today. What began as a craft and art industry has continued to contribute to the welfare of human being over the years. Indeed, at just about every turning point that the industrial world encountered, the textile industry was there to spark it, create it or contribute heavily to it. This great industry sparked the industrial revolution in the late 18th and into the 19th centuries.[1] In the early 1700s, one manual loom required four spinners and ten persons to prepare yarn to keep up with its slow production rate; weavers had to remain idle for lack of yarn. In 1733, this dilemma reached its peak when John Kay, a Lancashire mechanic, invented the first flying shuttle, speeding up the weaving process and imposing more pressure on the spinners to keep up with the speed. It took about 40 years to solve this problem when James Hargreaves invented his spinning jenny and Richard Arkwright introduced his 'water spinning frame' in the 1770s. These machines were capable of producing multiple threads simultaneously and in quantities. The increase in spinning production imposed pressure on the speed of fiber production.

This pressure was soon lifted by the invention of the cotton gin by the American Eli Whitney in 1793; it was an invention that not only sparked the industrial revolution but also forever changed consumer appetite from the traditional woolen clothing to cotton textiles. By the early 19th century, the cost of making cotton yarn had dropped dramatically and the labor cost of making fabric had fallen by at least 50%. Today, spinning speeds have reached over 400 m min^{-1}, fibers are rotating in air before consolidation at a rate reaching millions of revolutions per minute and yarns are inserted into the fabric via air and water since the shuttle loom was put to rest.

In the face of limited resources and constrained properties, the textile and fiber industry had a momentous vision for new fibers that has continued over the years, from mainly flax and wool in the 17th century, to cotton in the late 18th century and into the 19th century, to the first man-made fiber, rayon, before the end of 19th century. In the first half of the 20th century, a research team headed by Wallace Carothers of EI du Pont de Nemours & Company proved that a purely synthetic fiber can be made by chemical synthesis from readily available resources such as air, water, and coal or petroleum. By the 1930s, this team introduced nylon to the world, a fiber that has contributed to numerous products and never ceased to make a difference to human life. This marked the beginning of the synthetic fiber revolution with many synthetic fibers such as polyester, acrylic, polypropylene, and a host of regenerated man-made fibers being developed in the same century. Today, these fibers are used in many traditional fibrous products such as apparel, furnishing and household products. They are also integrated in many function-focus product categories such as speciality sports wear, agro-fiber products, geotextiles and medical products.

Realizing that the unique characteristics of fibers drive the developments of a wide range of fibrous products, the industry has developed high-performance fibers that can be used for high strength and high temperature applications.[2] These include: aramid fibers, gel-spun polyethylene fibers, carbon fibers, glass fibers, metallic fibers and ceramic fibers. These fibers can be consolidated into different types of fibrous assemblies so that not only are their original properties efficiently translated into the desired performance characteristics of the end product, but also they can be enhanced and perhaps modified to accommodate special needs and high-tech applications. The industry developed different forms of yarn from continuous to spun yarns, flat to texturized, twisted to twistless, and plain to compound or fancy yarns. Numerous fabric types were developed within the three major categories of fabric, namely woven, knit and nonwoven. Many speciality fabrics were also developed including crepe woven, dobby, piqué, Jacquard, pile-woven, double-woven, braided and multiaxial-woven structures. Finally, when fibrous assemblies needed additional performance enhancement or modification of some form, the industry has always been ready to offer numerous types of chemical and mechanical finishing treatments or special coating and lamination.

1.2   The 1990s: a turning point in the textile and fiber industry

Despite the great contributions that the textile and fiber industry has made over the years, it has generally been perceived as a commodity industry that relies on massive manpower and conventional technology to manufacture products that are essential for human needs. This perception has been a direct result of the fact that the industry has primarily been manufacturing-focused. The term 'manufacturing' is commonly used to describe operations that utilize well-known, often systematic, approaches to make product components in a highly consistent manner. Accordingly, basic tasks performed in a manufacturing environment include:[3]

- performing systematic analysis of technological and cost factors required to produce product items
- selecting of appropriate raw materials required to meet the specifications of intended products at the lowest cost possible
- setting and adjusting machines so that product units can be produced according to the desired specifications and at the highest efficiency possible
- monitoring and testing intermediate and final products
- implementing quality control and statistical process control (SPC) techniques to detect quality problems and diagnose their causes and effects

These tasks have represented the primary activities of the textile industry for many years and thousands of companies around the world are still performing these tasks in their daily operations. Indeed, the internal structure of a conventional spinning or a weaving mill is basically the same worldwide; a general manager, an assistant manager, operations supervisors, maintenance personnel, testing and quality control personnel, machine operators, business and human resources, and shipping and warehouse personnel.

From a business viewpoint, the manufacturing-focused approach of the traditional textile industry can indeed yield a significant profit provided that demands for products are continuous, massive units of products are manufactured, and maximum production efficiency is fulfilled. In other words, the profit made in the traditional textile industry has been primarily driven by the quantities it produces and the rate of production; a spinner must sell more yarns and a weaver must sell more fabrics to make profit. The significant importance of quantity stems from two main variable factors facing the traditional textile industry: (1) the cost of raw material and (2) the labor cost. The cost of raw material contributes significantly to the overall manufacturing cost of fibrous products. For example, the cost of cotton can be as high as 65% of the overall cost of yarn manufacturing and as high as 45% of the overall cost of finished fabric manufacturing.[4] This makes the profit margin highly vulnerable to changes in raw material price. Labor cost also represents a significant cost factor, particularly in industrial countries. Indeed, a general manager of a mid-size spinning mill in the USA in the 1980s was making a 15 to 30 times higher average salary than a person holding the same job in China or India, and the gap between machine operators in the USA and in these two countries was substantially much higher than that.

Traditionally, the high cost of raw material was handled through creative approaches to fiber blending and appropriate methods of waste reduction.[3] In addition, fiber price, particularly natural fibers, is quite competitive worldwide. It was the labor cost gap that truly represented a challenge to the USA and European countries in order to continue to compete in the traditional textile market. This cost differential sparked a major turning point in the textile and fiber industry in the early 1990s in which a rapid migration of the industry from the USA and Europe to Asia, South America and Africa was witnessed after a short period of attempting to outsource some of the textile operations which turned out to be too little too late to save the traditional industry.

The turning point that took place in the 1990s has resulted in a significant concentration of production of traditional fibrous products (apparel, furnishing and household items) in countries such as China, India, Pakistan, and in a number of South American and African countries, with China having the lion share of traditional textiles exports. In the USA, few spinning and weaving companies are still in business, attempting to compete in certain traditional niches such as high-end products and fashionable items. Other companies have switched to speciality products and new developments.

The turning point during the 1990s has left us with many lessons to learn and many challenges to face in the years ahead. These are summarized as follows:[4,5]

- The driving force of profit making in the traditional textile business is high quantity and high production rate at the lowest cost possible.
- In today's global market, products that can be duplicated easily will be duplicated rapidly, leading to a price-differential market controlled and dominated by a few retailers.
- A raw material that does not reflect a true added value in the market-place will remain a commodity and will yield a commodity product.

Indeed, the share of profit for fiber, yarn and fabric producers collectively in the total profit of a traditional textile product has historically been below 15%, and over 80% of the profit has been the retailer's share.[6]

- Ultimately, labor cost will reach a universal plateau as a result of globalization. Energy being a commodity will remain an important contributor to the overall cost of manufacturing. The efficiency with which the energy is utilized and the ways in which its use can be minimized will make a difference in any competitive advantages. The world population will continue to grow and the traditional textile industry will make a comeback in industrial countries. In what shape or form? It remains to be seen.

1.3   Function-focus fibrous products

The term function-focus fibrous products (FFFP) will be used throughout this book to describe non-traditional fibrous products such as those used in protective systems, transportation, construction, agricultural applications and health-related applications[5,7–9] (see examples in Table 1.1). Other terms used in the literature to describe non-traditional fibrous products include 'industrial textiles', popularly used in the USA, and 'technical textiles', popularly used in Europe. The use of the term industrial textiles seems inappropriate on the grounds that non-traditional fibrous products are utilized in many applications of a non-industrial nature,[5] which makes industrial textiles only a subcategory of non-traditional fibrous products. The use of the term technical textiles may seem more appropriate, except it implies technological applications, which makes it similar to the term industrial textiles. Indeed, one of the synonyms of technical is industrial. In the author's opinion, the common problem associated with both terms is that neither reflects the development and design efforts put into these products that stems from focusing on their specific functions.

The term 'fibrous products' is also deliberately used instead of the term 'textile products'. One of the reasons for this choice is that it reflects a more generic meaning of products that are fibrous-based including, of course, what are traditionally known as textile products. In this regard, traditional textile products will be recognized as traditional fibrous products (TFP). It should be noted that the term textile is derived from the Latin verb *texere*, meaning 'to weave'. Indeed, terms such as knit, lace, netting, felt, braid and cord were originally excluded from being given to textiles until it was necessary to use textiles as a generic term only for convenience.[1] In Chapter 2, the importance of selecting the appropriate terms in reflecting the nature and the tasks involved in product development will be elaborated further.

Table 1.1 Examples of function-focus fibrous products[5–9]

Application category	Function-focus fibrous products
Construction fibrous products	Fiber-reinforced concrete Lightweight roofing materials Insulation materials Retaining walls reinforcements Fibrous cables for bridges Fibrous-based piping and canalization
Protective fibrous products	Body armors Fire fighters uniforms Diving suit uniforms Particulate protection clothing (clean room) Chemical protection clothing
Human hygiene and medical practice	Wound care products Diapers Braces Protheses and orthoses Wipes Breathing masks Bedding and covers Fibrous-based implants Artificial tissues Joints and ligaments
Fibrous products for transportation	Safety air bags Safety seatbelts Tires Automobiles and airplane seats Parachutes and balloons Sails Automobiles and airplane interiors Aircraft wing and body structures Boat rumps (fiber composite structures) Inflatable components of satellites or other spacecraft Railroad foundations
Industrial fibrous products	Filters Ropes Conveyor belts Abrasive substrates Knitted nets and brushes Flexible seals and diaphragms
Agro-fiber and geo-fiber products	Artificial turfs Animal and insect-resistant fibrous products Solar radiation-resistant fibrous products Weather protection products Bridges, dams, roads, railways Embankment reinforcements
Smart fibrous products	Smart uniforms or shirts for wound detection and diagnosis Smart bras for breast tumor detection Thermoregulatory uniforms E-textiles

1.4   The move to function-focus fibrous products

A brief review of the major historical contributions of the traditional textile and fiber industry to human welfare over the years has been presented earlier in this chapter. The industry has also contributed significantly to the development of function-focus fibrous products and still represents the main test-bed for many of the current developments in this exciting market. A glimpse into history can provide much evidence that function-focus fibrous products have touched people's life in the air, in the sea and on the roads.[1] As early as the 1790s, the first soft parachute was developed from silk fibers. During World War II, nylon replaced silk in parachutes as a durable woven structure enhanced by silicone coating to prolong its durability. The introduction of nylon also resulted in further development of other speciality products such as tires, tents, ropes and many military supplies. Early boat sails were made from animal skins until the Ancient Egyptians wove cloth sails as early as 3300 BC. Cloth sails were made from flax fibers until 1851 when cotton sails were crowned as supreme after a victory of the racing yacht America, which defeated 14 British vessels in a sailing race around the Isle of Wight off the south coast of England. Later, polyester fibers were used instead of cotton as it became evident that they can offer lightweight, minimum surface friction, and they can be melted in the fabric to provide a closer porous fabric structure to prevent wind escape. Fibrous materials have also been used for many years in fiber-reinforced structures. As early as 1926, fiber-reinforced constructions were built by the South Carolina highway departments using a series of woven cotton fabrics to help reduce cracking and failures in road construction. This marked an increase in use of fibrous structures in pavement and embankment reinforcement and in other civil applications such as drainage, filtration and separation.

In recent years, particularly since the early 1980s, function-focus fibrous products have drawn a great deal of attention particularly in industrial countries such as the USA, Europe and Japan. Now, fibrous structures represent integral components in numerous areas and applications. The sparking reason for this revival of interest in these products was the growing need to create new market strategies in which innovations and value-added products can provide robustness against volatile market changes.

The move toward function-focus fibrous products does not imply a complete departure from traditional fibrous products, as the two markets must coexist, support one another and grow together in a profitable environment. Indeed, their coexistence is essential for the following reasons:[4]

• The traditional textile and fiber industry offers products that are essential for the existence of the human being and the market for products of this industry will always grow with the increase in population growth around the world.

- Traditional industry will always serve as a major supplier of fibrous structures to the function-focus industry.
- The function-focus industry can provide the traditional industry with many innovative concepts that can change the shape of traditional fibrous products and improve their marketing images through value-added features, such as antibacterial, odorless, UV resistance, flame retardant, soil resistant, and so on. Figure 1.1 shows examples of functional characteristics (the outer circle) that have been added to the basic performance characteristics (inner circle) of many traditional fibrous products in recent years.
- High-tech products often lack attractiveness, fashion and esthetic features; this is an area where the traditional industry has historical experience and can provide a great assistance.

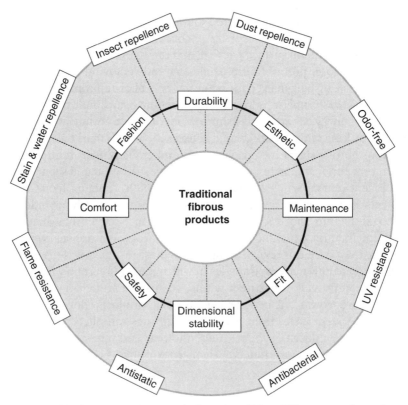

1.1 Performance characteristics of traditional fibrous products from basic to functionality.

1.5   Key differences between traditional and function-focus fibrous products

In the marketplace, the classification of traditional and function-focus fibrous products is often blurred by the inevitable overlap between these two categories of products and by the common components that can be used in both categories. However, three key factors can clearly distinguish between them: volume, value per unit and design complexity. The levels of these factors will vary depending on the extent of speciality of the intended product, which is a direct function of the levels of the desired performance characteristics. Figure 1.2 demonstrates this point conceptually using a percentage scale to express the three factors. The percentage values assigned for these factors are only used for demonstration of general trends as they will obviously vary from one product to another. The percentage of design complexity implies the extent of innovation. In this case, a 100% degree of complexity would conceptually mean very high innovation that cannot be duplicated. Indeed, a mirror image of this factor can be the degree of potential duplication, an issue that has faced the traditional industry for many years. A 100% degree of potential duplication would mean a common product that can be produced massively using basic technology.

As illustrated in Fig. 1.2, the performance characteristics associated with traditional fibrous products may range from conventional or basic performance characteristics to specified characteristics that require further development to improve their functionality with respect to some specific aspects, and to add value to the product that can be realized by the consumer. Examples of products reaching the upper limit of this range are numerous, including antibacterial clothing, flame-retardant clothing for children, functional athletic socks (e.g. Thro-Lo® foot equipment) and thermal clothing (e.g. Hydroweave®). The more specificity required from a traditional fibrous product, the more it approaches the status of a function-focus product with the ultimate limit of specificity being speciality. This point will be clearly demonstrated in the discussion of sportswear products in Chapter 12.

With regard to volume or quantity, traditional fibrous products will always be in demand and will be produced in huge quantities by virtue of their necessity to human life. Factors such as change in lifestyle, fashion and generation differences will always create dynamic changes in the demand for some traditional fibrous products. The design complexity of these products has normally been low and largely dependent on experience, design-by-duplication, or imitation of existing products. However, the recent trends of adding specific functionality to some of these products have been a result of creative design approaches. The value per unit of

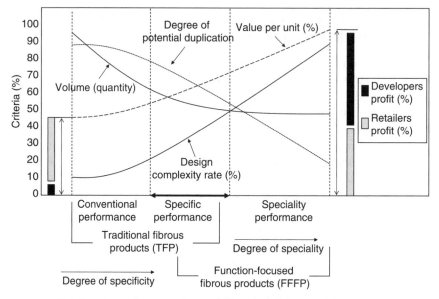

1.2 Conceptual comparison of fibrous products with respect to performance characteristics.

these products will also be relatively low, particularly at the lower end of their range of performance characteristics. However, the addition of functionality has resulted in a significant increase in market value. The problem, however, is that the retailers of these products seem to always have the lion share of their profit margins while the producers get a small percentage of this profit.[4]

As the degree of specificity of the performance of fibrous products increases, design complexity will increase and the value per unit will also increase. In general, the volume will be likely to decrease as a result of the narrower market niches associated with function-focus products. However, some markets have witnessed a steady growth in product demand including transportation, medical and health care, and military. In addition, exciting new markets such as E-textiles can indeed have substantial growth, particularly when fibrous products can be fully equipped with audio and video electronics.

Figure 1.2 also illustrates the overlap between traditional fibrous products and function-focus fibrous products by the line with a double arrow. Many sports products fall into this intermediate category as will be discussed in Chapter 12. As more specificity is added to the product or more functional performance characteristics are specified, the product approaches the speciality category, or become a truly function-focus product. In the marketplace, the most important measure of this status is the value per unit of the product. This value is largely influenced by the extent of design complexity (the degree of innovation and the possibility of duplication). Conceptually, a true speciality or function-focus fibrous product is the one that has a high level of design complexity, minimum potential for duplication or imitation, high value per unit and a high demand by potential users. The high value per unit of speciality products stems from the choice of high-performance fibrous materials and the optimum trade-off between different attributes including those that exhibit conflict in their effects. A good example of this trade-off is that between protection and comfort, as discussed in Chapter 15. The developers of function-focus products are likely to have a bigger share of profit than the developers of traditional fibrous products. However, this particular advantage will largely depend on the product type, the extent of innovation, market competitiveness and product lifecycle.

1.6   Fiber-to-fabric engineering

Historically, fibers have been made into various products through the use of well-established technological approaches in which a wealth of practical knowledge and a great deal of know-how have resulted in the production of billions of fibrous product items. The desired performance specifications of these items have been met through a great deal of work by the artist and designer, trial and error and design-by-reference to existing products. The involvement of science and engineering in making these products was manifested in specific areas such as new fiber development and machine design. Outside these areas, most approaches were largely technologically based. The role of science and engineering in this regard was mainly to explore the huge practical knowledge generated by the industry using fundamental scientific methods and to make some scientific sense of practices that were based on art and the experience of hundreds of years; in other words, the wagon was driving the horse not the other way around.

As will be discussed in Chapter 2, the term engineering has hardly been utilized in the traditional textile industry as it has been almost entirely driven by manufacturing technology. The fiber-to-fabric system was largely described as a linear system with the input being fibers and the output being fabric or fibrous assemblies, as shown in Fig. 1.3. Most tasks were manufacturing-oriented tasks, as mentioned earlier. These factors have made the industry largely predictable and competitively vulnerable. As industry in highly developed countries began to realize that innovations and creative ideas are the key to robust competitiveness and successful market strategies, it was almost 'too little too late' to make that transition, as other countries were also prepared to compete in these areas,

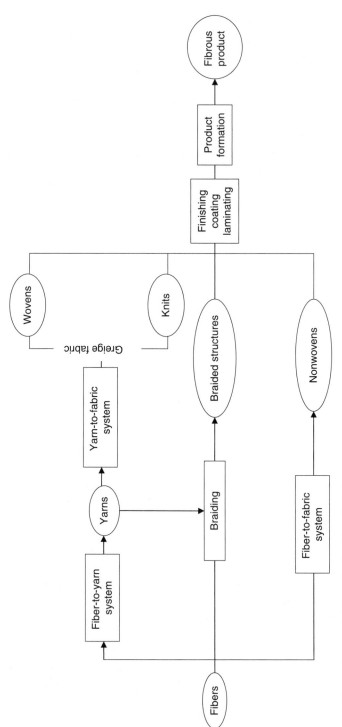

1.3 General material flow in the fiber-to-fabric system.

and scientific and engineering institutes were once again exploring the transition instead of contributing to it. Indeed, many textile institutes in the USA have decided to change their names to polymer and fiber engineering institutes years after the industry has made that transition. Today, the common realization is that product development via creative engineering design is the way to make new products, open new market channels and remain competitive. This realization largely inspired the concepts presented in this book.

Fiber-to-fabric engineering is the science and art of optimizing the utilization of fibers (natural or synthetic, conventional or high-performance) in various applications (existing or potential) using solid engineering concepts so that their ultimate benefits can be realized by all humanity. In contrast to the traditional technological approach, fiber-to-fabric engineering aims to optimize the entire fiber-to-fabric process and not just the fiber-to-fabric system. In the engineering world, while a system is defined by an entity with inputs and outputs, a process is defined by one or more of seven basic elements:[10–12] machine, material, people, money, energy, method and environment. A fiber-to-fabric engineering program is essentially a product development program based on reliable scientific methods in which all elements of the fiber-to-fabric process are taken into consideration.

In order to achieve a successful fiber-to-fabric engineering program, a number of key requirements must be fulfilled (see Fig. 1.4). These requirements will be discussed in detail in different chapters of this book. They are as follows:

1. understanding engineering basics – Chapter 2
2. understanding basic concepts and critical factors of product development – Chapter 3
3. understanding the basic aspects of product design – Chapter 4
4. understanding basic elements and tools of design conceptualization – Chapter 5
5. understanding the analytical tools required to perform design analysis – Chapter 6
6. understanding the concepts and the methods associated with material selection – Chapter 7
7. establishing good knowledge of the design-oriented aspects of fibers and fibrous assemblies such as yarns, fabrics and finished products – Chapters 8 to 11
8. understanding basic development and design concepts of traditional fibrous products – Chapter 12
9. understanding basic development and design concepts of function-focus fibrous products – Chapters 13 to 15.

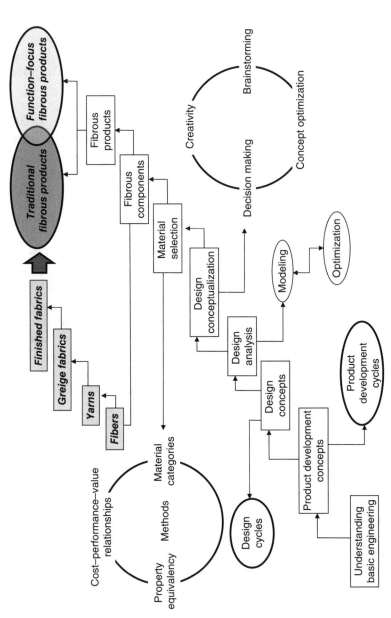

1.4 Fiber-to-fabric process.

It is important to point out that this book is not intended just for engineers who work in the textile and fiber industry but rather for engineers of all disciplines. Indeed, Chapters 2 to 7 are entirely focused on engineering concepts that can be useful for all engineers. It is with this approach that the author wishes to stimulate joint, cooperative and supportive thinking among engineers of different disciplines and to inspire engineers in the polymer and fiber field to expand their knowledge to other engineering areas.

1.7   References

1. TEXTILE, ENCYCLOPEDIA BRITANNICA ONLINE, http://www.britannica.com/, 2005.
2. HEARLE J W S, *High-performance Fibers*, Woodhead Publishing & The Textile Institute, Cambridge, UK, 2001.
3. EL MOGAHZY Y E and CHEWNING C, *Fiber to Yarn Manufacturing Technology*, Cotton Incorporated, Cary, NC, USA, 2001.
4. EL MOGAHZY Y E, '*Challenges Facing the US Textile Industry Today*', an article published in Quality Business Consulting (QBC) newsletter (http://www.qualitybc.com/), 1999.
5. BYRNE C, 'Technical textiles market – an overview', *Handbook of Technical Textiles*, A R Horrocks and S C Anand (eds), Woodhead Publishing & The Textile Institute, Cambridge, UK, 2000.
6. BONDURANT J and ETHRIDGE D, 'Proportions of the retail dollar received by Cotton Industry Segments: selected consumer goods', *Proceedings of the Cotton Beltwide Conference*, US Cotton Council, New Orleans, LA, 1998.
7. ANAND S C, 'Technical fabric structures: knitted fabrics', *Handbook of Technical Textiles*, A R Horrocks and S C Anand (eds), Woodhead Publishing & The Textile Institute, Cambridge, UK, 2000.
8. SHISHOO R, 'Safety and protective textiles: the opportunities and challenges ahead', *Proceedings of the 4th Conference on Safety and Protective Fabrics*, Industrial Fabrics Association International, Roseville, MN, 2004.
9. SONDHELM W S, 'Technical fabric structures: woven fabrics', *Handbook of Technical Textiles*, A R Horrocks and S C Anand (eds), Woodhead Publishing & The Textile Institute, Cambridge, UK, 2000.
10. DIETER G, *Engineering Design: A Material and Processing Approach*, McGraw-Hill Series in Mechanical Engineering, New York, 1983.
11. BEAKLEY G C and LEACH H W, *Engineering – An Introduction to a Creative Profession*, 3rd edition, Macmillan, New York, 1977.
12. GEE E A, and TYLER C, *Managing Innovation*, John Wiley & Sons, New York, 1976.

2

Textile engineering principles and concepts

Abstract: A fiber or a fabric engineer in today's competitive market does not work in a vacuum or perform in isolation. He/she must be fully aware of the surrounding engineering world and must be able to communicate and cooperate efficiently and effectively with engineers from various fields. This chapter discusses the evolution of engineering and the basic engineering concepts that can assist engineers of all fields in understanding what it takes to be good engineers. These concepts are often overlooked in the midst of detailed and highly technical engineering analysis associated with specific products or systems. Key engineering concepts discussed are: (a) knowledge gain and problem solving, (b) design, the foundation of engineering, (c) invention, innovation, dissemination and patenting and (d) natural resources. Many examples of fibrous and textile products are used to demonstrate engineering concepts. These include invention and innovation of synthetic fibers, and invention and innovation of spinning machines. The chapter also discusses the importance of natural resources in engineering applications.

Key words: fiber engineering, fabric engineering, textile engineering, problem solving, design, invention, innovation, dissemination, patenting, natural resources.

2.1   The evolution of engineering

According to *Encyclopedia Britannica*, the word 'engineering' originates from words such as 'engine' and 'ingenious'. These two words were derived from the same Latin root 'ingenerare', which means to create. Thus, engineering implies creation and creativity and for this reason it was a work of art before it became a scientific discipline. Indeed, the evolution of the engineering field began with a man who is recognized today among historians as the first engineer, the Egyptian, Imhotep, who built the step pyramid at Saqqarah, Egypt, probably in about 2550 BC. This may suggest that civil engineering was also the first engineering field recognized by humans. However, the reasoning used for the name civil engineering may reveal otherwise. The term 'engine' was originally used to describe the engines of war, or devices such as catapults, floating bridges and assault towers.[1] The designer of these engines was called a 'military engineer'. The counterpart of the military engineer was the civil engineer, named later, who applied essentially the same knowledge and skills to designing buildings, streets, water supplies, sewage systems and other projects.

In the 18th century, civil engineering emerged as the first engineering discipline to define the profession of designing and executing structural works that serve the general public. The same century also witnessed the emergence of mechanical engineering. This is defined as the branch of engineering concerned with the design, manufacture, installation, and operation of engines and machines.[2] Historically, much evidence indicated that the initiation of mechanical engineering as a discipline in the 18th century was largely driven by the evolution of the textile industry and its role in sparking the industrial revolution. According to the *Encyclopedia Britannica*:[1]

> The first unmistakable examples of manufacturing operations carefully designed to reduce production costs by specialized labor and the use of machines appeared in the 18th century in England. They were signaled by five important inventions in the textile industry: (1) John Kay's flying shuttle in 1733, which permitted the weaving of larger widths of cloth and significantly increased weaving speed; (2) Edmund Cartwright's power loom in 1785, which increased weaving speed still further; (3) James Hargreaves' spinning jenny in 1764; (4) Richard Arkwright's water frame in 1769; and (5) Samuel Crompton's spinning mule in 1779. The last three inventions improved the speed and quality of thread-spinning operations . . . A sixth invention, the steam engine, perfected by James Watt, 1785, was the key to further rapid development.

In the 19th century, other engineering fields emerged as a result of new discoveries and the need for more specialized and more focused disciplines.[2] The realization and growing knowledge of electricity and its impact on mankind from 1800 to 1870 led to the initiation of the discipline of electrical and electronic engineering, which grew rapidly as a result of discoveries by well-known scientists including James Clerk Maxwell of Britain and Heinrich Hertz of Germany in the late 19th century. The chemical engineering discipline grew out of the 19th century proliferation of industrial processes involving chemical reactions in metallurgy, food, textiles and many other areas. By 1880 the use of chemicals in manufacturing had created an industry whose function was the mass production of chemicals and the design and operation of chemical plants.

Today, engineering disciplines such as civil, mechanical, electrical and chemical are considered as basic engineering disciplines. In addition, the 20th century witnessed the emergence of new engineering disciplines, many of which were derivatives of basic engineering disciplines.[2-5] These can be described collectively as service and support branches of engineering in specialized areas and they represent a logical evolution of the engineering field as engineers began to touch upon every aspect of life and reach out to different areas and various applications. Examples of these derivative engineering disciplines and their functions are listed in Table 2.1.

Table 2.1 Examples of derivative engineering disciplines and their functions[1-5]

Engineering discipline	Definition or function
Architectural engineering	A discipline associated largely with civil engineering. It deals with the technological aspects of buildings, including foundation design, structural analysis, construction management and building operations
Highway engineering	A derivative of civil engineering that includes planning, design, construction, operation and maintenance of roads, bridges, and related infrastructure to ensure effective movement of people and goods
Environmental engineering	A derivative of civil engineering concerned with the development of processes and infrastructure for the supply of water, the disposal of waste and the control of pollution of all kinds. It is a field of broad scope that draws on such disciplines as chemistry, ecology, geology, hydraulics, hydrology, microbiology, economics and mathematics.
Marine engineering	A derivative of mechanical engineering concerned with the machinery and systems of ships and other marine vehicles and structures
Industrial engineering	An interdisciplinary branch of engineering dealing with the design, development and implementation of integrated systems of humans, machines and information resources to provide products and services
Material engineering	A derivative of mechanical engineering that focuses entirely on material characterization, selection and improvement
Petroleum engineering	A derivative of chemical engineering comprising the technologies used for the exploitation of crude oil and natural gas reservoirs
Biochemical engineering	A derivative of chemical engineering focusing on the application of engineering principles to conceive, design, develop, operate or utilize processes and products based on biological and biochemical phenomena. It has an impact on a broad range of industries, including health care, agriculture, food, enzymes, chemicals, waste treatment and energy
Agricultural engineering	An interdisciplinary field initiated to accommodate the expansion of the use of mechanized power and machinery on the farm. It utilizes appropriate areas of mechanical, electrical, environmental and civil engineering, construction technology, hydraulics and soil mechanics
Biomedical engineering	An interdisciplinary field in which the principles, laws and techniques of engineering, physics, chemistry and other physical sciences are applied to facilitate progress in medicine, biology and other life sciences. It encompasses both engineering science and applied engineering in order to define and solve problems in medical research and clinical medicine for the improvement of health care

Table 2.1 Continued

Engineering discipline	Definition or function
Computer-aided and software engineering	Computer-aided engineering: a discipline of engineering focusing on using computer software to solve engineering problems
	Software engineering: a discipline of engineering focusing on the process of manufacturing software systems (i.e. executable computer code and the supporting documents needed to manufacture, use and maintain the code)
Nuclear engineering	A branch of engineering dealing with the production and use of nuclear energy and nuclear radiation
Forensic engineering	A relatively more recent discipline of engineering that has gained more popularity after recent terrorist attacks. It is applied toward the purposes of law through using various engineering techniques to solve problems associated with criminal or terrorism situations
Genetic engineering	A specialized engineering branch that uses the techniques of molecular cloning and transformation in many areas including improving crop technology and manufacture of synthetic and human insulin through the use of modified bacteria

As the author, as a reader, was anxious to see disciplines such as 'fiber engineering', 'fabric engineering' and 'textile engineering' being mentioned, at least as derivative engineering disciplines, it was disappointing to find no mention of these critical disciplines in encyclopedias or in the records of professional engineering associations. Indeed, these terms were found to be largely unrecognized outside the textile field. Among workers in the field, the term 'fiber engineering' was occasionally used by synthetic fiber producers to describe the process of polymer manipulation to produce fibers of different and diversified performance characteristics. The term 'fiber engineering' was also used by cotton producers to refer to improvement in cotton fiber quality via molecular breeding or transgenic approaches, which may be more suitably called 'genetic engineering'. The term 'fabric engineering' was occasionally used in recent years to refer to the use of fabrics as membranes in technical applications such as architects and composite structures.

The term 'textile engineering', which has been used since World War II, as the name of textile programs in many engineering schools around the world has not been recognized by many engineering education accreditation programs in the USA or Europe as an independent engineering discipline. Only in the last 20 years, has it begun to be recognized by the US

Accreditation Board for Engineering and Technology (ABET), as perhaps a derivative of mechanical engineering.

Typically, the recognition of an engineering discipline is undertaken and communicated to societies by professional associations belonging to this discipline. Commonly, experts belonging to these associations are gathered to establish a definition of the discipline, document it and report it to engineering societies. To the author's knowledge, such effort was never undertaken by textile professional associations in the USA. despite the extensive efforts made by these associations to define other numerous textile-related terms.[6]

In general, textile engineering may be defined as an interdisciplinary field in which scientific principles, mathematical tools and techniques of engineering, physics, chemistry, and other physical sciences are utilized in a variety of creative textile applications including the development of pure fibrous structures, the innovation of fibrous elements that can be combined with other non-fibrous materials and the design of fiber-to-fabric systems that aim to optimize machine–fiber interaction and produce value-added fibrous products.

It is the author's hope that the term 'fiber-to-fabric engineering' introduced in this book and the concepts associated with it will represent a modest step toward a complete recognition of this critical field by engineers of other disciplines. It is also the author's belief that the recent developments in this great industry, particularly in the area of function-focus fibrous products, will inevitably earn this well-deserving recognition.

2.2   Engineering attributes and concepts

All definitions of engineering disciplines that are established by professional associations include a critical aspect of engineering, that is the knowledge of mathematical and scientific principles. To a great extent, this aspect reflects the pride of engineers. It takes a great deal of effort, knowledge gain, and practice through the educational process to become an engineer. In addition, an engineer must develop his/her own creative ability to become a good engineer.

In the practical world, most engineers have many common attributes. They are typically people of few words but lots of thinking and careful action. Unlike scientists, engineers cannot be very aggressive, as it is often less costly and less risky to research than to re-engineer. A scientist is always searching for knowledge, while an engineer is searching for a specific outcome through knowledge. Some scientists see things that relate to things that do not exist; these are called visionaries. As a result, many scientific breakthroughs are the result of picking and choosing pieces that were created for an unintended puzzle. Most scientists focus on identification and verification of physical phenomena; engineers benefit a great deal from these efforts to bear on practical problems. All engineers typically work and perform under some sort of pressure; they are not free to select the problem that interests them; scientists are. An engineer must learn to react efficiently to a predictable or an unpredictable problem, as problem-solving represents an essential engineering task. Scientists, on the other hand, spend more effort exploring the problem than solving it. Yet, an engineer should be appreciative of the earlier work by outstanding scientists and should always be aware of current research progress. As John Hearle[7] put it so appropriately, 'it is important to remember that yesterday's academic research is today's common knowledge and not to neglect today's academic research because we cannot see how it will be used tomorrow'.

In summary, to be an engineer, a person has to develop a number of skills to the point that they become second nature. The most important skills are to conceive, to create and to solve. These three skills summarize the basic functions of engineering. To conceive is to imagine, envision and visualize different inputs and outcomes of a system or a process. To create is to make things or generate concepts that stem from the nature of the problem in hand. To solve is to disentangle a problem and produce an optimum solution.

In the engineering world, a number of key concepts should be understood. These concepts are:[8,9]

- knowledge gain and problem solving
- design, the foundation of engineering
- invention, innovation, dissemination and patenting
- natural resources.

2.2.1   Knowledge gain and problem solving

As indicated above, engineering is based principally on physics, chemistry and mathematics and their extensions into materials science, solid and fluid mechanics, thermodynamics, transfer and rate processes, and systems analysis. Accordingly, all engineers should have good knowledge of the principles of mathematical and natural sciences. The need for this knowledge stems from their merits in solving specific design problems. The extent of depth of this knowledge will ultimately depend on the nature of the problem in hand and the specific tools required for solving it. In other words, understanding the basic principles often assists in identifying the appropriate mathematical or scientific tools to be used; gaining an in-depth knowledge, on the other hand, is typically a task that is proportional to the magnitude of the problem. Engineers may also rely on expert scientists for further support, particularly in the exploratory aspects of the problem.

Some problems are unique in nature and require special handling, but most engineering problems deal with common issues such as material appropriateness, bulk and weight, durability, reliability, energy saving, efficiency, ease-to-use, safety and environmental issues.[3,4,8,9] In addition, a host of other unique problems may also be encountered in finalizing the design and in the manufacturing process. These problems essentially stem from the need to satisfy unique criteria such as consumer appeal, human aspects (how they relate to humans during use), and multiple functionality. The establishment of a problem definition and the underlying concepts of problem solving will be discussed in Chapters 4 and 5 using examples of fibrous products.

It is important to emphasize that problem solving in the engineering world is essentially a design issue. This means that the solution to the problem should provide optimum outcomes with respect to all the factors associated with the problem.[9] Without an optimum solution to a problem, the outcome may be seemingly satisfactory, but the problem will ultimately reoccur. As a result, the engineering effort does not cease with a partial solution; instead a multiplicity of options is entertained, bounded by functional and economical constraints. It is also important to point out that engineering problems are different from quality problems often witnessed during manufacturing or in end use, but they may represent partial causes of these problems. For example, many textile quality problems including weight variation, yarn imperfections, fabric defects, machine stops and end breakage have existed for many years. The reoccurrences of these problems are partially due to inherent design problems that have not been resolved, either because of cost or due to their high complexity as a result of their dynamic nature.[10]

Dealing with engineering problems may vary in scope and complexity depending on the nature and the magnitude of the problem in hand. However, most engineers follow a general scheme in problem solving. This scheme typically consists of the following steps:[8,9]

- analysis of the situation and the factors contributing to the problem
- preliminary decision on a plan of attack in which the problem is reduced to a more categorical question that can be clearly stated
- deductive reasoning from known principles or by creative synthesis to address the stated categorical question
- answering the stated question (or designing the alternative)
- verifying the accuracy and adequacy of the answer
- interpreting the results for the simplified problem in terms of the original problem.

These points will be discussed in detail in Chapters 3 and 4.

2.2.2 Design: the foundation of engineering

The term 'design' is one of the most commonly used terms in human experience. It has been used to imply creation of things that have never been or to indicate rearrangement of things that have always existed. Engineers believe that they are the only people that design things and that design is the essence of engineering. As a result they generally prefer the first meaning of design, the creation of things that have never been. Accordingly, they are likely to protest about the Webster's dictionary definition of design as 'to fashion after a plan', since it excludes the creation aspect. As indicated earlier, creation is one of the basic engineering attributes. People in the fashion clothing business would, of course, prefer this definition and they would not see it as a creativity-free definition. The *Encyclopedia Britannica* has an interesting view of design, as it makes a simple analogy between design and flower arrangement:[1]

> the term flower arrangement presupposes the word design, . . . when flowers are placed in containers without thought of design, they remain a bunch of flowers, beautiful in themselves but not making up an arrangement. Line, form, color, and texture are the basic design elements that are selected, then composed into a harmonious unit based on the principles of design – balance, contrast, rhythm, scale, proportion, harmony, and dominance.

Over the years, the process of engineering design has evolved significantly to meet the changing needs of humanity. This evolution can easily be realized by comparing movies produced in the 1950s or 1960s with the movies made today. In the former period, all units of each product category (i.e. cars, phones, shoes, etc.) looked very similar in shape and geometry with few color options and minimal style differences. This period marked the end of a 200-year era of mass production that was based on unified static design, more popularly known as 'one-size-fits-all'. In that era, engineers viewed design as an internal proactive analysis in which customer needs were assumed to be invariable. They primarily focused on solving problems associated with the basic functional attributes of a product or a system and the durability of products as consumers purchased products to keep for life and not to dispose of after a short period of time.

In today's markets, virtually all consumer products become obsolete before they depreciate. This is a direct result of the rapidly changing technology in all fields driven by the power of computers and artificial intelligence. From a marketing viewpoint, this represents unlimited renewal of business opportunities that can lead to an exponential economical prosperity. From an engineering viewpoint, this trend represents a continuous pressure to develop new product ideas and innovative concepts to meet the highly unpredictable behavior of today's consumer. Although engineers are the people who started this market revolution, they are now the victims of their own success.

The textile and fiber industry was largely a part of that massive production era, but fashion design, as a historical cultural aspect, was ahead of the game in reacting to consumer needs and desires and in producing garments and household products that have always varied in style and color from year to year, season to season and from culture to culture. Although some historians like to mark the beginning of the fashion industry by the activities of George Brummell, who dominated the fashion scene in England in the early 1800s, the concept of fashion began many years before Brummell's time. Some historical records[1,2] clearly indicate that fashion was initially driven by environmental differences between different regions in the world. Arabs, who typically lived in a hot and dry climate, wore clothing that covered them from head to toe. Their loose white wool robes reflected the sun's rays and shielded them from the hot winds. They also provided insulation against the night-time cold. Eskimos wore two layers of clothing, usually sealskin or caribou furs. The inner layer consisted of undergarments and socks. The hairy side of Eskimo undergarments was worn against the skin. The outer garments (trousers, a hooded parka, mittens and boots) were worn with the hairy side out. Eventually, these varieties became integral aspects of cultures and desires and this created what we have come to know as fashion and clothing styling.[2]

In light of the above historical view, it follows that the fashion industry, supported by the textile and fiber industry, realized the concept of dynamic design many years before other industries. Indeed, and as described above, fashion design was not only about styling and look, it was about meeting highly technical consumer specifications and needs as dictated by environmental and cultural differences. One of the reasons that the fashion industry was not given enough credit in the evolution of engineering design is the historical perception of this industry as being an art and craftsmanship based. Now, concepts such as body scan and mass customization are among considerable evidence that the fashion industry is continuing to have a leading role in the consumer-driven market.

In today's market, it is inevitable that the design process emphasizes the user–product interaction. This may be called 'consumer engineering', or the establishment of design concepts that truly reflect both the objective and the subjective aspects of product uses. In this regard, the design process should account for most possible outcomes and leads to a cost-efficient product, which can perform according to its pre-intended function(s) and, in addition, accommodate most possible varying conditions including some which may fall outside its traditional functional boundaries.[10]

The key to achieving consumer-appealing design is to understand human-related factors associated with normal use as well as the potential abuse of a product. Traditionally, human-related factors, particularly those that are directly related to the intended functions of the product, have been accounted for in the initial design analysis. A classic case demonstrating this point is the automobile airbag, one of the fastest growing function-focus fibrous products. It is essentially an automatic safety restraint system that is built into the steering wheel or the side door of a car with instrumental panel. The initial idea was that upon crash, sensors set off an igniter in the center of the airbag inflator. A material such as sodium azide located in the inflator ignites and releases nitrogen gases, which passes through a filter into the airbag, causing it to inflate. In this regard, key design criteria included deployment mechanism, deployment time, material type, coating, fabric structure and fabric weight. Some of these design criteria indirectly account for human-related factors. For example, an optimum combination of nylon 66 (the fiber material used), tight continuous filament woven structure and silicon coating was selected primarily on the basis of preventing chemicals from penetrating through the fabric and burning the skin of the car occupant.

The merits of using automobile airbags to save lives were evident in the statistics reported during the 1990s:[11–13]

- Among belted occupants in frontal crashes, deaths in frontal airbag-equipped cars were 26% lower among drivers and 14% lower among passengers compared to vehicles without frontal airbags.
- Unbelted occupant deaths in frontal crashes were reduced by 32% for drivers and 23% for passengers.
- The US National Highway Traffic Safety Administration, NHTSA, estimated that the combination of an airbag plus lap/shoulder belt reduced the risk of serious head injury by 85% compared with a 60% reduction for belts alone.
- Side airbags have reduced deaths among passenger car drivers involved in driver-side collisions by about 45% when the side airbag included head protection and by 11% when the side airbag was designed to protect only the torso.

The above merits were the results of engineering design that totally focused on the functional aspects of safety. Unfortunately, just as airbags were the primary reasons for saving many lives, they also caused injuries and resulted in death. The energy required to inflate frontal airbags quickly and protect people in a crash was also found to cause injuries at high impacts. By January 2004, NHTSA reported that since 1990, deaths attributable to airbag inflation in low speed crashes amounted to about 230 cases.

Using purely deterministic design approaches, a deficiency such as that described above can easily go unnoticed. It is important, therefore, to account for the consumer use of the product. In this regard, deterministic design should be accompanied by probabilistic design[10] in which most random factors that can possibly be associated with the use of a product are incorporated in design conceptualization and design analysis. In case of airbags, these may include: occupant size or weight (children are more vulnerable) and occupant seating behavior.

The fiber and textile industry has had a long history of dealing with products and systems that require probabilistic design and accommodation of variable factors. In the fashion sector of the industry, dynamic changes in consumer desires and needs have always been a factor in fashion design. On the technical side, transferring the inevitable variability in incoming materials (e.g. natural fibers) into consistent average characteristics of yarns and fabrics has been achieved through implementation of probabilistic design concepts such as optimum fiber selection, optimum fiber blending and dynamic autoleveling.[14]

2.2.3 Invention, innovation, dissemination and patenting

Invention, innovation and dissemination are common terms that are often used by companies, organizations and nations as determinants of growth, progress, or state of development. Over the years, the words invention and innovation have been used alternatively to imply creative and original discovery. In the engineering world, invention and innovation are distinctly different; invention is to conceive a new concept or idea and innovation is to convert the concept or the idea into a real product. Dissemination, on the other hand, is an integrated phase of the innovation process but at a much larger scale or level of development; it implies a massive, efficient and cost-effective use of an innovation that originated from an invention.[15]

Patenting is a concept that can be traced back to the 18th century. It is the traditional legal way to assure the novelty or the authenticity of an invention. A patent is a certificate of grant by a government of an exclusive right with respect to an invention for a limited period of time.[15] In the USA, for example, a US patent confers the right to exclude others from making, using or selling the patented subject matter in the United States and its territories. An essential substantive condition which must be satisfied before a patent is granted is the presence of patentable invention or discovery. To be patentable, an invention or discovery must relate to a prescribed category of contribution, such as process, machine, manufacture, composition of matter, plant or design.

The meaning of invention has evolved over the years. The early human invented the wheel, one of the greatest inventions of all time. Typically, an invention is driven by a problem that must be realized first to motivate a solution. This realization may come from experience of a problem, personal impact, or a piece of knowledge that was not initially available. In this regard, an invention is primarily initiated by the creative thinking of a talented individual. Today, it is common to hear the sentence 'do not reinvent the wheel', meaning, do not come up with something that has been thought of before, or avoid duplicative thinking. In the corporate or business environment, we often hear the sentence 'think outside the box', which again means be creative, not duplicative; it also means bringing in outside viewpoints and starting from scratch to avoid reusing ideas. These common statements mainly refer to the concept of innovation. Invention is an idealization of a discovery; to invent is to conceive an idea that is totally original or new. Innovation, on the other hand, does not prohibit the use of an invention or an original idea for the benefit of mankind. The wheel is an invention, but the wheeled vehicle (viewed only on the basis of motion) is an innovation that stemmed from an invention called the wheel. In the textile industry, one can find numerous examples that illustrate the difference between invention and innovation. Two of these examples are reviewed below.

Invention and innovation of man-made fibers

The first idea leading to the development of a man-made fiber was proposed in the 17th century by Robert Hooke, an English physicist who is more popular for his discovery of the law of elasticity, known as Hooke's law. He suggested that it might be possible to imitate the process by which a silkworm produces silk.[1,2] Today, this approach is known among engineers and scientists as 'mimicking nature'. He proposed forcing a liquid through a small opening and letting it harden into a fiber. This was creative thinking that amounts to an invention. It took two centuries to revisit Hooke's idea. In the 19th century, the idea was tested with melted glass, but the resulting fibers could not be spun or woven into a useful fabric; it was the material not the concept that led to this disappointment. Today, glass fiber has become a reality, manufactured into fireproof curtains and reinforcing material. In addition, extremely pure glass can be made into fibers that transmit light, which carries information over long distances.

The glass trial stimulated other approaches to implement Hooke's first idea. Two scientists: Chardonnet, a French chemist, and Joseph Swan, an English physicist and chemist, attempted separately but during the same period to invent a man-made fiber but with a new material; this time it was cellulose.[1,2] The difference made at that point in history was more knowledge gain in the area of polymerization and molecular structures.[16] As Chardonnet and Swan came to realize, cellulose was not the easiest component to deal with; it did not easily liquidify and it was not meltable or easily dissolvable in any solvent. These challenges drove the idea of combining cellulose with nitric acid to form cellulose nitrate, or nitrocellulose, which does dissolve in a mixture of alcohol and ether. When the resulting solution was forced through small holes and the alcohol and ether evaporated, a fiber was formed.

In 1884, Swan exhibited fibers made of nitrocellulose that had been treated with chemicals in order to change the material to non-flammable cellulose. Unfortunately, Swan did not follow up the demonstration of his invention. Chardonnet, who began his work in 1878, found that nitrocellulose fibers could be chemically changed back to fibers of cellulose, which were much smoother and shinier than the original cotton or wood pulp from which they were made. He developed the rayon fiber by extruding a solution of cellulose nitrate through a spinneret, hardening the emerging jets in warm air and then reconverting them to cellulose by chemical treatment. The product, first called Chardonnet silk, was later renamed rayon and Chardonnet was named the father of the rayon industry.

The aspect of innovation came after the invention of rayon fibers. In this case, and with the help of the industry, the inventor also became the innovator. Chardonnet realized a problem associated with the highly flammable nitrocellulose, produced by treating cotton cellulose with nitric acid. This was one of the challenges that perhaps discouraged Swan from continuing his work. Chardonnet worked for several years on the problem of reducing the flammability of the new substance and at the Paris Exposition of 1889 he showed rayon products to the public for the first time. Soon after (1891), the first commercially man-made fiber was produced at a factory in Besançon.

Today, high tenacity and high wet modulus viscose fibers are common innovations, competing with cotton fibers in many products. In addition, hollow viscose fibers were innovated to give a more cotton-like feel through enhanced bulk and good moisture absorbency. The latest developments of viscose rayon involved the introduction of the environmentally friendly Lyocell or Tencel.[17]

Invention and innovation of the spinning machine

The evolution of the spinning machine can provide many design lessons for today's engineers. The early human realized ways to hand spin fibers into yarns and weave yarns into fabrics. The basic ideas underlying these techniques still represent the principles of today's spinning and weaving machinery, but with modern features such as high speed, automation and process control. The early machine used to turn fiber into a thread or a yarn was the so-called spinning wheel; it has an obscure origin (some believe it originated in India). At the beginning of the 16th century, the first semi-machined spinning system called the Saxon, or Saxony, wheel was introduced in Europe.[1] It incorporated a bobbin on which the yarn was wound continuously and the wheel was actuated by a foot treadle. Ironically, at that time the loom, which weaves the yarn into a fabric, was somewhat ahead in the process of development. This is a classic case of how invention and innovation can be independent of time sequence by virtue of consumer and market demands.

The shortage of yarn supply resulting from the improvement in the loom in the 18th century stimulated a series of inventions that converted the spinning wheel into a powered, mechanized component. This development began in about 1764 with the invention of the spinning jenny by James Hargreaves, an uneducated English spinner and weaver, who witnessed his daughter Jenny accidentally overturning his hand-powered multiple spinning machine. As the spindle continued to revolve in an upright rather than a horizontal position, he began to think in a different design direction that led to the invention of the spinning jenny with which one individual could spin several threads at one time.

Ten years after the introduction of the spinning jenny, an English spinner by the name Samuel Crompton developed the so-called spinning mule. This machine permitted large-scale manufacture of high quality thread and yarn. By our earlier definitions of invention and innovation, the spinning mule was an innovation, not an invention. It was motivated by the problem of excessive defects produced on the spinning jenny and by weaver demand for a defect-free yarn, but it did not add to the basic idea of the spinning jenny.

The ring spinning machine as we know it today (excluding automation and process control) was invented by the American John Thorp in 1828. It was an invention because the idea of using a ring and traveler was introduced for the first time. By the 1860s, ring spinning had largely replaced Samuel Crompton's spinning mule in the world's textile mills because of its greater productivity and simplicity.

The two examples presented above, in addition to their illustration of the difference between an invention and an innovation, also demonstrate an era where inventions and innovations were achieved by talented individuals who spent their own money, protected their inventions through costly patenting and followed their ideas until they became a practical reality. Today, the situation is quite different as individual efforts are now integrated into organized and well-structured institutes. Revolutionary inventions and innovations such as nylon and polyester fibers, rotor and air-jet spinning, rapier and air-jet looms are a few examples of many innovations that have resulted from organizational support. In the United States, the principal sponsor of invention is the federal government and the principal sponsor of innovation is the industry. Obviously, this trend shifts the focus from individuals to organizations and nations, with the results being more from team work and coordinated efforts. In today's world, the difference between highly developed countries and developing countries is primarily measured by the extent of invention and innovation in a country and the ability to protect these achievements through patenting and intelligent dissemination.

2.2.4 Natural resources

Natural resources are critical for all engineering applications. They are the basic goods and services that sustain human societies. Most engineering applications employ two categories of natural resources: materials and energy sources. Materials represent the basic component of most engineering designs. Common engineering tasks related to materials include:[15,18,20]

- selecting the type of material suitable for the product, system, or process in hand
- determining material properties
- manipulating material properties to meet the optimum requirements of the intended design.

Engineers often follow the classic scheme of categorizing different materials. In this regard, they consider four major categories:[19] metals, polymers, semiconductors and ceramics. Metals are generally defined as a class of highly crystalline substance with a relatively simple crystal structure distinguished by close packing of atoms and a high degree of symmetry. They are characterized by high electrical and thermal conductivity as well as by malleability, ductility and high reflectivity of light. They also represent almost 75% of all known chemical elements on earth. Examples of metal include aluminum, iron, calcium, sodium, potassium and magnesium. The vast majority of metals are found in ores (mineral-bearing substances), but a few such as copper, gold, platinum and silver frequently occur in the free state because they do not readily react with other elements.

Polymers are the class of natural or synthetic substances composed of very large molecules, called macromolecules that are multiples of simpler chemical units called monomers.[16] Polymers make up many of the materials in living organisms, including, for example, proteins, cellulose and nucleic acids. Moreover, they constitute the basis of such minerals as diamond, quartz, and feldspar and such man-made materials as fibers, concrete, glass, paper, plastics and rubbers. Polymeric materials may vary in their degree of crystallinity from highly crystalline to highly amorphous structures.

Semiconductors are a class of crystalline solids intermediate in electrical conductivity between a conductor and an insulator.[20] The popularity of semiconductors stems from the fact that they represent key elements in the majority of electronic systems, serving communications, signal processing, computing and control applications in both consumer and industrial markets. The study of semiconductor materials began in the early 19th century. Elemental semiconductors are those composed of a single species of atoms, such as silicon (Si), germanium (Ge) and tin (Sn) in group IV and selenium (Se) and tellurium (Te) in group VI of the Periodic Table. There are, however, numerous compound semiconductors, which are composed of two or more elements.

Ceramics are as old as the human race. It is a unique class of inorganic, non-metallic materials that exhibit many useful properties such as high strength and hardness, high melting temperatures, chemical inertness, and low thermal and electrical conductivity. Ceramics are also brittle and sensitive to flaws and crack propagation. They are typically categorized with respect to the type of product made from them. In this regard, there are two product categories:[20] traditional and advanced ceramics. Traditional ceramic products are made from common, naturally occurring minerals such as clay and sand. Advanced ceramics are often produced under specific conditions to serve intended functions or yield performance products ranging from common floor tiles to nuclear fuel pellets.

The revolutionary expansion of material applications and the continuous development of new materials have altered the world. Today, material type and material characteristics represent the major aspect of any engineering application. Many engineers categorize materials not by the classic categories discussed above but rather by their functional performance. Table 2.2 provides a list of categories of some of the functional materials available today.

Energy is the second major natural resource used by engineers. In simple words, energy is the capacity for doing work. Energy may exist in many different forms.[20] The most common form of energy is mechanical energy or the sum of kinetic (energy of motion) and potential (positional stored energy). Thermal energy is another form in which internal energy is manifested in a system in a state of thermodynamic equilibrium by virtue of its temperature. Nuclear energy is another form that manifests itself when the nuclei of atoms are either split (fission) or united (fusion). Solar energy results from thermonuclear fusion reactions deep within the sun. It provides the warmth necessary for plants and animals to survive. The heat from the sun causes water on the Earth's surface to evaporate and form clouds that eventually provide fresh rainwater.

The engineering interest in natural resources stems from three critical aspects:

Table 2.2 Examples of material categories and their definitions[2,9,16,20]

Material system	Definition
Fibrous material	A fiber-based material; mostly anisotropic (but can be isotropic), light weight and flexible
Composite	A mixture or combination of two or more macro-constituents that differ in form or material composition and are essentially insoluble in each other. A composite derivative is fiber composite, which is a material in which a fibrous phase that retains its physical identity is dispersed in a continuous matrix phase
Isotropic material	A material whose properties are not dependent on the direction along which they are measured
High-temperature material	A material with high-temperature capability, including some fibrous materials, superalloys, refractory alloys and ceramics; used in structures that are subjected to extreme thermal environments (e.g. spacecraft)
Thermoelectric material	A material that can be used to convert thermal energy into electric energy or provide refrigeration directly from electric energy; good thermoelectric materials include lead telluride, germanium telluride, bismuth telluride and cesium sulfide
Thermoplastic materials	A material that when heated (to the vicinity of the glass transition temperature), softens and flows controllably, enabling it to be processed at high speeds and on a large scale in the manufacture of molded products
Acousto-optical material	A material in which the refractive index or some other optical property can be changed by an acoustic wave
Active material	An energy-storing and energy-conversing material
Hazardous material	A poison, corrosive agent, flammable substance, explosive, radioactive chemical, or any other material which can endanger human health or well-being if handled improperly
Infrared-transparent material	An optical material that transmits infrared radiation; examples include sodium chloride (0.25 to 16 µm), cesium iodide (1 to 50 µm) and high-density polyethylene (16 to 300 µm)
Acoustical material	Any natural or synthetic material that absorbs sound
Contact material	A material having high electrical and thermal conductivity
Electrorheological material	A material possessing rheological properties that are controlled by an imposed electric field
Electro-optic material	A material in which the indices of refraction are changed by an applied electric field
Dielectric material	A material which is an electrical insulator or in which an electric field can be sustained with a minimum dissipation of power
Depleted material	Material in which the amount of one or more isotopes of a constituent has been reduced by an isotope separation process or by a nuclear reaction

Table 2.2 Continued

Material system	Definition
Bridging material	A fibrous, flaky, or granular substance added to a cement slurry or drilling fluid to seal a formation in which lost circulation has occurred. Also known as lost-circulation material
Magneto-optic material	A material whose optical properties are changed by an applied magnetic field
Nano-structured material	A material whose composition is modulated over nanometer length scales in zero, one, two, or three dimensions. Also known as nano-composite material
Nano-phase material	A material made up of phases that have dimensions of the order of nanometers. An ultrafine single solid phase where at least one dimension is in the nanometer range, and typically dimensions are in the 1–20 nm range
Lossy material	A material that dissipates energy of electromagnetic or acoustic energy passing through it
Lossless material	An ideal material that dissipates none of the energy of electromagnetic or acoustic waves passing through it
Non-linear material	A material in which some specified influence (such as stress, electric field, or magnetic field) produces a response (such as strain, electric polarization, or magnetization) which is not proportional to the influence
Barrier material	Packing material impervious to moisture, vapor, or other liquids and gases. It can also be an inert material placed in an explosive charge to shape the detonation wave
Optical material	A material which is transparent to light or to infrared, ultraviolet, or X-ray radiation, such as glass and certain single crystals, polycrystalline materials (chiefly for the infrared) and plastics
Paramagnetic material	A material within which an applied magnetic field is increased by the alignment of electron orbits
Phase-change material	A material which is used to store the latent heat absorbed in the material during a phase transition
Piezoresistive material	A metal or semiconductor in which a change in electrical resistance occurs in response to changes in the applied stress
Electrochromic material	An organic or inorganic substance that can interconvert between two or more colored states upon oxidation or reduction, that is, upon electrolytic loss or gain of electrons
Infrared optical material	A material which is transparent to infrared radiation
Radioactive material	A material having one or more constituents that exhibit significant radioactivity
Radar-absorbing material	A material that is designed to reduce the reflection of electromagnetic radiation by a conducting surface in the frequency range from approximately 100 MHz to 100 GHz
Phase-change material	A material which is used to store the latent heat absorbed in the material during a phase transition.

- Natural resources represent an absolute necessity for making products and operating systems.
- Natural resources do not represent a static target or readily available resources that can be used whenever they are needed; many natural resources are likely to be depleted.
- Natural resources represent an essential aspect of optimum design, particularly when material and energy are design factors of multiple levels or sources.

In recent years, the issue of natural resources has become of a worldwide concern. This has resulted in dividing earth's natural resources into two main categories:[20]

- non-renewable – minerals, oil, gas, and coal
- renewable – water, timber, fisheries and agricultural crops.

Without renewable resources, accessibility to air, water and food will not be possible. Non-renewable resources, on the other hand, provide the energy essential for industrial economies and are the source of important products ranging from iron tools to silicon chips. Since most resources are limited, engineers must be concerned with the development of new resources as well as efficient utilization of existing ones. These two requirements often represent the essence of optimal design. As indicated earlier, engineering is a practice that serves mankind in an optimal fashion. Accordingly, any engineering task must eventually be integrated into the overall national and global interest. The world population is not static and natural resources are not static either. This means that the outcomes of all engineering tasks must be judged and evaluated on the basis of meeting an optimum combination meeting the challenge of growing populations and increasing levels of resource consumption.

One common misconception regarding natural resources is that the biggest challenge of resource conservation will involve non-renewable resources. This may seem logical on the basis that renewable resources can reproduce themselves under the appropriate conditions. However, history indicates that in fact the opposite is the case as a result of significant engineering and scientific developments. When non-renewable resources have been depleted, new technologies have been developed that effectively substitute for the depleted resources. In many situations, new technologies have often reduced pressure on these resources even before they were fully depleted. Historians use examples such as fiber optics that has substituted for copper in many electrical applications and renewable sources of energy, such as photovoltaic cells, wind power, and hydropower that may ultimately take the place of fossil fuels when stocks are depleted.

Nations and large organizations should be concerned about the management of the human use of natural resources to provide the optimum benefit for current generations while maintaining capacity to meet the needs of future generations. This issue is particularly essential for non-renewable natural resources. In this regard, many engineering tasks can be implemented as summarized below:

1. Beneficiation – the use of technological improvements in upgrading of a resource that was once too uneconomical to develop (e.g. utilization of fiber wastes and fiber by-products particularly in the natural fiber and fiber blend processes)
2. Optimization – the use of engineering design to prevent initial waste, to minimize material weight and to maximize integrated benefits
3. Substitution – any opportunity in which a readily available natural resource is used in place of rare or depleting resource (e.g. the use of aluminum in place of less abundant copper for a variety of products and the use of fibrous materials instead of metals for light weight and reduction of energy consumption)
4. Recycling – this is the most commonly used approach for optimizing natural resources. It involves the concentration of used or waste materials, their reprocessing and their subsequent reutilization in place of new materials. Examples of recycling are numerous in the textile industry, from chemical recycling to solid material reutilization.

2.3   Conclusions

The main purpose of this chapter was to review some of the commonly known engineering concepts. These concepts are often overlooked in the midst of detailed and highly technical engineering analysis associated with specific products or systems. A fiber or a fabric engineer in today's competitive market does not work in a vacuum or perform in isolation. He/she must be fully aware of the surrounding engineering world and must be able to communicate and cooperate efficiently and effectively with engineers from various fields.

2.4   References

1. *Textile*, Encyclopedia Britannica Online, http://www.britannica.com/, 2005.
2. AccessScience@McGraw-Hill, http://www.accessscience.com, 2004.
3. BEAKLEY G C and LEACH H W, *Engineering – An Introduction to a Creative Profession*, 3rd edition, Macmillan, New York, 1977.
4. GEE E A and TYLER C, *Managing Innovation*, John Wiley & Sons, New York, 1976.

5. FRENCH M J, *Conceptual Design for Engineers*, The Design Council, London, Springer-Verlag, Berlin, 1985.

6. *The Textile Institute, Textile Terms and Definitions*, 10th edition, Textile Institute, Manchester, 1994.

7. HEARLE J W S, 'The contribution of academic research to industrial innovation in fibers and textiles', *World Textiles: Investment, Innovation, and Invention, Annual World Conference*, The Textile Institute, London, May, 1985.

8. JAMES E, *Human Factors in Engineering and Design*, 4th edition, McGraw-Hill, New York, 1976.

9. DIETER G, *Engineering Design: A Material and Processing Approach*, McGraw-Hill Series in Mechanical Engineering, New York, 1983.

10. EL MOGAHZY Y E, *Statistics and Quality Control for Engineers and Manufacturers: from Basic to Advanced Topics*, 2nd edition, Quality Press, Atlanta, USA, 2002.

11. FERGUSON S A, LUND A K and GREENE M A, *Driver Fatalities in 1985–1994 Air Bag Cars*, Insurance Institute for Highway Safety, Arlington, VA, 1995.

12. NATIONAL HIGHWAY TRAFFIC SAFETY ADMINISTRATION. *Fifth/Sixth report to Congress: Effectiveness of Occupant Protection Systems and Their Use*, US Department of Transportation, Washington, DC, 2001.

13. BRAVER E R and KYRYCHENKO S Y, *Efficacy of Side Airbags in Reducing Driver Deaths in Driver-side Collisions.* Insurance Institute for Highway Safety, Arlington, VA, 2003.

14. EL MOGAHZY Y E and CHEWNING C, *Fiber to Yarn Manufacturing Technology*, Cotton Incorporated, USA, 2001.

15. LEECH D J and TURNER B T, *Engineering Design for Profit*, Ellis Horwood, New York, 1985.

16. BILLMEYER, JR. F W, *Textbook of Polymer Science*, 3rd edition, John Wiley & Sons, New York, 1984.

17. COLE D J, 'A new cellulosic fiber – *Tencel*', *Advances in Fiber Science*, The Textile Institute, Manchester, 1992.

18. *Handbook of Product Design for Manufacturing: a Practical Guide to Low-cost Production*, Bralla J G (editor), McGraw-Hill, New York, 1986.

19. BARKER J W, *Engineering*, in AccessScience@McGraw-Hill, http://www.accessscience.com, 2000.

3

Textile product development: basic concepts and critical factors

Abstract: Product development is the primary recipe for success in today's competitive market. In order to remain competitive, an organization must sell more products, create new products and find new market opportunities for existing intellectual properties. This traditional view of product development implies that it is largely an engineering exercise. However, in today's competitive market, the concept of product development has largely been expanded to accommodate and integrate critical product-related aspects such as consumer perception, product attractiveness, value appreciation, market niche and anticipated performance over a product's lifecycle. Textile and fiber engineers must expand their knowledge base beyond the traditional concepts and should work with other groups within their organization such as manufacturers and marketing personnel to meet the overall objectives of a product development project. In this chapter, basic concepts and various elements of product development are discussed.

Key words: product development cycle, product lifecycle, competitor-focus organization, competition models, time-to-market pressure, oligopoly, monopoly, market segmentation, market shift, product-focus development, user-focus development.

3.1   Introduction

In Chapter 1, fiber-to-fabric engineering was defined as the science and art of optimizing the utilization of fibers (natural or synthetic, conventional or high-performance) in various applications (traditional or function-focus, and existing or potential) using sound engineering concepts so that their ultimate benefits can be realized by all humanity. A fiber-to-fabric engineering program is essentially a product development program based on reliable scientific methods in which all elements of the fiber-to-fabric process are taken into consideration. It is important therefore to begin our discussion by introducing the basic concepts and critical factors associated with product development.

As indicated in Chapter 1, fibrous products can be generally divided into two major classes: traditional fibrous products (TFP) such as apparel, furnishing and household products, and function-focus fibrous products (FFFP) such as protective and safety fabric systems, automotives, aerospace, medical and hygiene and geosynthetics. From a product development viewpoint, it is important to distinguish between these two categories of products as each will typically involve a uniquely different development strategy by virtue of the differences in the associated technologies, the types of application and the customer's perception. It is also important to keep in mind the common performance criteria that bridge these two categories of product. These include light weight, high flexibility and ease of manipulation of dimensional characteristics (e.g. length, fineness, cross-sectional shape and density).

Product development is the primary recipe for success in today's competitive market. In order to remain competitive, an organization must sell more products, create new products and find new market opportunities for existing intellectual properties. In the traditional sense, product development is the process of improving existing products, or converting new ideas into innovative products that are acceptable to potential users. This traditional view implies that product development is largely an engineering problem. However, in today's competitive market, the concept of product development has largely been expanded to accommodate and integrate critical product-related aspects such as consumer perception, product attractiveness, value appreciation, market niches and anticipated performance over a product's lifecycle. This means that engineers must expand their knowledge base beyond the traditional concepts and should work with other groups in the organization such as manufacturers and marketing personnel to meet the overall objectives of a product development project.

In this chapter, basic concepts and various elements of product development will be discussed. First, a simplified view of product development will be presented to familiarize the reader with the basic tasks constituting most product development programs. Second, a brief review of the evolution of product development cycle will be presented. Attention will then be turned to a critical aspect of product development, namely product lifecycle. Finally, critical business and marketing aspects related to product development will be discussed.

3.2   Simplified view of product development

As indicated above, the concept of product development has undergone a major transition in recent years; from merely a conversion of an idea to a physical product, to a process in which all product-related aspects, from the inception of an idea to the final phase of a product lifecycle, are carefully integrated. In addition, the use of powerful computer-aided tools has provided a significant dimension to the process of product development, not only in the design phase of this process but also in other important areas including information gathering and marketing strategies.

The basic steps of any product development project are as follows (see Fig. 3.1):

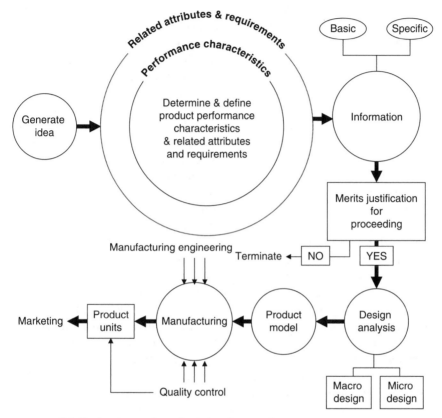

3.1 Basic steps of product development.

- Generate or identify an idea that is worthy of further consideration;
- Define and determine product performance characteristics and associated product attributes and requirements;
- Gather relevant information about the product idea (basic information, patents, and market information);
- Justify the merits of the idea before you proceed further;
- For justified ideas, perform design analysis;
- Develop a product model or prototype;
- Manufacture the product;
- Establish a strategy for marketing the product.

The sparking point of a product development project is idea discovery or the identification of an idea that is worthy of further consideration. This may be brought about by a need to improve the performance of an existing product such as the need for a bulletproof vest that is both highly protective, yet reasonably comfortable, or the desire to develop a suture product that is both flexible and tissue-compatible. The idea may also come from a totally different product concept via creative thinking such as a fiber composite structure that can assist in resisting or breaking winds at speeds exceeding 100 miles per hour (160 km per hour), or a reusable diaper that is environmentally friendly. In this case, significant efforts must be made to evaluate the idea and verify its feasibility.

Once a product idea is conceived and tentatively accepted, the next step in product development is to determine and define the anticipated performance characteristics of the intended product. These are the characteristics that describe the target function(s) of a product and reflect product expectations in the marketplace. A performance characteristic is typically a function of a set of attributes and requirements that must be satisfied in the product to meet its intended performance. For example, the main performance characteristic of a fibrous product may be determined as 'durability'. This may be defined as the ability of the product to withstand external stresses resulting from possible harsh actions and environmental exposure (e.g. military uniforms or mine working outfits), or from great physical activities (e.g. sportswear). In this regard, it is important to point out that durability *per se* is not a measurable attribute with well-defined or specified levels. Instead, it is a function of many attributes that can collectively lead to a durable product. In the case of fibrous products, these attributes are those of the components constituting the product (i.e. polymer, fiber, yarn, fabric and final assembly). They may include fiber strength, yarn strength, yarn structure, fabric tear, bursting strength, UV resistance, or heat resistance. These are well-defined and measurable parameters with values that can be considered in the design analysis of the fibrous product.

In general, product performance characteristics and corresponding attributes will depend on the complexity of the intended product and whether it is a modified version of a pre-existing product or a totally new product. For example, a fibrous surgical implant may be described in terms of two basic characteristics: durability and biocompatibility. The former is typically related to a set of attributes such as filament tensile strength, load-bearing capacity, corrosion resistance and fiber weight. The latter is commonly related to another set of attributes such as polymer type, chemical compatibility and fiber surface morphology. Other performance characteristics such as allergic risk and X-ray transparency may also be considered. These are more medically oriented, yet they must be accounted for in the choice of fibrous material. On the other hand, a traditional product such as table cloth may be described in terms of simple attributes that directly reflect its anticipated performance characteristics. These include weight, dimensions, water absorption or hydrophilicity and stain resistance.

Performance characteristics are also influenced by product requirements, particularly those that can affect the consumer's view of the product. These include natural resources, quantity of material required to produce a product unit, expected cost per unit, anticipated customer perception of the product, anticipated appreciation of product value, time allocated to complete the product development cycle, product safety, anticipated product lifecycle and recycling or reclamation potential.

The next step in product development is information gathering. In this regard, two types of information should be available: basic information pertaining to the intended product and specific information associated with the desired performance characteristics. In the case of fibrous products, basic information includes fiber type, fiber attributes, yarn structure, yarn attributes, fabric construction and fabric attributes. Specific information is typically driven by the target functions of the product as discussed above. Primary sources of this type of information will typically include existing patents, specific technical information of similar products, some research literature and expert opinions.

In connection with product requirements, information typically represents market research results such as outcomes of competitive analysis, information about market availability, information about potential customer sectors or segments, and costs of product delivery and distribution. As expected, this type of information will be less expensive and easier to obtain to modify a pre-existing product than to develop a totally new product.

After gathering all relevant information, it will be important to pause and ask key questions such as Should the product development project proceed further? Should it be taken different directions? or Should it be terminated? These questions are often difficult to answer prior to collecting all the relevant information about the intended product. The reason for this critical step is that many product development projects go forward under time-to-market pressure only to face unexpected failure at the final phase and after a huge amount has been spent. In this regard, the common wisdom should be 'fail fast and bleed less'. Obviously, proceeding with the product development project will not always be a guarantee of a successful product; other criteria must be met such as reliable design, good product model and initial market appeal.

The core of any product development project is product design. This is the process of transforming a product idea into a product model with optimum performance. An engineering design begins by defining a design problem, the solution of which results in a product model. Based on the outcomes of the previous steps in product development, a combination of creative thinking and analytical approaches is implemented to seek an optimum solution of the design problem. In Chapter 4, the focus will be entirely on product design and common design cycles. In general, a design process should consist of two main tasks: design conceptualization and in-depth design analysis. These two tasks are discussed in detail in Chapters 5 and 6, respectively. Design conceptualization is essentially the task of dealing with the design problem at a macro level through exploring all solution ideas and evaluating the effectiveness of the selected few. In-depth design analysis deals with the problem at a micro level through focusing on the selected approaches for solving the design problem and seeking optimum analytical solutions that yield an acceptable product model.

The outcome of a design process is a product model that can be validated, finalized and presented to the manufacturer to consider for mass production. Obviously, a great deal of communication and coordination should be carried out between design engineers and manufacturing engineers to finalize the design process and reach compromising solutions. In this regard, specific issues such as detailed configuration of the product, product specifications, assembly planning, cost and manufacturing specifications should be addressed.

The result of the steps above is a product that can be tested in the marketplace. Again, the task of marketing a product will largely depend on whether the product is a modified version of a pre-existing product or a totally new one. The success of marketing the former will be determined largely by the status of the current competing product in the marketplace and its stage in product lifecycle (i.e. growth, maturity, or decline stage). This point will be discussed later in this chapter. Marketing a modified product will also be determined by whether it was a result of a deficiency in the current product discovered by the consumer, or an internal desire by the company, inspired by consumer interest, to enhance the current product and add new features. For totally new products, the marketing phase should overlap with all the previous steps of product development. In this regard, the product idea should be market tested, information should be partially market-oriented and the final product model should undergo a great deal of market evaluation via consultation with product distributors and consumer surveys.

The issue of marketing is critical for all fibrous products. The fact that consumers are fully aware of traditional fibrous products makes it truly challenging to develop new features that continue to attract buyers to these products. In this regard, the most active sector producing this type of product is the fashion industry. In recent years, other sectors of the industry have made significant strives to develop traditional products with new attractive features, such as sportswear that are uniquely suitable for specific types of athletic applications and medical socks. With function-focus fibrous products, the challenge is even greater, particularly in situations where fibrous materials are either being combined with non-fibrous structures or substituting traditional non-fibrous materials (e.g. metals or gravels). This challenge obviously stems from the long stand of traditional materials in the marketplace. For example, the use of fibrous materials to develop fabric-reinforced concrete has been associated with a struggling market in which gravel or metal reinforcements have been compared with fiber reinforcement, particularly with respect to reliability issues. Although this product idea has been around since the Ancient Egyptians built homes with straw in bricks, conceptualizing this idea to suit modern applications has been difficult. As a result, features such as greater resistance to cracking and thermal changes, thinner design sections, less maintenance and longer life had to be emphasized in marketing this type of product.

The above discussion, intended to simplify the basic steps in product development, may give the impression that product development is a systematic step-by-step approach the end of which is a guaranteed developed product. This could not be further from the truth, as a great deal of judgment and critical decision making must be involved in each project, depending on its complexity. Indeed, a standard product development scheme does not exist because of the following factors:[1-3]

- The extent of success of a product development approach will not only depend on following the common basic steps of product development but also on the effectiveness and the efficiency of coordinating and integrating these steps.
- Product development involves a great deal of creative and imaginative effort that is likely to vary from one company to another and from one product to another.
- Reaching an optimum solution to a design problem is not only a question of putting together the most talented design team, or utilizing the most reliable tools; it is often a question of resources available, time-to-market pressure and knowledge capacity. As a result, an optimum solution will typically represent the most compromising solution within these constraints.
- Developing a good product does not necessarily guarantee that it will be a market success. Some products that enjoyed great initial market success did not make it beyond the growth phase of the product lifecycle.

3.3   The product development cycle

The basic steps in product development discussed above, with exception of the marketing aspects, follow closely the so-called multiple serial design–build–test cycle[1] (see Fig. 3.2a). In this approach, a product idea is conceived, a design procedure is implemented to provide a product model, and the model is validated, built and tested. The outcome of this cycle is a

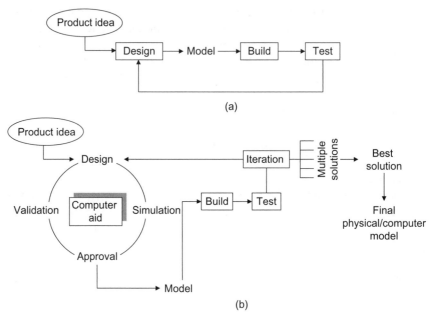

3.2 Examples of product development cycles.[1-4] (a) Multiple serial design–build–test cycle. (b) Product development with computer aid.

physical prototype that is ready for manufacturing planning and mass manufacturing.

In this information era, product development has emerged to a new paradigm for utilizing effective and efficient computer analyses in performing critical tasks such as design analysis, product modeling, performance simulation and validation, and development of a final model for building and testing.[2-4] This paradigm, shown in Fig. 3.2b, allows an iterative process leading to multiple solutions from which the most compromising one is selected and tested.

The product development cycles illustrated in Fig. 3.2 represent classic approaches to product development that typically begin with a product idea and end with a product model based on optimum solutions to the design problem. When fibrous products, traditional or function-focus, are the target of product development, it will be important to establish a product development program that accounts for the coordination between different personnel involved in product development, as manufacturing and marketing represent key aspects in the development of these types of product. In addition, analysis of product lifecycle should be a part of product development so that companies can be prepared to make timely decisions about modifying existing products or introducing new products to the market.

Accordingly, a more appropriate product development program should be based on a dynamic product development cycle in which both coordination and lifecycle analysis are integral parts. Figure 3.3 shows a generalized outline of a dynamic product development cycle.

Some of the merits of implementing a dynamic product development cycle include:

- insuring that the initial reception of a new product is successful;
- monitoring and testing a new product at the growth stage of the product lifecycle and as it reaches the maturity stage to avoid costly rejects or liability issues that were not anticipated in the design process;
- modifying current products to enhance their performance (responding to user's concern and feedbacks);
- seeking alternative product options that may be less costly, energy efficient, performance efficient and more attractive.

3.3.1   Coordination in product development

Figure 3.3 illustrates the basic components of a dynamic product development cycle. As can be seen in this figure, the core functions of a dynamic product development cycle are similar to those discussed above. In addition, the cycle involves continuous evaluation of market needs, continuous entertainment of new ideas, design conceptualization, in-depth design analysis, continuous coordination between different teams involved in product development, and monitoring of product performance in the marketplace or by product lifecycle analysis. In this section, the focus will be on the coordination aspect.

In order to meet the ultimate objectives of a product development program, a great deal of coordination should be achieved between the design team and the manufacturing team. Through this coordination, many practical aspects associated with making a product can be explored and some design limitations can be overcome during the manufacturing phase. This coordination is particularly important in developing fibrous products as it is well known that a fabric construction designed on paper may face a number of obstacles during manufacturing that can either make it very costly to produce (in the absence of readily available technology), or result in high cost that cannot be justified in determining the value of the final product. Similarly, a yarn structure designed on paper may have nominal structural features that may not be easily achievable during manufacturing. It is important, therefore, that design engineers and manufacturing personnel come to agreement on how the final model of the product will be handled during manufacturing.

In the context of coordination between design and manufacturing teams, it is important to distinguish between two groups of manufacturing

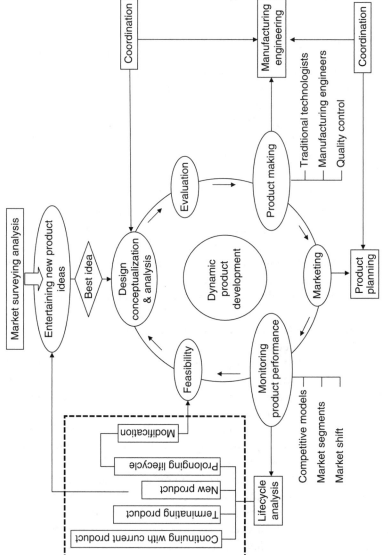

3.3 Dynamic product development cycle.

personnel:[5] traditional technologists and manufacturing engineers. A traditional technologist is a person with long manufacturing experience who simply follows and implements instructions regarding the manufacturing of massive product units. A traditional technologist often operates on the basis of a 'product-in' approach.[5] This means that his/her task ends with the manufacturing of a part required for a product assembly. For example, a spinning room supervisor has traditionally assumed the role of a technologist with his/her task being to produce a yarn that can be used for weaving or knitting. Typically, the main concern of this technologist has been to meet the required production efficiency of the spinning units under his/her supervision and satisfy the specifications required in the yarn. Once the yarn is packaged and delivered, this technologist shifts his/her attention to the next lot of yarns to be manufactured and would have little concern about the impact that the yarn may have on the efficiency of the subsequent stages of processing or the quality of the end product.

A manufacturing engineer, on the other hand, operates on the basis of 'market-out' approach, in which every component in a product assembly is manufactured with a clear vision of its impact on the cost and performance of subsequent processes or on the end product. Accordingly, he/she will be involved in many engineering activities such as the creation and operation of the technical and economic processes that convert raw materials, energy and purchased items into product. In comparison with the spinning technologist discussed above, a spinning manufacturing engineer will be involved in critical tasks such as determining the type and the amount of fibers required to make the yarn, estimating the operational cost per pound (or kilogram) of yarn, optimizing the various parameters of the spinning process, evaluating the extent of meeting design specifications and quality requirements, and forecasting the impact of the yarn produced on the performance of end product (woven or knit fabrics). It is through these activities that a manufacturing engineer can effectively assist a design engineer solving problems associated with product development.

In a dynamic product development cycle, coordination between the manufacturing team and marketing personnel (sales representatives, distributors and retailer representatives) should be a continuous process. The purpose of this coordination is to make certain that the product will meet with the customer's approval and that quality or performance problems are being solved in a timely fashion. This type of coordination has become critical in today's competitive market as a result of the increasing consumer awareness and user demand for more consistent and high quality products. Three-way coordination between manufacturers, design engineers and marketing personnel is also important. From a product development viewpoint, the key factor in this coordination is the performance status of the product in the marketplace or the stage in the product lifecycle, as discussed in the following section. Based on this coordination, critical decisions regarding the product may be made in a timely fashion. As shown in Fig. 3.3, these may include termination of an existing product, development of a new product, or prolonging the product lifecycle through revisiting the product development cycle and seeking modifications that can revive product usefulness and attraction.

3.3.2 Product lifecycle

In the above section, the term 'lifecycle' was used repeatedly to describe the progressive performance of a product in the marketplace. Product lifecycle analysis has become an essential task of product development in today's global market. In general, a product lifecycle consists of four common phases of product performance:[5–8] initiation, growth, maturity and decline. A generalized product performance profile over a complete lifecycle is shown in Fig. 3.4. Common market changes associated with product lifecycle are demonstrated in Fig. 3.5. Although some features of each stage of a product lifecycle may vary depending on the product type, most products will exhibit common features such as those described in Table 3.1.

The initiation stage in a product lifecycle encompasses the generation of product idea and the different tasks of product development discussed earlier. Toward the end of this stage, a product undergoes rapid changes

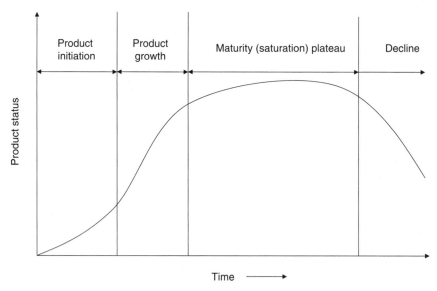

3.4 Basic stages of product lifecycle.

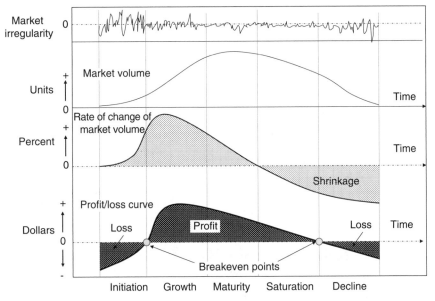

3.5 Market evolution pattern: market lifecycle.[5]

and adjustment to optimize its performance in the marketplace. In some situations, the product undergoes extensive testing by selected customers or potential users of the product. User testing of a product not only reveals the extent of success of the design process but also evaluates the outcome of the assembly process of the product during manufacturing. More importantly, product testing by potential users may also reveal other concerns or interests that have been overlooked during the design or the manufacturing process. When intense 'time-to-market' pressure increases, some organizations rush into introducing their new products without extensive user testing. As a result, many new products may be recalled shortly after their introduction to the market for safety and liability reasons or due to some functional flaws.

From a market viewpoint, the initiation stage of a product lifecycle is typically associated with erratic patterns of growth and instability in the market. This a direct result of slow customer awareness of the product at the initiation stage, particularly when promotion of the new product is limited owing to cost reasons. Market volume typically expands slowly during this period as a result of high market resistance. The rate of growth increases faster than the market volume itself, because each additional dollar represents a higher growth percentage than at any later stage of the cycle. Assuming that the company initiating the market is a pioneer, and recognizing that the initial outlay for product and market development is

Table 3.1 Product and market attributes associated with the product lifecycle[5-8]

Lifecycle phase	Product evolutionary attributes	Market evolutionary attributes
Initiation	Invention to innovation Rapid changes and adjustment to optimize product performance Possible instability in technology	Customer or consumer awareness and acceptance of the product is slow Erratic pattern of growth in market structure, and competitive status Market volume typically expands slowly as a result of potential market resistance Possible initial net loss as a result of costs exceeding revenues
Growth	Better knowledge of product capabilities and flaws Further attempts to reduce manufacturing cost as mass production continues More customization of product performance to meet customer requirements	Customer awareness increases rapidly during this period until it reaches a maximum at the end of the growth period Steady or rapid increase in market volume Revenues exceed cost (net gain)
Maturity to saturation	Product development largely ceases New ideas may be introduced to re-design, rejuvenate the product or introduce better products	Wide acceptance of the product Customer expectation of lower price as sales volume increases Slow rate of market volume
Decline	Production rate and volume declines rapidly Product support declines rapidly	Market volume shrinks Customers shift to other alternatives

often quite substantial, the product initiation stage is generally characterized by costs exceeding revenues (or net loss). As a company approaches the end of this phase, it should reach a breakeven point for the innovation. This means that its total revenue should at least be equal to its total cost. From a competitive viewpoint, if two companies share the pioneering stage of a certain product, the company that reaches the breakeven point first will be likely to have more flexibility and competitive advantage in the next stage of the product lifecycle.

Passing the initiation stage is critical for any product, as failure to achieve this can only mean substantial losses for the organization developing the product. The period of time that an initiation stage can take will obviously vary depending on many factors including the type of product, resources, time-to-market pressure, competitive pressure and customer acceptance. Most traditional fibrous products exhibit a short initiation stage. Function-focus fibrous products on the other hand are likely to be associated with longer initiation stages by virtue of the design complexity and marketing challenges.

After the initial stage, the product enters the growth stage of the product lifecycle. In this stage, the product should exhibit booming sales and the growth rate should also increase. The profile in Fig. 3.5 shows a peak of growth rate at about the middle of this stage. Further growth occurs at a decreasing rate, often because of a steadily expanding base. Although profit reaches its highest point during the growth stage, the trend may reverse in the middle of this phase as declining prices and rising costs prevail.

Products that exhibit steady growth are likely to continue their growth, normally at smaller rates to the next phase of product lifecycle, which is the maturity stage. In this stage, the market volume continues to grow at a slower rate until it reaches the stagnation level at the end of this stage. Profits diminish but can still be healthy. Markets that have grown to maturity may reach a saturation stage in which volume and profit/loss all go through a negative change rate. Costs and competitive pressure reduce profits further until they cross the breakeven point again at the end of this stage.

Maturity may continue for many years depending on the extent of customer satisfaction with the product (as indicated by a steady market volume) and the willingness of the organization to continue supplying the product. However, it is at this stage and before a product enters the saturation stage that an organization should revisit the product concept and make important decisions, such as (a) improving the product model so that a new generation of the product can be initiated, (b) terminating the product concept and turning into another product idea, or (c) prolonging the maturity stage through promotional activities or price reduction.

Product decline is an inevitable stage of the lifecycle of virtually any product that must be expected and anticipated by any organization. This is a fact that has been historically realized by many organizations particularly those that produce fashionable fibrous products. Indeed, these particular products are associated with short product lifecycles as a result of a self-imposed decline so that new fashions associated with higher profits can be introduced. Decline may also occur for other products as a result of the introduction of new products that exhibit better performance or meet new demands. This will result in shrinkage of market volume as a result of booming substitute markets.

In the fiber-to-fabric system, one can find numerous examples of products at various stages of their lifecycle. For example, the development of high-speed shuttleless looms has rapidly forced conventional shuttle looms into the decline phase. On the spinning side, new yarn-forming technologies such as rotor spinning, air jet spinning and friction spinning with their superior production rates have largely slowed the growth of conventional ring spinning. However, they could not fully eliminate ring spinning in niche markets such as fine woven fabrics and soft knit goods. Products such as cloth diapers have suffered a major decline as a result of the development of disposable diapers. Now, there are new trends toward the use of cloth diapers again as a result of environmental concerns (un-recyclable) and some adverse effects including the outbreak of diaper rash and skin irritation.

3.4   Business and marketing aspects related to product development

In today's competitive market, engineers in various fields should be aware of business and marketing issues that can influence their approaches to product development and design methodologies. Obviously, there are many business and marketing issues that may surround the process of product development. However, the most influential factors are (1) the competitive status of an organization, (2) market segmentation and (3) market shifts. These factors are discussed below.

3.4.1   The competitive status of an organization: competition models

Product developers should be aware of the competitive status of their organization. If an organization is leading in the competitive race, the pressure imposed on the product development process will be to maintain or improve the organization image and reputation through developing products that are superior to the existing ones or totally innovative. This pressure is typically offset by sufficient financial and human resources that leading organizations typically allocate to create and develop new product ideas. Organizations trailing in the competitive race typically face a different type of pressure, which is gaining or regaining consumer confidence in their products, and intense 'time-to-market' pressure of new products that can assist them in catching up with leading organizations. This pressure may be further complicated by other factors, such as the feasibility of risk taking and limited resources. New organizations that wish to enter the competitive race should have completely different approaches to product development depending on the size of investment and the organization experience. For newcomers with moderate investment, niche markets and speciality products may represent the best options.

As engineers often like to discuss matters using some modeling techniques, it will be beneficial to discuss the different competition models that are likely to exist in the marketplace. Commonly, two main competition models can exist in the market:[9-13] pure competition and monopolistic competition. Pure competition is a market scenario in which many autonomous and knowledgeable suppliers and buyers of an identical product are present, yet none are capable of gaining price advantage or changing the price. Typical conditions of a pure competition model are as follows:

1. No single supplier or buyer is large enough or powerful enough to affect the price of the product.
2. Products are identical and they exhibit no disparity in quality, and no brand name or image issues exist. As a result, customers are less likely to prefer one supplier over another.
3. No collusion (conspiracy or hidden agreements); as a result, each supplier and buyer act autonomously. This means suppliers would compete against one another for the consumer's dollar. Buyers would also compete against each other and against the supplier to obtain the best price.
4. No mutual loyalty is necessary between the buyers and the suppliers as all products are identical giving little reason for loyalty to a particular supplier on the basis of product merits.
5. Non-regulated market environment; this means that suppliers and buyers are free to get into, conduct, and get out of business. This makes it difficult for a single supplier to dominate the market and dictate unfavorable prices as other suppliers can freely enter the market and stabilize the price.

In today's globally competitive market, pure competition hardly exists and the more common competition model is the so-called monopolistic competition. This model typically exists at any time the conditions of pure competition listed above are not met. Under a monopolistic competition model, products cannot be considered as identical even if they are seemingly perceived to be the same. As a result, product developers must continually strive to establish features in their products that are uniquely different from those of competitor products. This may require design emphasis on performance and quality differentiation rather than price differentiation.

Monopolistic competition may be represented by many familiar models in the market place. These include oligopoly, pure monopoly and near monopoly. Oligopoly implies a market of few large firms that dominate the market and have the ability to influence prices. These firms tend to have common strategies and one may follow another in product lines, price change, or customer service procedures. They also tend to keep their prices and their product values far from reach by smaller competitors. Despite their independent management, they always seem to act in unity particularly in price fixing and in dividing the market so that each is guaranteed to sell a certain quantity. Familiar organizations that operate largely under oligopoly model include Coca Cola and Pepsi Cola, and some giant airlines.

Pure monopoly is a model which hardly exists in the real world as it implies domination by one organization of a specific product that has no substitutes. Even if this model holds for some time, other competitors will certainly come into play and attempt to develop similar or better products. A more realistic model is the so-called 'near monopoly'. This is best illustrated by its derivatives such as natural monopoly, government monopoly, geographic monopoly and technological monopoly.

Natural monopoly is when society is best served by the existence of monopoly. Organizations that enjoy natural monopoly are typically subject to many regulations to serve the interest of the consumer. As they grow larger, they are likely to bring about lower prices as a result of lower costs (the economy of scale). Government monopoly is a common condition in all countries, but mostly in the third world countries. Under this condition, the government owns, operates and develops products for the public. These are mostly of a service nature such as healthcare, public transportation and utility. Geographic monopoly is a historical type of near monopoly that some organizations have enjoyed as a result of their geographic locations and consumer demands in these locations. With the increasing trend toward a global competitive market, this type of monopoly is likely to decline as many organizations will find their ways through different cultures and different areas in the world. Technological monopoly is perhaps the most accepted type of near monopoly and the most difficult one to compete against. From an engineering perspective, technological monopoly lies in the heart of product development. It reflects the special rights that an organization can enjoy for developing new products. In most countries including the USA, these organizations are given exclusive rights to manufacture, use, or sell new products that were invented by them.

In light of the above discussion, it is critical that competitive analysis be a part of the product development process. Unfortunately, there is no universal approach for performing competitive analysis at the product development stage. As a result, different organizations seek different approaches to determine their competitive status. Some organizations rely on their market share statistics to determine their competitive status. Historically, this approach has been useful in creating organizational confidence and investors' trust. However, at the product development stage, more intimate knowledge of the competitive status of an organization in view of the products being developed can be critical.

In a competitive environment in which monopolistic competition models are likely to prevail, an organization should always be aware of the external forces influencing the market. These can be represented by a competitive loop showing the various market forces, which can serve as a guideline for possible action toward achieving competitive advantages.[5] A simple competitive loop will consist of a number of organizations competing for the same consumer or potential buyer. In this environment, each organization attempts to attract the consumer to its product through offering better function performance, lower price, better quality and more reliable service than the other organizations. This type of competitive loop (Fig. 3.6) represents a simple case of competition that is normally experienced in domestic markets.

In a globally competitive market, an organization is typically subject to many market forces, some of which can have a significant impact on the progress of product development and the competitive status of the organization. In addition to local competitors, the organization may be subject to foreign competitors. This imposes additional market challenges such as cultural gaps, coordination, logistics and communication issues. Trade, standards and consumer organizations will also play a larger role in the global market. These organizations are likely to coordinate their efforts to achieve fair trade, standardized measures, common product specifications

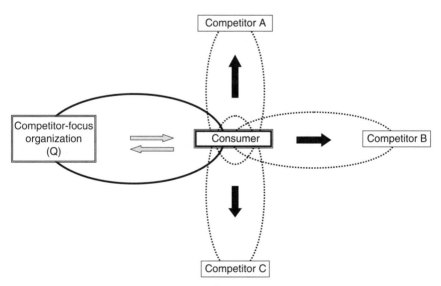

3.6 Simple competitive loop.[5]

and more knowledgeable consumers. These various forces are illustrated in the competitive loop shown in Fig. 3.7.

3.4.2 Market segmentation

In general, there are two major marketing segments: mass market and target market. In a mass market, a product is developed for a large population of consumers. Traditional fibrous products such as bed products, curtains, towels and cleaning items are developed for mass markets. These types of product are typically diverse in order to satisfy a wide range of consumers. They may also be associated with standard matching applications such as standard size beds or windows of standard dimensions. Most other traditional fibrous products are typically developed for target markets such as fashion clothing, children's clothing, sportswear and uniforms. As the market becomes more global, most products will be developed with target segments of consumers in mind. Even with similar products, geographic and cultural differences may dictate changes in product development to meet target consumers in different areas in the world.

In the face of high competition in the mass market segment, smaller organizations may find target markets as the best way to compete, grow and achieve profits through developing value-added products. Many producers of fibrous products followed this approach. A good example was the Thor-Lo, a manufacturer of athletic socks, which has turned an unglamorous commodity into a truly performance-differential product for particular target markets. In this approach, the term athletic socks, which was a

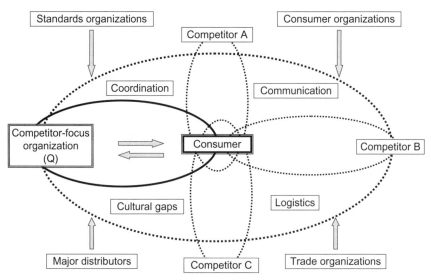

3.7 Complex competitive loop.[5]

development in itself, was replaced by the term 'foot equipment' to reflect the new design in which sport socks were made with as much terry cloth as could be packed into the sole without making it too thick in the arch. The dense packing was skillfully placed in areas where the foot takes the most bruising. In addition, Thor-Lo further segmented the market on the basis of lifestyle criteria and special needs, providing specialty socks for senior citizens, and for health care needs. Another example was the Goretex organization founded by Bob Gore (1969), who discovered that PTFE (polytetrafluoroethylene) can be expanded mechanically to form a micro-porous membrane without chemical reactions. Furthermore, only low amounts of this polymer were needed to create an airy, lattice-like struc-ture. This opened up the possibility of applying the particular properties of this polymer to many fibrous products with the idea being to transform some traditional fibrous products into function-focus products, taking advantage of the many special properties of PTFE such as extremely high resistance to chemicals and ultraviolet radiation, thermal stability (between $-250°C$ and $+280°C$) and recyclability. Now, Goretex (W. L. Gore & Asso-ciates) continues to develop and produce specific PTFE membranes for different target markets.

Companies developing function-focus fibrous products for target markets should always evaluate the market growth of these products via reliable forecasting analysis, as sudden changes are likely to occur and long-term steady growth may be highly unlikely. For example, the global market for electrically enabled smart fabrics and interactive textile (SFIT) technol-ogies was worth US $248.0 million in 2004. Forecasting analysis revealed that the growth of this market is expected to continue at an annual rate approaching 18–20% for many years to come.[14] If this trend holds, product developers should continue to seek novel polymers and innovative designs in order to develop products with interactive performance functions such as electrical conductivity, ballistic resistance and biological protection. On the cautionary side, product developers should be aware of the factors that could inhibit the growth of these products including the high price of fin-ished articles, shortage of research and development (R&D) funding and the lack of industry standards.

Obviously, the challenge associated with developing products for target markets typically stems from the need for a great deal of product custom-ization to meet specific performance needs. This typically comes with higher cost and significant effort in the product design phase. On the other hand, the main advantage of developing products for a target markets is the value-added nature of these products, which primarily stems from the intelligence consumed in product design. Organizations developing prod-ucts for a target market can also become leaders in these markets, enjoying technological monopoly for many years.

3.4.3 Market shifts

In today's global market, producers and potential customers of certain products may change locations as a result of cost considerations or increase in demand for a product in some areas in the world. Although this is largely a manufacturing and marketing issue, changes in location are typically associated with changes in many product development factors including different customer interests imposed by cultural or social differences, communications and coordination between product developers and manufacturing engineers, domestic regulations, trade regulations and environmental regulations. These factors should be considered in the product development process.

In the fiber and textile industry, market shifts have represented a common trend in recent years as evidenced by the substantial changes in the global roadmap of fiber production and textile manufacturing. As a result, the stability of product development programs in this industry has been significantly disturbed by the need to redeploy resources both geographically and strategically. For example, the USA was producing about 3.7 million metric tonnes of cotton in 2000. Over 60% of this cotton was used by US mills to produce traditional fibrous products in the local market and only 30% was exported, with the remainder being stocked. In 2006, the USA produced about 5.2 million metric tonnes of cotton, yet only 25% of this amount was used by US mills and over 70% was exported. On the other hand, China produced about 4.4 million metric tonnes of cotton in 2000, which was used entirely by Chinese textile mills to produce textiles. In the same year, China imported about 0.7 million metric tonnes of foreign cotton, which was used by local mills. In 2006, China produced about 5.7 million metric tonnes of cotton that was also used entirely by Chinese textile mills. In addition, China imported about 4.1 million metric tonnes of foreign cotton, which was used by local mills for exporting textile goods to the world. These market shifts in manufacturing locations are likely to continue owing to many factors including cheap labor and low operating costs.[15,16]

The significant shift in the production and consumption of fibrous products has resulted in noticeable changes in approaches to developing textile machinery, particularly in the spinning sector. In 1980s and early 1990s, when industrial countries such as the USA and Europe were very active in textile manufacturing, the design of this equipment focused on critical features such as automation, process control and automated material handling and transportation. As the manufacturing locations shifted to Asia and other developing countries, new lines of machinery were developed with non-automated or semi-automated features and more reliance on human handling. From a product development viewpoint, these trends may seem to represent a development setback. However, market demands and economical benefits often dictate the nature and the progress of product development.

3.5   Product-focus versus user-focus product development

Over the years, the process of product development has evolved from a pure product-focus approach to a user-focus approach. In a product-focus approach, an organization develops a product that it sees some need for in the marketplace. This need may stem from conventional use of similar products, uniqueness or from a deficiency in existing products. Input from potential users is typically considered after a new product has been developed. This approach follows closely the classic product development cycle of generating product ideas, defining performance characteristics (based on internal specifications and design tolerances), gathering relevant information, justification, performing design analysis and developing a product model or prototype. The marketing and sales division then attempts to persuade potential consumers to purchase the product through common promotional activities.

The user-focus approach to product development is primarily based on the realization that an organization selling a product to a consumer is essentially offering a service in the form of a product. As a result, one may consider the user-focus approach as a service approach in which the user or consumer dictates the merits of the outcome of product development. In this regard, an organization develops a product that it has discovered some need for through market research in which the consumer expresses a true desire for the product, perhaps through comparison with existing products in the market. This approach follows closely the dynamic product cycle discussed earlier. The product development division works closely with the manufacturing division and the marketing and sales division throughout the entire process of the making of the product. In addition, potential consumers are continuously consulted as the product progresses from the development stage to the manufacturing and finally to the marketing stage.

Another aspect of the user-focus approach, which has been largely implemented in recent years, is what can be crudely called 'playing on consumer feelings'. This is typically achieved through providing cosmetic or functional excitement to the product that attract special consumer groups (e.g. children and youth) and make them switch from an existing product to a new one. Similarly, some organizations develop products partly on the basis of creating artificial accelerated obsolescence by making design changes that intentionally tempt users to replace goods with new purchases more frequently than would be necessary as a result of normal wear and tear. This approach was originally invented by the fashion industry but now it has become a common approach in all sorts of products including electronics, automobiles and toys.

It should be pointed out that the user-focus approach to product development should not imply a consumer-defined product, as this can hinder the creativity and the imaginative ability of design engineers. It mainly implies a consumer-inspired product. This may be called 'aligning user voice and design intelligence'. A successful user-focus approach should integrate consumer interests into the product development process in such a way that preserves and highlights the inherent intelligence of the product design. For example, police officers may want a bulletproof vest that is both protective and comfortable. To convert this desire into reality, design engineers will realize the conflicting aspects of these two performance characteristics. With protection being the primary purpose of the product, the emphasis should initially be on meeting this objective and reaching the highest level of protection. Upon meeting this objective, efforts should then be made to reduce the burden of wearing a bulletproof vest for long hours and under harsh physical or environmental conditions. In other words, while user's priorities may be simultaneous in nature, developer's priorities should be functionally ordered.

Traditionally, user-focus organizations have drawn a clear distinction between consumer needs and consumer wants, with the former being products that are classed as absolutely necessary for the consumer way of life and the latter being products that are desired by consumers for complementary purposes. In today's market, this distinction is no longer appropriate for classifying products as it has become blurred by wants turning into needs as a result of the dynamic progressive change of people's lifestyle and the increase in prosperous societies. A more precise categorization of products commonly used in today's market is 'convenience goods' versus 'shopping goods'. Another less commonly used categorization is 'specialty goods' versus 'unsought goods'. These classifications largely reflect the consumer's perception of the products and not necessarily the inherent nature of the product as often described in the product development phase.[5] Indeed, some products may surprise the product developers themselves in the way they are received by consumers in the marketplace. This issue is likely to become more critical as products increasingly cross the global map through different consumer interests and different cultures.

The majority of traditional fibrous products can be classified as shopping goods rather than convenience goods. Typically, shopping goods are associated with price and quality comparison and style preference. Even with common products such as baby diapers, consumers typically go in a cycle of search for the right brand, price, size and quality. In general, shopping goods typically fall into two categories:[5] similar products and dissimilar products. For example, most home appliances may be perceived as similar shopping products. On the other hand, clothing and furniture are considered to be dissimilar shopping goods. The key, however, is making these products with quality and performance differential rather than price differential. Convenience fiber products include some uniforms, kitchen cleaning items, mops and utility towels.

3.6   Role of research in product development

The basic difference between a design approach and a scientific research approach is that the former is always driven by identification of needs, while the latter is often driven by scientific curiosity. The question then is how scientific research can fit into the product development process? To answer this question, engineers should realize that from the flood of scientific research in which scientific principles are utilized to meet research goals, many ideas can indeed be utilized in product development. A quick avenue to these ideas would be the patented research ideas, which represent a very small portion of the numerous research activities conducted daily by research institutes and universities around the globe. Many research patents have some validity for practical implementation.

The effectiveness of research in product development will largely depend on the objectives established and the duration over which research projects need to be completed. Typically, organizations that utilize effective research work will have more patents and creative ideas than they can possibly implement or transform for the real world. In addition, dynamic market changes and prohibitive cost factors normally dictate the extent of utilizing new ideas. It is important, therefore, that research outcomes be implemented carefully and within the overall objectives of product development.

3.7   References

1. DIETER G, *Engineering Design: A Material and Processing Approach*, McGraw-Hill Series in Mechanical Engineering, New York, 1983.
2. ETTLIE J E, *Idea Generation in New Product Development: Japanese Ways and Western Ways*, University of Michigan Business School, Ann Arbor, Michigan, January, 1993.
3. WASTI S N, 'Design outsourcing and product development: a study of the US and Japan Automotive Industries', *Proceedings of the 9th World Productivity Congress*, Vol. 1, Istanbul, Turkey, June 4–7, 1995.
4. SOBEK D K, *Principles that Shape Product Development Systems: A Toyota–Chrysler Comparison*, PhD Dissertation, University of Michigan, 1997.

5. EL MOGAHZY Y E, *Statistics and Quality Control for Engineers and Manufacturers: From Basic to Advanced Topics*, 2nd edition, Quality Press, Atlanta, GA, 2002.
6. BOX J, 'Extending product lifetime: prospects and opportunities', *European Journal of Marketing*, 1983, **17**, 34–49.
7. DAY G, 'The product life cycle: analysis and applications issues', *Journal of Marketing*, 1981, **45** (Autumn), 60–7.
8. LEVITT T, 'Exploit the product life cycle', *Harvard Business Review*, 1965, **43**, 81–94.
9. BLANCHARD O J and KIYOTAKI N, 'Monopolistic competition and the effects of aggregate demand', *American Economic Review*, 1987, **77**, 647–66.
10. BRAKMAN S and HEIJDRA B J, *The Monopolistic Competition Revolution in Retrospect*. Cambridge University Press, Cambridge, MA, 2004.
11. CHAMBERLIN E H, *The Theory of Monopolistic Competition*. Harvard University Press, Cambridge, MA, 1933.
12. DIXIT A K and STIGLITZ J E, 'Monopolistic competition and optimum product diversity', *American Economic Review*, 1977, **67**, 297–308.
13. DIXIT A K and STIGLITZ J E, 'Monopolistic competition and optimum product diversity: reply', *American Economic Review*, 1993, **83**, 302–4.
14. *Global Market for Smart Fabrics and Interactive Textiles*, report published by Textile Intelligence, http://www.mindbranch.com/products/R674-214.html, March, 2006.
15. EL MOGAHZY Y E, 'Cotton utilization efficiency: new ways to deal with cotton in the global competitive market', paper presented at the *8th International Conference of High Technologies in Textiles*, Cracow, Poland, Institute of Textile Architecture, Lodz, Poland, 18–20 September, 2005.
16. EL MOGAHZY Y E, 'Competitive US cotton & textile industry via engineering & scientific approaches', paper presented at the *19th EFS® Conference*, Greenville, SC, Cotton Incorporated, Cary, NC, June, 2006.

4

Textile product design

Abstract: The basic elements of product design are introduced including: (a) the product design cycle, (b) design conceptualization and (c) design analysis. When design is targeted at developing fibrous products, one of the key challenges faced is the trade-off between the fundamental design aspects of functionality and styling. The former implies meeting the intended functional purpose of a product at an optimum performance level and the latter implies satisfying a combination of appealing factors such as appearance, color and comfort. In traditional fibrous products, functionality and styling have always been considered simultaneously in the design process as a result of the expected intimacy of such products with humans. With function-focus fibrous products, fulfilling both functionality and styling requirements often represents a design challenge and an optimum trade-off between these two basic criteria may be required. This chapter discusses these critical issues.

Key words: functionality, styling, green design, product design cycle, manufacturing planning, mass manufacturing, design conceptualization, design analysis.

4.1 Product design: the core task in product development

In Chapter 3, basic concepts and tasks of product development were introduced. Among these tasks, product design was considered to be the core task in product development cycles. The importance of product design stems from the fact that it involves a great deal of imaginative and creative effort which often distinguishes one product development program from another and differentiates between products in the same category in the marketplace.[1] In addition, reoccurring problems during manufacturing or during product use are often a result of an inherent design deficiency that was not discovered in the product development stage or was overlooked owing to high cost or limited resources. It is important, therefore, to devote a great deal of discussion to the issue of product design. This discussion begins in this chapter by providing a review of basic aspects of the design process. In Chapter 5, the discussion will shift to the basic tools of design conceptualization and in Chapter 6, specific design analyses will be reviewed.

When design is targeted at developing fibrous products, one of the key challenges faced is the trade-off between two fundamental design aspects, namely functionality and styling. The former implies meeting the intended functional purpose of a product at an optimum performance level and the latter implies satisfying a combination of appealing factors such as appearance, color and comfort. With traditional fibrous products, functionality and styling have always been considered simultaneously in the design process as a result of the expected intimacy of such products with humans. Indeed, the distinction between functionality and styling in these products is often difficult as some characteristics, such as comfort and appearance, can be considered both functional and styling features of these products.

With function-focused fibrous products, fulfilling both functionality and styling requirements often represents a design challenge and an optimum trade-off between these two basic criteria may be required. For example, consumers expect a seat cover to last the lifetime of a car with a minimum or no maintenance required. This is a functional aspect that requires fabrics of high abrasion resistance and low UV degradation. On the other hand, an uncomfortable seat may result in an unsatisfactory perception of the entire car. From a design viewpoint, achieving high abrasion resistance may result in excluding fiber types that can provide good comfort and appearance (e.g. cotton, viscose rayon and lyocell) owing to their poor abrasion resistance. When cost is added to the design formulae, wool fiber may also be excluded despite its high abrasion resistance and great thermal comfort, except for up-market cars. When the focus is shifted to low UV degradation, acrylic fibers may seem to be the best choice. However, its abrasion resistance is not as good as other synthetic materials.[2] This leaves the design engineer with fewer fiber options with which simultaneously to satisfy functionality and styling performance. In this regard, the common option is to use polyester fibers owing to their high abrasion resistance and moderate UV degradation resistance, which can be further improved by using UV light-absorbing chemicals. The design engineer may also have to rely on the glass in car windows to provide further support by filtering out the UV light radiation.[2] In addition to the selection of an appropriate fiber material, styling features can also be enhanced through the design of a fabric structure that exhibits appealing patterns and colors and provides good fabric hand.[3]

The styling aspect of design has become a major selling point in modern products, from automobiles with different looks and styles, to elegantly looking computers, mobile telephones and other household products. In addition, a wide variety of design attractions, from the use of recycled materials (green design) to the spread of organic products, is often introduced to reflect the need for optimum utilization of natural resources and a healthier environment. Moreover, nanotechnology and wireless engineering came as a timely contribution to both functionality and elegance, as products such as hand-held electronics and tiny audio equipment have become a commercial reality. Ultimately, nanotechnology will result in the continued miniaturization of new technology and wireless engineering will transform the world into human network systems in which the workplace will be connected to the home and people will be wirelessly connected to each other.

Although the objective of design is to develop a product model, the ultimate goal is to transform this model into a commercial product through massive manufacturing of product units. This means that a design process should account not only for product structure and optimum performance but also for product manufacturability.[4–6] As a result, manufacturing engineering should be implemented in order to reach the most acceptable product in the marketplace. As indicated in Chapter 3, manufacturing engineering represents a host of activities that are involved in the creation and operation of the technical and economic processes that convert raw materials, energy and purchased items into components for sale to other manufacturers or into end products for sale to the public.[4] Manufacturing engineers coordinate their efforts with design engineers in the process of modifying product designs for the assurance of product manufacturability.

4.2   The product design cycle

The use of cycles to illustrate engineering tasks is a common approach that stems from the need for an iterative procedure to reach compromising solutions to engineering problems. In performing design analysis, a product design cycle will be necessary particularly in view of the fact that most design problems are open ended in nature.[4] A universal or standard product design cycle cannot be established since the design approach may vary depending on many factors including the company's philosophy, the product type, the availability of resources, the accessibility of relevant information and personnel qualification. However, common requirements and tasks used in all product design cycles are shown in Fig. 4.1.

As can be seen in Fig. 4.1, the foundation of a product design cycle consists of two key aspects: design planning and design conceptualization. Design planning implies making the basic requirements of a design project available prior to entering a design cycle. These include qualified personnel (e.g. design engineers, representatives of manufacturing engineers and representative of marketing personnel), hardware and software, information sources and natural resources. The second foundation of a product design cycle is design conceptualization. This is the process of generating ideas for an optimum solution to the design problem. As will be discussed later, design conceptualization is the most critical aspect of design as it determines whether a product idea is justifiable in view of functional merit,

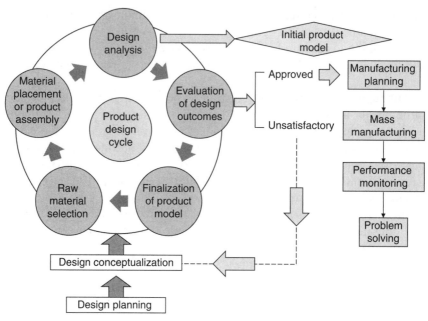

4.1 Key tasks in a product design cycle.

cost and value, with the results being to proceed with the design project, modify the product idea or terminate the project. An introduction to design conceptualization will be presented later in this chapter and the different tools required to perform design conceptualization will be discussed in Chapter 5. If the outcome of design conceptualization is to proceed with the design project, the product design cycle is then implemented, as shown in Fig. 4.1.

The common task sequence of a product design cycle is as follows: raw material selection, material placement or product assembly, design analysis, evaluation of design outcomes, finalization of product model and communication with manufacturing engineers regarding the final product model. The task of raw material selection and placement will largely depend on whether the intended product is a modification of an existing one or a totally new product. Typically, the latter will involve a greater effort than the former. In Chapter 7, the subject of raw material selection will be discussed in detail. Material placement or product assembly is first initiated on the basis of previous experience and finalized after performing design analysis.

The key aspect of design analysis is the choice of the appropriate analytical and modeling tools required to explore the various factors influencing product performance and to obtain an optimum solution leading to a product model. In Chapter 6, many design analytical tools will be discussed. Obviously, the type of analytical tool required will vary with the type of product and its intended functional performance. A design analysis will yield an initial product model that should be tested and evaluated to determine whether it should be finalized or modified through an iterative analysis in which the design cycle is revisited, perhaps many times. The evaluation of an initial product model can be performed either by the original design team or by external experts representing other design engineers, manufacturing engineers and marketing personnel. If this evaluation reveals an acceptable product model, the design team will then proceed by finalizing the product model. If the results are unsatisfactory, further analysis should be made in which the design cycle is repeated, perhaps with consideration of alternative raw materials, different product assemblies or different design approaches. This cycle may continue until a satisfactory model is developed. In Chapter 5, detailed discussion of the methods of evaluating design outcomes will be presented in the context of design conceptualization.

For simple products (e.g. simple constructions, or pre-existing products that require slight modifications), the design cycle will be typically short and it may be completed in one or two iterative loops at the end of which the product model is finalized and handed to the manufacturing process. For newly developed products, the design cycle can be long and many loops of the iterative procedure may be required to reach optimum product performance at a minimum cost.

The points above indicate that product design cycles should be carefully implemented in view of the anticipated complexity of the intended product. In this regard, simple design cycles should be handled by a design team that operates on the basis of having a limited time frame to complete the cycle in a minimum period of time. This team will obviously have long experience of the nature of the intended product (e.g. strengths and weaknesses, specific alterations, customer complaints and competitive advantages or disadvantages). When a design cycle is implemented for a product that has not been developed by the same organization in the past, or for a totally innovative idea, a special design team should be formed to undertake this task. Typically, this team operates on the basis of a longer time frame, dictated by a slower learning curve associated with the new innovation. Later in this chapter, this point will be illustrated further using the concept of 'resource–time elapse profile'.

In all situations, the final product model is handed to manufacturing engineers to determine the key manufacturing aspects associated with the product. These include manufacturing planning, mass manufacturing, performance monitoring and problem solving of quality issues. These tasks are performed through continuous coordination between the design team and manufacturing engineers.[5,6]

4.3   Design conceptualization

As indicated above, design conceptualization is the process of generating ideas for an optimum solution to the design problem. These ideas should stem originally from the product idea and stated definitions of the design problem. In most situations, product ideas are inspired by some needs or wants expressed by people, often in non-specific terms, as a sought-for goal. For example, the continuous rise in oil price has always inspired ideas for alternative energy sources and the limitation in natural resources such as fibers and soils have always inspired ideas for developing alternative synthetic materials. Sometimes the need is inspired by a deficiency in an existing product. For example, a bulletproof vest may be highly protective but it may also lack comfort as a result of heavy weight imposed by fabric construction and additional inserts. In this case, the design of an alternative product will be associated with a set of problems that reflect both the product idea, being a protective and comfortable bulletproof vest, and the need to overcome the discomfort problem. This is where design conceptualization becomes a necessity since a trade-off must be achieved between protection and comfort to obtain an optimum solution rather than an absolute best solution.

The process of design conceptualization requires a great deal of brainstorming to examine the merits and the feasibility of a product idea against physical, economic and functional factors. For example, if alternative resources to the existing ones impose high cost, design conceptualization will aim at ways to reduce the cost or to justify the added cost in view of long-term benefits. In this regard, issues such as value, cost, price and liability should be accounted for in design conceptualization.[7]

In light of the above discussion, a number of key steps should be taken before reaching the conceptualization stage of design. These steps (illustrated in Fig. 4.2 and discussed below) are justification, problem definition and information gathering.

4.3.1   Is the product justifiable?

Product justification is typically a function of four key factors: (a) consumer-added value reflected in the specific needs or wants for the product, (b) potential users or customers for the product (user category, size, income, culture, location), (c) producer-added value (profit, image, reputation, etc.) and (d) regulations and liability (marketing, safety and environmental aspects). These factors must be evaluated carefully to justify further effort toward designing a product.

The consumer-added value of any product can be determined via scientific research that aims to explore technical factors or market-oriented

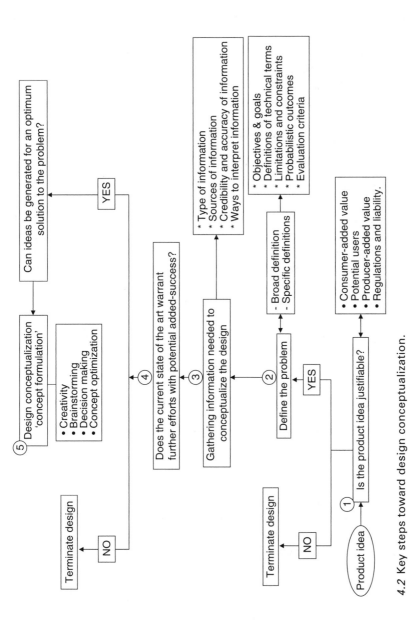

⑤ Design conceptualization 'concept formulation'
- Creativity
- Brainstorming
- Decision making
- Concept optimization

Can ideas be generated for an optimum solution to the problem?

YES

NO → Terminate design

④ Does the current state of the art warrant further efforts with potential added-success?

* Type of information
* Sources of information
* Credibility and accuracy of information
* Ways to interpret information

③ Gathering information needed to conceptualize the design

- Broad definition
- Specific definitions

* Objectives & goals
* Definitions of technical terms
* Limitations and constraints
* Probabilistic outcomes
* Evaluation criteria

② Define the problem

YES

NO → Terminate design

① Is the product idea justifiable?
- Consumer-added value
- Potential users
- Producer-added value
- Regulations and liability.

Product idea

4.2 Key steps toward design conceptualization.

factors associated with the intended product. Indeed, product justification represents a great opportunity for linking design analysis with scientific research in a product development program. In some situations, it will be useful to conduct scientific research to test potential materials and examine the effects of some factors on the intended product. When existing products are used as a reference for developing new products, scientific research can explore the capabilities of existing products using system identification analysis.[8] Market research is conducted to evaluate the dynamic nature of consumer behavior and product acceptability. As indicated in Chapter 3, addressing marketing issues such as competitiveness, market segmentation and market shifts can provide useful guidelines in justifying product ideas.

Any product idea should be partly justified in view of its potential user. In this regard, the key issue is whether the product represents a common consumer product that will potentially be used by the vast majority of people, or it may be directed toward a specialty application or a market niche. The justification aspect here stems from the fact that depending on the potential user of a product, the cost of developing and supporting a product may vary substantially. In general, different users may exhibit different qualifications that are reflected in their handling of the product. For common products, the wide range of user qualification may result in a wide range of product handling; from perfect use to misuse or even accidental abuse of the product. This point has typically been overlooked in traditional design philosophies owing to the focus of the design process strictly on the specific functions of the product. In today's market, a misuse or accidental abuse of a product by a consumer could be blamed on the product developer, on the ground that an appropriately designed product should be easy to use and should be robust against potential misuse.[8]

The producer-added value represents the entitlement of the producer to make a profit for the effort made in developing the product. When a product is used by a large number of consumers, this value is reflected in the sale of a large number of product items that make a large profit as a result of massive sales. This is typically the case for traditional fibrous products such as apparel and furnishing. Products developed for specialty applications (e.g. function-focus fibrous products) are normally made in small quantities, but they are justified on the basis of their high value in the marketplace.

Regulations and liability are often considered in product planning or marketing. However, some regulations can have a direct impact on the design concept. Certainly, a product design that does not account for market regulations will not find its way into the market. Different products may be associated with different types of regulation depending on the type of product and its potential users. Most traditional fibrous products are associated with regulations dealing with their maintenance aspects such as washing, drying and handling. Companies that design clothing for children must be fully aware of the safety precautionary rules that are established to assure safe and appropriate use of this product. For example, the US Consumer Product Safety Commission sets national safety standards for children's sleepwear flammability. These standards are designed to protect children from burn injuries if they come into contact with an open flame, such as a match or stove burner. Under amended federal safety rules, garments sold as children's sleepwear for sizes larger than nine months must be flame resistant or snug-fitting garments. The latter need not be flame resistant because they are made to fit closely against a child's body. Snug-fitting sleepwear does not ignite easily and, even if ignited, it does not burn readily because there is little oxygen to feed a fire.

For function-focused fibrous products, numerous rules and regulations are established to insure reliability, durability and safety. These are updated and published regularly by many organizations. For example, fabric requirements for car interior safety are regulated by the Federal Motor Vehicle Safety Standards and Regulations established by the US Department of Transportation or the National Highway Traffic Safety Assurance-Office of Vehicle Safety Compliance. Bulletproof vests are regulated by the Justice Department Federal Safety Guidelines Body Armor Safety Initiatives. Protective clothing systems are regulated by many organizations including the Occupational Safety and Health Administration (OSHA) and the Environmental Protection Agency (EPA). Medical products are divided into broad categories such as hygienic absorbent products, hospital and healthcare products, implants and scaffolds. Because of their significant impact on human health and safety, they are produced according to strict regulations and legislation. These should be designed-in and not just considered via product instructions. Products involving toxic chemicals or fire hazards are also regulated. In this case, the regulations are clearly highlighted on the product package.

Sometimes regulations can restrict some potential users from accessing the product for their own safety or for the safety of others. Many medicines, for example, are only accessed via prescription. In the case of bulletproof vests, regulations and laws make it unlawful for criminals or people that are convicted of a felony to acquire or to use bulletproof vests (United States Law 18USC931).

4.3.2 Define the design problem

The term 'problem' is commonly used to imply a negative situation, caused by some unpredictable, or predictable factors, which requires immediate solution. In engineering design, this term reflects a question raised for inquiry, consideration or solution. Accordingly, every design project is associated with a problem or perhaps a set of problems that must be solved to reach a product with optimum performance, with 'optimum' meaning the amount or degree of something that is most favorable to some end, or the greatest degree attained or attainable under implied or specified conditions.

In general, a design process is often viewed as a series of discrete problems that must be solved, one after the other, to reach an optimum or semi-optimum solution. Unlike systematic problems that are typically experienced in manufacturing or in handling customer complaints, a design problem is often not 'a problem to react to', but rather 'a problem to anticipate and attack'. In other words, a design problem represents a statement of challenging aspects associated with a proposed product idea.

The key to reaching an optimum solution to a design problem is to define the problem clearly and specifically. In a typical design process, one can divide problem definitions into two main categories: broad definitions and specific definitions. The order of these two categories is not important. A design team may commence with a broad definition of the problem and break it down into more specific definitions of the problem. On the other hand, a design team with long experience of the product type may begin by establishing specific definitions of a number of problems associated with the product design and then integrate these to form a broader or a collective definition.

A broad definition of a design problem may be initiated by establishing a general description of the proposed product. For example, a bulletproof vest is generally described as an article of protective clothing that works as a form of armor to minimize injury from projectiles fired from handguns, shotguns and rifles. This broad description serves as a generalized concept of the product which may result in hundreds of product options and alternatives. A specific definition of the product is typically a derivative of the broad definition that deals with more specific aspects. For example, a more specific definition of the design problem of the bulletproof vest would be to specify the level of protection expected; say, against 9 mm full metal jacketed round nose bullets, with nominal masses of 8.0 g or 124 gr (grain), impacting at a maximum velocity of 427 m s^{-1}. This is a performance-related definition. Another specific definition may be stated as 'a three-layer garment construction, with protective inserts that exceed the overall fabric weight by no more than 20%, with garment layers being reducible to accommodate less risky situations'. This is a structural-related definition.

Under time-to-market intense pressure, the phase of defining the problem is often minimized, reduced to one or two problems, or strictly applied to few specific performance aspects of the product. As a result, some products find their way to the marketplace with problems that have not been fully defined at the design stage. On the other hand, dwelling on the broad definition for so long may slow down the design process and result in an unnecessary or costly over design. The key to overcoming these issues is to make all relevant information available to the design team.

Failure to establish a broad definition of the problem can be fatal in some situations. The design of safety automobile airbags discussed in Chapter 2 clearly demonstrates this point. In this product, specific definitions of the problem were well established and optimum solutions were provided. However, surrounding factors that are outside the specific functional boundaries of the product were largely neglected. As a result, safety problems occurred after the product had been used in the marketplace for quite some time.

With regard to the specific definition of the problem, this should also be established with full awareness of the surrounding and probabilistic factors that may come into play even in the most remote situations. One example that best demonstrates this point is the supersonic Concorde airplane. This giant product was commercially introduced in the mid-1970s. It was a result of a joint effort by companies representing major industrial countries in France and the UK, with the idea being a very fast (under four hours flight time between New York and Paris) and comfortable (no atmospheric turbulence at the very high altitude, 18 000 m, at which it flies) flight. Concorde's take off speed was 397 km h^{-1} compared to a typical subsonic aircraft takeoff speed of 300 km h^{-1}. This aircraft typically accelerated to supersonic speed over the ocean in order to avoid a sonic boom over populated areas. It also flew at 2179 km h^{-1} (2.04 mach), which was just over twice the speed of sound, and the range of the aircraft was about 6580 km.

The Concorde was associated with many specific problems that required accurate definitions and specific solutions. Two of these problems were (a) the ability of the fuel tank to withstand an external impact hit without rupture and (b) the puncture resistance of the tires by sharp metal objects. These two problems represented unusual situations that rarely occur under normal take-off, landing or flying conditions. The relationship between these two problems is that the tires on the Concorde were in dangerously close proximity to the engines and engine inlets as well as to the fuel tanks. Unusual as they may seem, the failure to define these specific problems in the design phase resulted in a major disaster that eventually led to the end of the Concorde aircraft business. On Tuesday, 25 July 2000 the very first fatal accident involving Concorde occurred involving Concorde 203, F-BTSC out bound from Paris to New York. It crashed 60 seconds after take off as a result of suffering a tire blow out that caused a fuel tank to rupture. All 100 passengers and nine crew on board were killed. Four people in a local hotel on the ground were also killed.

One of the reported causes of the Concorde crash was that as the plane was on its take-off run, a piece of metal (a titanium strip that fell from another aircraft) punctured the tires which then burst, puncturing the fuel tanks and leading to the aircraft flying in flames. After the crash, some design efforts were made to correct these problems. These included more secure electrical controls, Kevlar lining for the fuel tanks and specially developed, burst-resistant tires. Unfortunately, these efforts failed to recover passengers' trust in the Concorde aircraft and in May 2003, the last Concorde flew from New York to Paris.

The examples mentioned above clearly demonstrate the importance of establishing broad and specific definitions of the design problem. They also demonstrate a key aspect of design, which is failure analysis, which will be discussed in Chapter 6. The design process of a justifiable product idea should involve a problem statement, which expresses what the design is intended to accomplish. Critical points in this statement include objectives, goals, definitions of technical terms, limitations and constraints associated with the design, possible probabilistic outcomes, and criteria that will be used to evaluate the design outcome. Initially, these points may be stated on the basis of the experience of the design team. As more information is gathered and perhaps after the second or third iteration, a more detailed problem statement should be made.

4.3.3 Gather relevant information

In Chapter 2, the general relationship between problem solving and knowledge gain was discussed. In Chapter 3, the importance of information gathering was also discussed in the context of product development. In general, a solution of a design problem requires special tools that may take their effectiveness from knowledge of basic aspects in physics, chemistry and mathematics and their extensions into materials science, solid and fluid mechanics, thermodynamics, transfer and rate processes, and systems analysis. In addition, each design process requires the gathering of specific information related to the problem in hand. The amount of information needed will depend on the degree of complexity of the problem and the extent of previous background knowledge of the nature of the intended product. Obviously, the more complex the problem is, the more information will be required to handle the problem.

When the design goal is to modify an existing product in order to achieve better performance, a great deal of information should already be available and the task in this case will be to seek the most relevant, most specific and most current information. Sources of this information may include technical reports of sponsored projects, company technical documents, patents, supplier catalogs and trade journals. Another critical source of information is investigative reports associated with failure or deficiency in pre-existing products. On the other hand, when the intended product is a result of a new innovation, information may be limited and expert inputs may have to compensate for the lack of information.

In seeking information relevant to the design problem, an 'information priority list' should be made to insure that the information gathered will be efficiently useful. The broad and specific definitions of the problem should represent the driving issues in this list. In other words, the problem definition statement should guide the design team to the type of information to be gathered and, in turn, to the sources of information. Other key points on this list will include the credibility and accuracy of information and ways to interpret the information.

4.3.4 Design concept formulation

As shown in Fig. 4.2, the previous steps lead to a key question which is 'Does the current state of the art warrant further efforts with potential added-success'? Addressing this question provides additional justification for the merits of the design project. Unlike the first justification step in which participants included engineers, manufacturers and marketing personnel, this question is addressed by the design engineers and supported by relevant information and a clear definition of the design problem. A 'no' answer to this question may result in terminating the design project. A 'yes' answer, on the other hand, would mean that the product idea is justifiable, clearly defined, and current information is available and indeed warrants further effort for potential added-success. This leads to the conceptualization step, in which the key question is 'Can ideas be generated for an optimum solution to the problem'? It is at this stage of product design that different design teams may follow different approaches depending on the level of creativity and imaginative talent of the team. The key difference will be how the design team forms the design concept. In this regard, four basic problem solving and mind tools can be used to assist engineers in design conceptualization projects. These are creativity, brainstorming, decision making and concept optimization methods. In Chapter 5, these tools will be discussed in detail.

It should be pointed out that forming a design concept does not necessarily mean finding a final solution; it is rather finding an idea for a solution or a direction of thought, as some design problems may be associated with numerous solution ideas and others may indeed have limited or no apparent solutions. For example, if the proposed idea is to design a military uniform that can be used under various environmental conditions (from hot to cold and from dry to wet), provided that a need exists for such a product and the multiplicity of problems associated with achieving these conflicting conditions are defined, the design concept may be formulated in many different ways. One concept would be perhaps to think about multiple fabric layers; another is to think in terms of impregnating smart polymeric components that can react to surrounding temperature changes; and another is to think about utilizing a special fiber blend supported by unique yarn and fabric structural features. These are not specific solutions as they only represent design concepts expressed in broad terms which can be followed so that analytical efforts can be directed toward fulfilling them in part or in full. Most design concepts are expressed in qualitative terms with minimum constraints associated with cost, resources availability or even absolute realism. As Robert Mann described it,[9] 'concepts should be considered as possible solutions to engineering challenges that arise initially as mental images which are recorded first as sketches or notes and then successively tested, refined, organized, and ultimately documented by using standardized formats'.

The process of forming a design concept often results in revisiting the preparatory tasks discussed earlier, as new justifications may surface and more important problems may arise, requiring the establishment of more specific definitions. It is also possible that through the brainstorming process of design conceptualization, the design team finds themselves lacking important pieces of information that must be acquired to continue with the process. The point to be emphasized here is that the design process is not a linear or one-way process. It is an iteration process that should be implemented in the most flexible fashion, yet constrained by time and financial resources.

4.4   Design analysis

In view of the definitions of engineering design presented in Chapters 2 and 3, engineering design is the creation of processes, systems and instruments that are useful for humanity. The term 'creation' here implies a combination of art and creative work (mostly implemented in design conceptualization), as well as standard analytical procedures to meet design goals. Most design analyses represent iterative procedures as the search for optimum solutions (e.g. the best functional performance at the lowest cost) represents the core of any design analysis. In today's computer era, numerous analytical tools can be used in design analysis. These include modeling and optimization techniques, statistical analyses, neural network, genetic or evolutionary algorithms, simulated annealing and finite element analysis. In Chapter 6, the principles underlying these tools will be reviewed and their merits in design applications will be highlighted.

Design iterative analysis typically reveals a sequence of progressively specific outcomes which, depending on the product complexity, may represent a series of discrete partial, temporary, or interim solutions. These outcomes are compared with one another to reach the best practical solution. This process involves geometrical sketches of product structure, material type and specifications, different parts that will be needed to assemble the product, machine type and settings to be used for manufacturing the product, technological information, and possibly logistics and production control information.

Again, it should be emphasized that design analysis aims to provide the best compromising solution and not necessarily the best optimum solution. In this regard, the limitations that typically face design engineers in reaching an ideal or near perfect solution include the costs associated with time and resources consumed in reaching optimum solutions and the ability to test and validate each solution. In some situations where physical knowledge of all factors associated with the best solution is inadequate, the analytical effort is reduced or supported by experimental analyses and empirical modeling to reveal actual performance outcomes of the product model.

The issue of reaching an optimum solution or a best compromising solution constrained by time and budgetary limitations brings about the issue of the relative time contribution of the design process to the overall product development process. For truly new and creative ideas, the design phase may indeed take a significant period of the time of the total duration period of product development. When only modification of an existing product is required to develop a new and improved version of the product, the design process may consume less time, as physical knowledge of the current product performance and realization of the necessary modifications are typically adequate to perform efficient analysis. In this regard, the so called 'resource–time elapse profile' should be used to evaluate the relative time contribution of the design process to the overall product development process.[9]

As shown in Fig. 4.3, a resource–time elapse profile clearly describes the elapse of time and expenditure of worker effort in the evolution of a product development project. In this regard, design time duration is divided into two periods: the period for design conceptualization and the period for design analysis. The vertical axis represents the available resources. In the conceptualization phase, these are mostly human resources[10] that are typically available within an organization. Over the years, the time required to perform design analysis has been reduced significantly owing to the availability of computing power and capable software programs. However, this process still requires many more tasks and significant resources than that of design conceptualization. Typical tasks include trying different materials, testing and validation, product model assembly, and so on, but

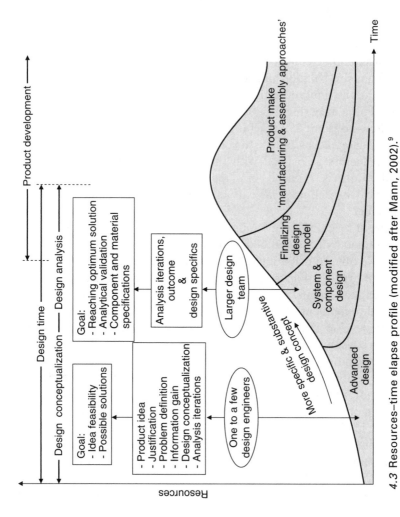

4.3 Resources–time elapse profile (modified after Mann, 2002).[9]

the most significant time is that consumed in finding the optimum solutions. Organizations with higher resource curves will be likely to meet the design objectives in less time than those with lower resource curves. By the time the product model is at its final stage and as the product enters the manufacturing phase, resources are typically at their maximum level. Thus, the merit of implementing a resource–time elapse profile is to establish a time frame for meeting the task of design with respect to the total time consumed in product development.

4.5   References

1. ULRICH K T and EPPINGER S D, *Product Design and Development*, 4th edition, McGraw-Hill Education–Europe, 1995.
2. FUNG W and HARDCASTLE M, *Textiles in Automotive Engineering*, Woodhead Publishing Limited, Cambridge, UK, 2001.
3. BEHERY H M (ed), *Effect of Mechanical and Physical Properties on Fabric Hand*, Woodhead Publishing Limited, Cambridge, UK, 2005.
4. DIETER G, *Engineering Design: A Material and Processing Approach*, McGraw-Hill Series in Mechanical Engineering, New York, 1983.
5. NEVINS J L, *Manufacturing Engineering*, in AccessScience@McGraw-Hill, http://www.accessscience.com, DOI 10.1036/1097-8542.404100, last modified: May 4, 2001.
6. BRALLA J G (ed), *Handbook of Product Design for Manufacturing: a Practical Guide to Low-cost Production*, McGraw-Hill, New York, 1986.
7. LEECH D J and TURNER B T, *Engineering Design for Profit*, Ellis Horwood, New York, 1985.
8. EL MOGAHZY Y E, *Statistics and Quality Control for Engineers and Manufacturers: From Basic to Advanced Topics*, 2nd edition, Quality Press, 2002.
9. MANN R W, *Engineering Design*, in AccessScience@McGraw-Hill, http://www.accessscience.com, DOI 10.1036/1097-8542.233800, last modified: January 11, 2002.
10. ERNEST J, *Human Factors in Engineering and Design*, 4th edition, McGraw-Hill, New York, 1976.

5

Textile product design conceptualization: basic elements and tools

Abstract: Design conceptualization is the process of generating ideas for an optimum solution to the design problem. These ideas should stem originally from the product idea and stated definitions of the design problem. In most situations, product ideas are inspired by some need or want expressed by people, often in non-specific terms, as a sought-for goal. This chapter focuses entirely on design conceptualization. It begins with general guidelines of design conceptualization including simplicity, support, familiarity, encouragement and safety. It then shifts to the discussion of basic tools of design conceptualization. These include (a) creativity tools, (b) brainstorming, (c) decision making and (d) concept optimization.

Key words: creativity, brainstorming, decision making, concept optimization, decision theory, state of nature, payoff table, decision tree, Pascal's theory, Bayes' theorem, resources–time elapse profile.

5.1 Basic differences between design conceptualization and design analysis

As indicated in Chapter 4, the design process consists of two basic tasks: design conceptualization and design analysis. The former is the process of generating ideas for an optimum solution to the design problem and the latter represents the specific analytical tools and procedures required to satisfy the design concepts and meet design objectives. A list of the differences between these two design tasks is presented in Table 5.1.

One of the key differences is the qualification required for engineers who formulate design concepts and those who perform design analysis. Typically, concept formulators are talented individuals distinguished by spontaneous creative thinking and problem-solving skills. A good background in mathematical and natural sciences can greatly enhance people's creativity. However, creativity is not a scientific discipline that can be taught in conventional education. It is rather a result of a combination of different thinking approaches supported by realization of possibilities and alternatives that is derived from experience, study and special talents. Design analysis, on the other hand, requires good knowledge of fundamental and applied mathematics. This qualification is earned in many scientific

Table 5.1 Basic differences between design conceptualization and design analysis

Aspect	Design conceptualization	Design analysis
Goal	Generating ideas for an optimum solution to the design problem	Performing appropriate analysis to satisfy the concepts established
Qualifications	1 Creativity 2 Great communication skills 3 Experience in problem-solving and decision-making approaches 4 Knowledge of mathematical and natural sciences	1 Good mathematical background 2 Great communication skills 3 Good computer background (particularly software programming)
Outcome	1 Idea(s) for solutions 2 Problem definitions 3 Justification points 4 Critical design parameters 5 Possible constraints	1 Product model 2 Optimum solution(s)
Tools	1 Creativity tools 2 Brainstorming tools 3 Decision-making tools 4 Concept optimization tools	1 Modeling analysis 2 Optimization techniques 3 Specific analytical tools
Conclusion	Mostly qualitative	Mostly quantitative

and engineering disciplines. In addition, good knowledge of powerful software programs and high skills in computer programming represent important qualifications in design analysis.

Engineers involved in design analysis and design conceptualization should have good communication skills. Any engineering subject (e.g. applied mathematics, stress analysis, thermodynamics, fluid mechanics, etc.) is typically associated with familiar terminologies and universal symbols that are familiar to most engineers. Engineers of the same discipline or of different disciplines typically communicate using these terminologies.

Different engineering fields may also use different terminologies that are useful for their unique applications. In the area of fiber-to-fabric engineering, many terminologies and unconventional units are used to describe some of the characteristics of fibers and fibrous assemblies that are difficult to measure using conventional methods. For example, fiber or yarn fineness is expressed in 'mass per unit length' instead of direct diameter units (e.g. millimeter or centimeter). This is a result of the difficulty of obtaining direct microscopic measures of fiber or yarn diameter and the high variability of these measures.[1] As a result, engineers dealing with fibers or yarns should get familiar with fineness measures such as the 'tex' expressed as mass in grams of 1 km of fiber or yarn, or 'denier' expressed as mass in grams of 9000 m. In characterizing the strength of these materials, the conventional engineering stress (e.g. force per unit area) is also replaced by the so-called specific stress, expressed in 'cN/tex' or 'cN/denier'. Another example of a characteristic that is difficult to measure directly is the bulk density. This is due to the significant amount of air entrapped in fibrous assemblies. As a result, conventional measures of bulk density such as 'g cm^{-3}' are not commonly used. Instead, area density (gram per square meter or oz per square yard) is used for fabric, assuming a two-dimensional structure. When the third dimension is considered, the thickness cannot be expressed with any degree of accuracy without consideration of the lateral pressure under which it was measured. In addition, the air entrapped inside a yarn or a fabric structure calls for unique terminologies such as yarn packing density and fabric cover factor (fiber/air ratio). In Chapters 9 and 10, common measures and terminologies associated with yarns and fabrics will be discussed.

It is important to understand that communication is not only an issue of appropriate information exchange but also a key to efficient design projects, as time-to-market pressure will always be a significant challenge. When fiber and polymer engineers work jointly with engineers of other disciplines, it is important to establish effective and efficient communication tools. This means that both conventional and unconventional terminologies should be clearly communicated and ways to transform one type of unit to another should be established.

Another key difference between design conceptualization and design analysis is the type of outcome revealed by each task. Typically, design conceptualization yields ideas for solutions supported by carefully stated problem definitions, justification points, critical design parameters and possible constraints. The role of design analysis is then to entertain these ideas using appropriate analytical approaches so that a product model of optimum performance can be developed.

The differences discussed above and listed in Table 5.1 require special tools to be used by concept formulators and design analysts. In this chapter, the supporting tools required for design conceptualization are discussed. These are mainly problem-solving and mind tools. In Chapter 6, basic tools for design analysis will be discussed. These include modeling, simulation and optimization tools.

5.2   General guidelines for design conceptualization

Before proceeding with the discussion of the subject of this chapter, it is important to point out some of the general guidelines that should be taken into consideration in the design conceptualization phase.[2,3] These are summarized below.

Simplicity: A key aspect of design is that no matter how complicated the design analysis may be it should ultimately result in great simplicity from a product's user viewpoint. This means that the final product model should be associated with simple and straightforward usability. The importance of this principle is that engineers often focus on the product function and pay minimum attention to the usability aspect. These are two different attributes of a product that should be addressed in design conceptualization. For example, the function of an E-fibrous jacket equipped with some nano-medical sensors is to monitor some body functions such as blood pressure, sweat rate or heartbeat. Reaching these functions would require a great deal of sophisticated work. However, from a user's viewpoint, easy handling and simple access to the output (signals or measures) should be achieved in the design process. This requires a well-organized interface that supports the user's benefits with the highest possible efficiency. In other words, the user should gain the benefits of the product with minimum interaction and minimum interference with the product function.

Support: Sometimes the notion of a 'correct' engineering sequence using a particular product can be different from the sequence that would naturally be followed by a typical user of the product. In this type of situation, it is important either to adopt the user's sequence to meet the product functions or clearly to guide the user to the appropriate sequence. In other words, the user should be in control of the product and any design assistance should take the shape of a proactive and not a reactive approach. This support is often critical, particularly in products that require easy utilization in a very short period of time; a classic example of this is the use of a parachute to jump off a helicopter.

Familiarity: One of the key aspects of design conceptualization is to build on the users' prior knowledge of a product, which was gained from experience with similar products. The user of a product should not have to learn too many new things to perform familiar tasks. This is particularly true when the product in question is an extension or a modified version of an existing one.

Encouragement: Predictability is a key aspect of design concepts. In other words, a user's interaction with a product should yield expected or predictable outcomes. This principle requires a design engineer to understand clearly the user's task experience. This will result in an increase in the user's confidence in using and exploring the product capabilities without fear of adverse consequences. In this regard, a lesson to be learned from the user–computer interface, is to attempt to design products in such a way that users' actions do not result in irreversible consequences. In the design of physical products, this may require building a system in which tasks are coordinated but that none are so dependent on one another that performing one task can lead to adverse side effects in performing another.

Safety: The concept of safety deals not only with the adverse impact that the normal use of a product may cause to the user as a result of safety-hazard design, but also with what some unintentional errors or confusion during the use of a product may cause. This is partly a simplicity issue and partly a design robustness issue. In other words, the product should be robust against misuse and accidental abuse by the potential users.

5.3   Basic tools of design conceptualization

In Chapter 4, the key steps toward design conceptualization were discussed (see Fig. 4.2). These steps addressed key points such as product justification, problem statement, knowledge gain through gathering relevant information and realization of the state of the art. These steps lead to the key conceptualization question: 'Can ideas be generated for an optimum solution to the problem?' This question leads us to the task of design conceptualization or the task of generating ideas or concepts for an optimum solution to the problem in hand. In general, there are no universal procedures to perform design conceptualization. However, there are some basic problem-solving and mind tools that can be used to assist engineers in formulating design concepts. These tools are creativity, brainstorming, decision making and concept optimization. Since these tools also reflect key interactive criteria for engineers involved in design conceptualization, the order of their implementation is not important. Indeed, the four tools can be considered simultaneously at any stage of design conceptualization. For this reason, they may be collectively called the design conceptualization circle as shown in Fig. 5.1.

5.3.1   Creativity: concepts and misconceptions

In general, there are two different types of creativity: artistic creativity and technical creativity. Artistic creativity is more born of individual talents and skills that are difficult, if not impossible, to be completely cloned into another individual. Technical creativity is the ability to develop new concepts and create new theories and new technologies. This is the type of creativity that is emphasized in this chapter.

Any company developing new products would like to have as many creative individuals on board as possible. Indeed, people are often surprised by questions in job interviews such as 'What is creativity?', 'Are you a creative individual?' or 'What creative work have you done?'. These are the types of question to which a company is likely to receive as many different answers as the number of people being interviewed. What truly constitutes

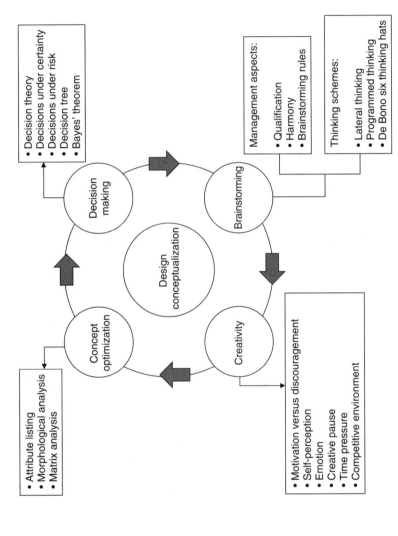

5.1 The design conceptualization circle.

creativity has been the subject of many learning studies over the years, particularly in the areas of critical thinking, epistemology and cognitive psychology.[4–6] These studies aimed to search for the meaning of creativity and ways to improve human creative ability. Some studies suggested that creative ideas arrive with flash-like spontaneity and others arrive after a slow and deliberate process.[3] In addition, creativity is not a static attribute; it is often a learning curve that is enhanced by study and experience. Even the most creative individuals are likely to be influenced by many of the surrounding factors (market and environment) that are dynamic in nature. Discussed below are some of the points that indicate the author's reflection on the numerous studies on creativity.

Motivated versus discouraged creativity

Research on creativity indicates that everyone with normal intelligence is capable of doing some creative work. However, a creative ability may be either decelerated or accelerated by many surrounding factors that can either discourage or motivate a person to think creatively.[7] Discouragement often results from fear of failure or risk taking. Product development projects should be associated with largely unconstrained and fear-free thinking to allow a creative environment. Motivation can result in a great deal of creativity. Indeed, many people are not fully aware of their creative potentials as a result of working in environments that impede intrinsic motivation.[8] Although financial reward may represent a significant motivation for creativity, it is not the main one. In some situations, the concept of creativity-for-money can be in conflict with the spirit of creativity that encourages risk taking in the brainstorming process. People can realize their creative potentials through a supportive environment in which the opportunity for creative thinking is often the best motivation.

Self-perception is a key creativity force

Creative individuals often enjoy a combination of attributes such as self-confidence of their creative ability, observability and freedom to create.[7–9] As a result, they seek every opportunity to be creative. Uncreative people are either inherently unconfident owing to growing problems, or made uncreative by an inappropriate educational or working environment.

Creative emotion

One of the interesting theories of human brain is that human is not a thinking machine, but rather a feeling machine that can think.[5] This is one of the fundamental differences between humans and computers. Some very creative ideas can come from emotional involvement in the problem which drives needs and wants and inspires solutions. On the other hand, extreme emotional involvement may result in creativity hindrance heightened by the emotional pressure. It is important therefore to have a balanced emotion to be creative, as the outcome of creative work should exhibit some level of objectivity.

Creative pause

One of the factors that can emotionally prepare a person to be more creative is the so called 'creative pause' proposed by Edward de Bono.[10,11] He suggests that people with creative potential should set aside a time needed to take a step back and allow themselves to ask if there is a better way of doing things. According to De Bono, this should be a short break of maybe only 30 seconds, but that should be a habitual part of thinking induced by self-discipline, as it is easy to overlook.

Time pressure

Some studies on creativity indicate that people were the least creative when they were fighting the clock.[7-9] This reaction is described as 'time-pressure hangover' and it indicates that the creativity of some people may be hindered by time pressure and by the fact that most people cannot deeply engage with the problem in a very short period of time. In contrast, history tells us about many creative ideas that were driven, at least partially by time pressure owing to ongoing wars or natural disasters. In today's competitive market, time has become a critical factor in all aspects of business, including product development. It is important therefore to be prepared for situations where creative ideas can be generated in a short period of time as a result of an urgent need for these ideas. The key factor in this regard is to avoid distractions. Indeed, many people can work under time pressure if they are allowed to focus on the work. It is also useful to create a diary of ideas against possibilities and unpredictable outcomes so that when the time comes they can provide the necessary preparation and guidelines for more specific ideas.

Competitive environment

This is another issue that studies have struggled with. On one hand, some studies suggest that creativity can be highly driven by a competitive atmosphere. On the other hand, competition within the same organization may not work in favor of creative work as it often hinders collaboration.[7] Indeed, some studies suggest that the most creative teams are those that have the confidence to share and debate ideas.[3,7] But when people compete for recognition, they stop sharing information.

In light of the above discussion, the key human-related factors that can greatly influence creativity are motivation versus discouragement, self-perception, emotion, creative pause, time pressure and competitive pressure. In any situation, these factors may have little effect without the support of other qualification factors such as experience, willingness to learn and study, and good knowledge. The role of creativity in design conceptualization is largely reflected in two aspects: (a) realization of deficiency in the current state of the art and (b) the ability to generate ideas for an optimum solution to the problem. Since these two aspects represent the core of design conceptualization, it is imperative that the team involved in the design process exhibits a great deal of creative ability. In the absence of a standard approach, the points made above may be formulated into some form of questionnaire to judge someone's creativity skills.

5.3.2 Brainstorming

Brainstorming is the process of focusing on ideas and thoughts that can result in feasible solutions. In a typical product development program, sources of ideas are generally inspired by the specific problem in hand. When the task is to modify an existing product, experience and in-depth analysis of the current product deficiency may drive new ideas to make modifications that can improve its performance. When a completely new product is proposed, fresh thinking with a higher level of creativity will be required and sources of ideas must be discovered. In both cases, the rule of thumb is to ask the right questions, stimulate creative thinking and capture the best idea or the best combination of ideas. These are the essences of brainstorming.

Brainstorming is a popular tool that has proven to be very useful in all product development projects. Typically, a brainstorming session involves a group of individuals who may exhibit different educational backgrounds, experience and creativity levels. These individuals come together to share ideas and exchange viewpoints with the common goal of creating a new product idea that is bounded only by the stated definition of the problem and the outcomes of the preparatory steps in design conceptualization.

One of the key aspects of brainstorming is management and rule setting. Having a group of creative individuals brainstorming together is often a difficult task unless the process is carefully managed to allow each individual to contribute to the enrichment of the solutions explored. A key management issue is the selection criteria of the brainstorming team. This team should exhibit the necessary technical qualification and experience that can allow them to participate effectively in the brainstorming process.

It is also useful to involve a few individuals with less experience, as brainstorming represents an environment to cultivate, teach and enhance individual creativity.[7-9]

Another critical management aspect of the brainstorming process is team harmony. By virtue of human nature, some may view a brainstorming session as an opportunity to express their dominant creativity talents. This attitude in itself is healthy unless it comes at the expense of the involvement of other individuals in the brainstorming process. Others may be reluctant to express their views or propose their ideas for the fear of being criticized by others who happened to have longer experience or higher positions in the organization. These are common management issues that are almost inevitable but must be resolved in advance to establish an effective and efficient brainstorming process.

Obviously, there are many ways to manage a brainstorming process. A good approach is to establish appropriate rules for the brainstorming process. Examples of these rules are illustrated in Fig. 5.2. In general, the process can be divided into three phases:[3,7] initial exploration, idea crunching and discovery. In the initial exploration phase, all suggested ideas are presented and no idea is criticized. The outcome of this phase is a document or a display of all proposed ideas. The second phase is idea crunching. This is the phase where each individual will have the opportunity to either defend his/her idea, shift toward someone else's idea, or combine his/her idea with other ideas presented in the initial phase. At this stage, the team structure may undergo some changes; individuals with the most accepted ideas will become more involved and more committed to the process and may take leadership positions in the team; others will remain on the team as key supporting elements. The final phase is discovery. In this phase, the most accepted ideas in the crunching phase are re-evaluated in view of pre-established criteria, and further narrowed down to the most feasible and most promising ones. Commonly, the brainstorming process yields a primary solution and one or more backup solutions. The final course of action will then require decision making about which concepts to follow and possible concept optimization in the presence of the many concepts presented. These aspects are discussed later in this chapter.

In addition to the management aspects discussed above, the brainstorming process should also be based on a thinking scheme that is selected in view of the nature of the problem in hand and the target product (modified or totally new). In general, there are two main approaches to thinking:[3,7,8] programmed thinking, which relies on logical or structured ways of creating a new product or service, and lateral thinking, which relies on the brain's ability to recognize patterns and possibilities. These approaches represent schools of thought supported by specific tools for problem solving. For example, programmed thinking uses familiar tools such as attribute

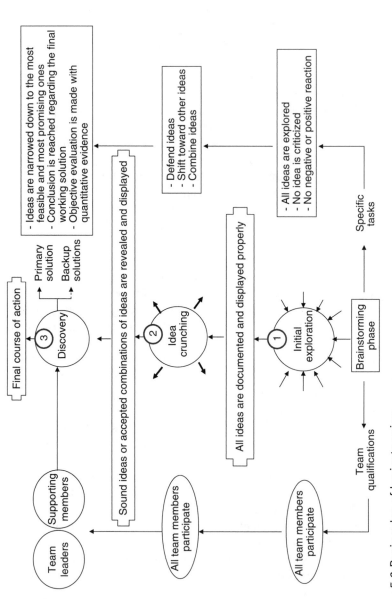

5.2 Basic rules of brainstorming.

listing, morphological analysis and the reframing matrix.[3,7] Lateral thinking was popularized by Edward de Bono in his famous 'six thinking hats' approach.[10,11]

Brainstorming should initially be based on the lateral thinking approach. This is due to the fact that it requires people to think individually and come up with ideas and thoughts some of which may initially seem shocking or unusual. As the process progresses, these ideas are modified and improved to yield final thoughts that can be substantially different from the initial ones. Different individuals will be likely to have different abilities for recognizing patterns and possibilities, an attribute that computers (despite the great progress in artificial intelligence) cannot do very well. Humans do not function like computers. It takes a long time to train people to do even simple arithmetic; computers are more suitable and more efficient for this task. On the other hand, it will take a great deal of modeling and complex algorithms to train computers to recognize patterns, objects and situations; these can be often recognized by humans instantaneously, although developments of computers have made a huge leap in this area. More importantly, humans can think spontaneously, in multi-directions, and with a great deal of creativity.

The main problem with human thinking, however, is getting excessively involved with particular familiar patterns that may prohibit or limit thinking outside their boundaries. As a result, some solutions typically represent reflections of previous experiences with similar problems or familiar solutions that happened to work instead of the original solutions. A successful brainstorming process should be based on breaking out the patterned way of thinking through looking into solutions in other unfamiliar patterns; this is the essence of lateral thinking, which is particularly useful when totally new product concepts are required. On the other hand, programmed thinking or logical disciplined thinking can be very effective in the final stage of design conceptualization.

The lateral thinking approach: De Bono six thinking hats

As indicated above, one of the greatest challenges associated with brainstorming is the ability of individuals to perform lateral thinking during brainstorming. In this regard, it will be useful to introduce one of the approaches that has been implemented with great success in recent years, the six thinking hats developed by Edward de Bono.[10,11] The idea underlying this thinking approach stems from some common behavior associated with most thinking processes. For instance, many successful outcomes happen as a result of rational and positive thinking approaches. These outcomes are typically associated with product development projects in which an organization's goal is to further excel in what it does already either by adding new features to enhance the current product performance or by developing new products that can take the overall performance upstream. On the other hand, many product development projects may be carried out under intense economical or competitive pressure. This often results in a thinking process that is unintentionally based on a mix of emotion, negative attitudes and even irrational views. In these situations, pessimists may be excessively defensive and more emotional people may fail to look at decisions calmly and rationally.

According to De Bono each thinking hat represents a different style of thinking. These are explained below:

- *The white hat* is a thinking mode in which the focus is on looking at the available data and converting it into useful information. The outcome of this thinking phase is a knowledge gain from available information and realization of gaps in knowledge, which have to be overcome either by trying to fill them or by taking account of them. Analytical approaches, such as past trend analysis and extrapolation from historical data, can be useful under this thinking hat.
- *The red hat* is a thinking mode in which individuals should look at problems using intuition, gut reaction and emotion. It will also help to think how other people will react to the problem emotionally, not just from your side of reasoning but more critically from their viewpoints.
- *The black hat* is a thinking mode in which individuals should look at all the points of the decision cautiously and defensively by evaluating why it might not work and highlighting potentially fatal flaws and risks before embarking on a course of action. The outcome of this phase is a list of all the weak points in a plan. This outcome often results in taking a variety of actions including elimination of some points, altering some, or preparing a contingency plan to counter them. This typically results in more resilient and robust plans.
- *The yellow hat* is the phase of positive thinking. All benefits associated with the decision are highlighted in an optimistic environment. This phase provides a psychological motivation to move to the next step under circumstances that may look gloomy or difficult.
- *The green hat* stands for creativity. This is the phase where creative ideas are invited with the attitude that no idea is so small or so unsound. In other words, it is a freewheeling way of thinking, in which there is little criticism of ideas.
- *The blue hat* stands for process control. According to De Bono, this is the hat worn by people chairing meetings with the task being to redirect the brainstorming process depending on the progress made. When the team is running into a stage of dry ideas, blue hat thinkers

may direct activity into green hat thinking, and when contingency plans are needed, they may ask for black hat thinking.

It is important to point out that the different approaches of thinking including the De Bono thinking hats should be considered primarily as guidelines and not as firm or systematic ways of thinking. Each product development team should consider different thinking approaches and perhaps develop its own scheme of thought. In any case, this scheme should be pre-defined before starting a brainstorming process.

5.3.3 Decision making

Decision making is an essential element of design conceptualization. It begins with the first preparatory step of design conceptualization, as we must decide whether the product idea is justifiable and continues all the way until a final solution or a product concept is reached. Unlike creativity, which is typically entertained without fear, instead, with joy and pleasure, and in contrast to brainstorming, which is again a fear-free individual input into a team establishment, decision making is often a stressful action that has to be made by one or two individuals. This is largely due to the fact that many decisions are associated with a great deal of uncertainty or risk; some may result in major gain and others may lead to a substantial loss. Indeed, some decision makers often risk their reputation and credibility, particularly when they make seemingly unfavorable or unpopular decisions. On the other hand, decision making shares some common attributes with creativity and brainstorming. They all involve some subjectivity and human judgment and they all are influenced by behavioral aspects. But the most common feature is that they all result in verifiable outcomes. In other words, no one can claim to be creative or a good decision maker without providing evidence and verifiable results.

From a design conceptualization viewpoint, a good decision is the one that yields optimum outcomes. Therefore, decisions should be made with the aid of some analytical approaches that support the decision-making process and assist in exploring all the options available. In this regard, all engineers should be familiar with decision theory.

Decision theory

Decision theory is an analytical approach in which a body of knowledge is used to assist in making choices from a set of alternatives in view of their possible consequences. Decisions are treated strictly against nature. In other words, the result, or return, from a decision will primarily depend on the action of nature. For example, if the decision is to coat a fabric with some type of abrasion-resistant finish (or not), then the return, which is failure against abrasive action (or not) will depend on what action nature takes (nature here being the contacting medium). In this model, the returns accrue only to the decision maker or the product under consideration and it does not influence nature, as the contacting medium does not care what the outcome is. This simple example illustrates the difference between the decision theory and the 'game theory' in which both players (the decision maker and an opponent, the nature in this case) have an interest in the outcome.

Using decision theory, three basic scenarios can be considered:[3,12,13] decision under certainty, decision under risk and decision under uncertainty. A decision under certainty implies that each action results in a known outcome (or each alternative leads to one and only one consequence). This means that a choice among alternatives is equivalent to a choice among consequences. Mathematically, an outcome resulting from each action will have a probability of 1. A decision under risk implies that each alternative will have one of several possible consequences, and the probability of occurrence of each consequence is known. In other words, each state of nature has an assigned probability of occurrence. Therefore, each alternative is associated with a probability distribution and a choice among probability distributions. The level of risk in this case is determined by the probability value.

In situations where probability distributions are unknown (totally new ideas), the decision is taken under uncertainty; that is each action can result in two or more outcomes with unknown probabilities of the states of nature. When the decision theory is expanded to encompass the game theory, the decision is then considered to be 'under conflict'. In this case, the states of nature are replaced by courses of action determined by an opponent (another player) who is trying to maximize his/her objective function.

Before giving some examples to demonstrate the use of decision theory, it is important to point out that the theory assumes that the ranking produced by using a criterion is consistent with the decision maker's objectives and preferences. Being a mathematical approach, decision theory will offer no assistance in defining objectives, establishing alternatives, or evaluating consequences. These are decision maker's concerns.

The basic model of the decision theory is represented by the so-called payoff Table 5.2[12-15] shown on p. 97, in which the entries r_{ij} are called payoffs for each possible combination of decision, d_i, and state of nature, N_j.

For example, consider an application in which the goal is to design a new fabric for ladies dresses. Suppose that one of the decisions or courses of actions is to choose one of three material types, namely: polyester (d_1), cotton (d_2) and cotton/polyester blend (d_3). The states of nature, N_1, $N_2, \ldots, N_m$, represent the consequences that can result from these decisions. These are out of the control of the decision maker. Suppose that the states of nature represent the reaction of potential customers to the new fabric expressed as N_1 = negative or weak (W), N_2 = average (A), N_3 = strong (S) and N_4 = very strong (VS).

The payoffs r_{ij} for different combinations of decision, d_i, and state of nature, N_j may be represented by management best estimates of percent increase in profit with reference to the profit made from a conventional fabric used for this application. As indicated earlier, decision theory assumes that these estimates are available, as they reflect the state of knowledge that the decision maker should have in order to be able to make an appropriate decision. In practice, estimating payoffs may require in-depth analysis of market factors such as potential customers, supply and demand, the performance status of current similar products, and the price of similar existing products. It should also be noted that nature, in this case, does not care about the decision, but it is affected by its consequences. For this example, suppose that the payoff table is as shown in Table 5.3:

Table 5.2 A 'pay-off' table: example 1

Decision	State of nature			
	N_1	N_2	. . .	N_m
d_1	r_{11}	r_{12}	. . .	r_{1m}
d_2	r_{21}	r_{22}	. . .	r_{2m}
. . .	. . .	. . .	. . .	. . .
d_n	r_{n1}	r_{n2}	. . .	r_{nm}

Table 5.3 A 'pay-off' table: example 2

Decision	State of nature			
	W	A	S	VS
Polyester (d_1)	5	15	20	25
Cotton (d_2)	−20	6	8	10
Cotton/polyester (d_3)	3	18	30	35

At this point, the decision maker is faced with three options or possible decisions, consequences or states of nature, and payoffs represented by estimates corresponding to various combinations of decision and state of nature. The additional piece of information, which must be available before a decision is made, is the degree of certainty associated with the states of nature. In our example, if the decision maker knows which states of nature will surely occur, he/she can then select the decision that yields the largest return for the known state of nature. For example, if a very strong market (VS) was surely anticipated, then the decision would be to select cotton/polyester blend as the primary material since it results in the largest payoff; this is the d_3/VS combination.

Obviously, the selection of which decision to make may involve some subjectivity or ranking of decisions, depending on the state of knowledge and other factors that may arise. In some situations, the decision makers may be faced with many possible decisions that are too close to select from. If these possible decisions are represented by a vector d and the return by the real-valued function $r(d)$, the decision problem can then be formulated as: maximize $r(d)$, subject to feasibility constraints on d, which requires linear programming analysis (to be discussed in Chapter 6).

Decision theory is best used when there is an inevitable risk involved in making the decision.[14,15] Mathematically, a risk is represented by a chance or a probability value associated with a state of nature. Accordingly, decisions under risk are handled under two basic assumptions. The first assumption is that there is more than one state of nature, N_j, and the second one is that the probability, p_j, of each state of nature is known. The approach to handling this situation is to implement the concept of 'expected value'. The idea here is that when faced with a number of actions, each of which could give rise to more than one possible outcome with different probabilities, the rational procedure is to identify all possible outcomes, determine their values (positive or negative) and probabilities that will result from each course of action, and multiply the two to give an expected value. The action to be chosen should be the one that gives rise to the highest total expected value in case of gain situations. Under risk, if the decision maker makes decision d_i, the payoff will be the expected value ER_i, determined by the following equation:[15]

$$ER_i = r_{i1}p_1 + r_{i2}p_2 + \ldots + r_{im}p_m \tag{5.1}$$

Accordingly, the goal will be to make the decision, d_i, which maximizes ER_i.

Suppose that for the example discussed above, the probabilities associated with the four states of nature are: $p(W) = 0.2$, $p(A) = 0.5$, $p(S) = 0.2$ and $p(VS) = 0.1$. In this case, the expected gains associated with the three decisions would be as follows:

$$E(d_1) = E(\text{polyester}) = 0.2(5) + 0.5(15) + 0.2(20) + 0.1(25) \qquad (5.2a)$$
$$= 15$$

$$E(d_2) = E(\text{cotton}) = 0.2(-20) + 0.5(6) + 0.2(8) + 0.1(10) = 1.6 \quad (5.2b)$$

$$E(d_1) = E(\text{cotton/polyesterblend}) \qquad\qquad\qquad (5.2c)$$
$$= 0.2(3) + 0.5(18) + 0.2(30) + 0.1(35) = 19.1$$

Judging by the expected value of payoff or percent increase in profit, the decision maker would select the cotton/polyester blend as the material that would maximize the increase in profit.

A convenient way to represent a decision making problem is by using so-called 'decision trees',[1,13,14] as in Fig. 5.3. A square node will represent a point at which a decision must be made and each line leading from a square will represent a possible decision. A circular node will represent situations where the outcome is uncertain and each line leading from a circle will represent a possible outcome. Using a decision tree to find the optimal decision is commonly known as 'solving the tree'.

It is critical to point out that decision theory is a mathematical concept and does not provide insights into the cause and effect of the decision to be made. It simply provides a quantitative measure or a value index that can be used as a guideline in making the decision. The effectiveness of this measure will mainly depend on the state of knowledge or the input information and the assumptions that the decision maker is willing to take in performing the analysis.

When decisions are made under uncertainty, the probabilities associated with different states of nature are unknown. This complex situation has become more common in today's global market as a result of the rapid changes in demand for some products and unpredictable consumer behavior or reaction towards new products. Behavioral thinkers and philosophers use the classic Pascal's Wager to illustrate thinking under uncertainty.[13–15] According to Pascal, uncertainty is at its utmost level when the question is whether or not God exists. In this case, personal belief or non-belief in God is the decision to be made. However, the reward for belief in God if God actually does exist is infinite. Therefore, this becomes a decision of weighing the unknown probability of God's existence against the expected value, which often results in the conclusion that the expected value of belief exceeds that of non-belief, so it is better to believe in God.

Following Pascal's theory, a decision in our example can be solely based on the best anticipated pay-off values for each decision. In this case, the only criterion would be the decision at the state of nature of very strong increase in profit, with a pay-off of 35 at the VS/d_3 combination leading to the decision to select the cotton/polyester blend. It is also useful to be

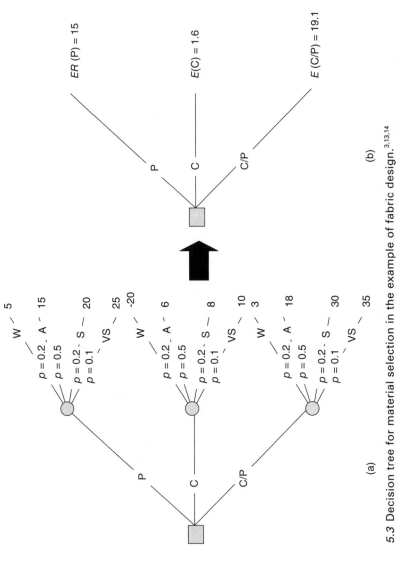

5.3 Decision tree for material selection in the example of fabric design.[3,13,14]

prepared with a backup decision as the market warrants changes. In this case, testing the market through market research can be very useful.

Bayes' theorem

A practical approach to dealing with decisions under uncertainty should be based on the realistic view that the actual probability of a state of nature may not be known with great confidence without some history of the product in question. Since newly developed products are typically associated with unknown probabilities, attention should be shifted to estimating the probabilities of the states of nature. One common approach to making these estimates is Bayes' theorem,[16,17] which provides ways of revising prior estimates of probabilities in view of new information that has become available. This approach sounds more appropriate as it resembles the need to upgrade the level of certainty in the face of dynamic changes in consumer's behavior. The obvious limitation of this approach is the time and cost associated with reaching more reliable probabilities and the changes that might be made as a result of learning new probabilities.

Bayes' theorem can be considered as a mathematical triviality that is directly derived from the probability theory.[18] The basis for this theorem is that rational belief is governed by the laws of probability, particularly conditional probabilities. The theorem relates the 'direct' probability of a hypothesis, H, conditional on a given body of data E, or $P_E(H)$, to the 'inverse' probability of the data conditional on the hypothesis, $P_H(E)$:

$$P_E(H) = \left[\frac{P(H)}{P(E)}\right] P_H(E) \qquad (5.3)$$

The inverse probability $P_H(E)$ is referred to as the 'likelihood' of H for E. It expresses the degree to which the hypothesis predicts the data given the background information codified in the probability P. In general, inverse probability makes the analysis more certain and less subjective than direct probability.

In the following discussions, the use of the Bayes' theorem in decision making is demonstrated using simple examples. Readers who will be involved largely in decision making under uncertain situations are encouraged to read more on this subject, particularly on the different forms that the theorem can take depending on the type of application considered.[14–19]

Example 1: body-armor Bayesian decision

Consider a product development process in which a new body armor (A_n) was designed to replace an existing one (A_c) for military applications. Both products were designed and produced by the same company. The current body armor (A_c) was designed in multi-layers to protect against fragments and low velocity bullets. It was characterized by low weight, low bulk, high flexibility and good thermo-physiological comfort. However, it was associated with an unacceptable rate of injuries and casualties as it was often defeated by flechettes (tiny needle-shaped objects) disseminated in large amounts by exploding warheads or shells. The new body armor (A_n) was designed to be more protective by careful reinforcment with shaped plates made from metals and ceramics in areas covering sensitive parts of the body. As a result, it was a heavier weight resulting in less flexibility and some discomfort. This is a classic trade-off in this type of product between protection and comfort.

Since the key performance of body armor is its protective ability, the company relied on statistics of injuries or casualties to determine whether the new product will provide significant advantage over the current one. Considering the massive amount of the current product being used by military personnel, the company decided to test the new product in small quantities and over a certain period of time to see whether a decision to replace the current product slowly by the new one can be made.

The above scenario follows one of the Bayesian decision making forms. In this case, data is available on a population that is stratified by current and new products, and better estimates of the probabilities of the states of nature are required. The way the Bayes' theorem is used in this case is summarized in the following steps:

- The analysis begins with a key hypothesis, H. In this example, it is:
  H: a body armor unit picked randomly will fail to protect from injuries
- Suppose that the company delivered a total of one million units (N) of body armor last year that were entirely used, and the number of injuries that were related directly to failures of this body armor were 8000. This data yields the unconditional probability, $P(H)$, that any unit picked randomly will fail to protect from injuries. Thus, $P(H) = H/N = 0.008$. This probability is not very impressive as it implies that approximately 1 (or exactly 0.8) of each 100 units will fail to protect.
- Now, suppose that of the one million units delivered, only 50000 units of the newly designed body armors were delivered and used. This is only a small percent (5%) of the total units used, but it represents the data, E, that is required for better probability estimates that can assist in making an appropriate decision about the new product. Suppose that 20 units of these failed to protect. This information represents a key input to the Bayesian decision making analysis.
- To find the probability of the hypothesis, H, conditional on the information, E, that the failing unit was the newly designed one, we divide the

probability that the unit was one of the newly designed and failed to protect, $P(H\&E) = 20/10^6 = 0.00002$, by the probability that the unit was a newly designed, $P(E) = 50000/10^6 = 0.05$. Thus, the probability of failure of a product unit given that it was a newly designed is $P_E(H) = P(H\ \&\ E)/P(E) = 0.00002/0.05 = 0.0004$.

- Notice how the size of the total population, N, factors out of the above equation, so that $P_E(H)$ is just the proportion of newly designed units that failed (20/50000).
- Now, if the above quantity, which gives the failure rate among the newly designed units, is contrasted with the 'inverse' probability of E conditional on H, $P_H(E) = P(H\&E)/P(H) = 0.00002/0.008 = 0.0025$; this is the proportion of failing units in the total population that occurred among the newly designed units.
- Thus, the condition that a product unit failed in this trial is a fairly strong predictor of newly designed products. Indeed, a value $P_H(E)$ of 0.0025 indicates that 0.25% of the total failure occurred in the newly designed product. Using Bayes' theorem, this information can be used to compute the 'direct' probability of the failure of a product given that the product was newly designed. This can be done by multiplying the 'prediction term' $P_H(E)$ by the ratio of the total number of failures in the population to the number of newly designed units in the population, $P(H)/P(E) = 8000/50000 = 0.16$. The result is $P_E(H) = 0.0025 \times 0.16 = 0.0004$, just as expected.

Example 2: flame-resistant finish Bayesian decision

Bayes' theorem can also be used in making decisions regarding purchasing of materials that are suitable for some particular applications on the basis of performance tests. For example, suppose that a distributor of flame-resistant uniforms has purchased a batch of garments from a manufacturer. Typically, this type of fabric is subjected to chemical finishes to enhance its flame resistance. Two types of finish are commonly used, A and B, with the former being known as superior to the latter in flame-resistance criteria such as fire spread and dimensional stability. The distributor does not know whether the fabric was treated with finish A or B. From previous experience, based on laboratory and actual performance testing, it was known that fabrics with finish A has a 97% chance of passing an ignition test, while fabric with finish B only has a 92% chance. Suppose that distributor suspects that on average manufacturers send batches consisting of about 80% of fabric rolls of finish A and 20% of finish B.

The above scenario reflects uncertainty that cannot be resolved without actual data, E, evaluated against some pre-established hypothesis, H. The data can only be generated from actual testing of a random sample, say five rolls. The results can then be evaluated against the hypothesis that the batch was entirely produced from finish A. Suppose that the results of testing the randomly selected rolls revealed no failure in the ignition test, E. In this case, Bayes' theorem can be used as follows:

- On the basis of the information known, the probabilities that the batch consists of fabrics made from both types of finish are: $P(A) = 0.80$ and $P(B) = 0.20$. These are roughly determined probabilities based on a guess work.
- The probability, for each finish, that a random sample of the fabric will pass the ignition test is: $P_A(E) = (1 - 0.03)^5 = 0.858734$ and $P_B(E) = (1 - 0.08)^5 = 0.65908$.
- In this case, the following form of Bayes' theorem is used:

$$P_E(A) = \frac{P(A) \cdot P_A(E)}{P(A) \cdot P_A(E) + P(B) \cdot P_B(E)} = \frac{0.8 \times 0.858734}{0.8 \times 0.858734 + 0.2 \times 0.65908} = 0.84$$

- The above result provides a better estimate of the probability that the batch was made from fabrics with finish A, given the information obtained from testing a random sample. An increase from the previous anticipated probability of 0.8 to 0.84 that the batch was produced from fabric with finish type A, is useful information that can assist in making better decisions.

The above examples are used only to demonstrate the usefulness of statistical tools and some well-established theories in making appropriate decisions, particularly in uncertain situations. Since cause and effect are not revealed by the outcomes of any of these tools, they should be used very carefully with full consideration of the purpose of the decision-making analysis.

5.3.4 Concept optimization

The basic elements of design conceptualization discussed in the previous sections (i.e. creativity, brainstorming and decision making) can ultimately assist in reaching design concepts that can be implemented for modifying existing products or creating totally new products. Typically, these efforts yield one or two ideas that are deemed feasible to implement. In this case, the choice may be easy and the process is narrowed to one of the two ideas, with the other acting as a backup plan. In some situations, the product development team may be faced with a host of ideas or options that seem to be equally beneficial. Obviously, this is a problem that any company would like to have rather than a scarcity of concepts. Under these favorable circumstances, the challenge is often called 'concept optimization', that is the process of determining which of the concepts should be chosen, or what ideas should be implemented in the presence of the many ideas proposed. Obvious options in this regard are:

- eliminating all ideas but one (typically the best)
- combining some ideas into one
- dividing different ideas into different models of the same product
- dividing different ideas into different products
- directing some ideas toward cost/value-oriented products

The first two of the options above represent opportunities for developing unique products. The last three options represent opportunities for making diverse products, a key criterion in today's competitive market.

Some approaches that can assist in concept optimization are attribute listing, morphological analysis and matrix analysis.[3,18,19] These tools are commonly used to find new combinations of products or services and they are sufficiently similar to be discussed together. The common steps in these tools are as follows:

- List all the attributes of the product in question. Attributes may constitute components, properties, quality parameters or design elements. For example, the attributes of a fabric gasborne dust collector are the gas source, the cleaning mechanism, gas parameters (temperature, moisture, acid content), dust particle size, material, fabric construction and finishing application.
- Construct a table or a matrix using these attributes as column headings and related levels or variants as entries. In other words, the matrix should show all possible variations of each attribute obtained from a brainstorming process or from other sources of information as shown in Table 5.4. Again, this example is presented only to illustrate the concepts.
- Perform morphological analysis by selecting one entry from each column to create various combinations. This can be done randomly or by selecting interesting combinations.
- Finally, evaluate and improve the mixture to see if it has performance, economical or market merits.

The morphological analysis should be performed with full realization of product requirements, drawbacks and advantages. For this example, the factors listed represent the critical performance criteria for a fabric dust collector. Depending on the problem definition stated, the design concept formulators may decide to focus on one product concept or option as listed below.

- *Product option 1:* a fabric dust collector that will accommodate dust particle sizes of small to medium size (0.1 to 0.15 μm), for gases at low to medium temperature (<100°) at low moisture content and low acid

Table 5.4 Illustration of concept optimization

Gas source	Cleaning	Temperature	Durability	Moisture	Acid content	Particle size	Material	Construction	Finishing
Pulverizing	Shake	Very high	Very high	Very high	Very high	Very large	Polyester	Woven	Heat setting
Combustion	Reverse air	High	High	High	High	Large	Polyaramid	Needlefelts	Singeing
Milling	Pulse jet	Medium Low	Medium Low	Medium Low	Medium Low	Medium Small Very small	Polyimid Cotton Fiberglass	Knitted	Raising Calendering Chemical

Table 5.5 Output matrix: product option 1

Gas source	Cleaning	Temperature	Durability	Moisture	Acid content	Particle size	Material	Construction	Finishing
Pulverizing	Shake								
Milling		Medium Low	High Medium Low	Medium Low	Low	Medium Small Very small	Cotton	Needlefelts	Singeing

content. The filter may use shake cleaning (i.e. by switching off the exhaust fan and flexing the filter sleeve mechanically), which requires moderate to high durability. As a result, the filter may be made from cotton fibers in a needlefelt construction. In addition, since staples cotton fibers will be used, a singeing treatment may be required to minimize or remove projecting fibers that often inhibit dust cake release by clinging to the dust particles. This product combination is directly obtained from the matrix above and highlighted in the output matrix in Table 5.5.

- *Product option 2:* a fabric dust collector that will accommodate dust particle sizes of a wide range (0.1 to 0.25 µm), for gases at medium to high temperature at low to medium moisture content and medium to high acid content. The filter may use shake cleaning. As a result, the filter may be made from chemically treated polyester fibers in a woven construction (perhaps of simple satin design). This product combination is directly obtained from the matrix above (Table 5.4) and highlighted in the output matrix below (Table 5.6).
- *Product combination options:* Using morphological analysis, different product combinations may also be detected. For example, a combination of shake and reverse air cleaning can be used for enhanced performance, or a needlefelt/woven construction combination with a woven basecloth or scrim sandwiched by two layers of needlefelt. In the latter combination, the woven structure provides more stability to the needlefelt structure as it allows it to withstand the tensile and flexing stresses imposed by the pulse cleaning mechanism, and the needlefelt provides good filtration efficiency and good protection for the basecloth from abrasion caused by constant flexing against the filter cage wires.

As can be seen from the above example, morphological analysis is largely a programmed thinking approach, which allows the concept formulator to seek different product options. The decision about what options to take will obviously depend on other factors such as customer need, market conditions and cost considerations. These can also be incorporated in the matrix and entertained using morphological analysis. Ultimately, all options will be exhausted or reach their limitations. This is when lateral thinking and ideas outside the traditional boundaries become critical. Just as in any sport, programmed thinking is the approach that maintains records at a competitive level; lateral thinking, on the other hand, is the approach that continuously breaks these records.

5.4   Conclusions

Lessons from history can assist a great deal in understanding how ideas are often discovered. Many ideas come as a result of a long and extensive

Table 5.6 Output matrix: product option 2

Gas source	Cleaning	Temperature	Durability	Moisture	Acid content	Particle size	Material	Construction	Finishing
Combustion	Shake	High Medium Low	Very high High Medium						
Milling				Medium Low	High Medium Low	Large Medium Small Very small	Polyester	Woven	Chemical

study of the subject matter supported by a great deal of knowledge. For example, the continuous research effort by Stephanie Kwolek in high performance chemical compounds for the DuPont Company led to the development of a synthetic fiber that is five times stronger than steel of the same weight. This is Kevlar, a now household name, which was patented by Kwolek in 1966 as a material that does not rust nor corrode and is extremely lightweight. Today, lives are saved by bullet proof vests made from this material, and many useful products are made from this fiber including underwater cables, brake linings, space vehicles, boats, parachutes, skis, and building materials are made from Kevlar.

Ideas may also come through outcomes in some areas that lead to discoveries in other areas. For example, the first waterproof coat was discovered by Charles Macintosh, a Scottish chemist, in 1823 as a result of working in an unrelated area. While he was trying to find uses for the waste products of gasworks, Macintosh discovered that coal tar naphtha dissolved rubber. He then implemented this idea in cementing two pieces of cloth together by taking wool cloth, painting one side with the dissolved rubber preparation and placing another layer of wool cloth on top. This created the first practical waterproof fabric. Obviously, the fabric was not perfect as it was easy to puncture when it was seamed and the natural oil in wool caused the rubber cement to deteriorate. As a result, the fabric became stiffer in cold weather and sticky in hot weather. When vulcanized rubber was invented in 1839, Macintosh's fabrics improved as a result of the ability of the new rubber to withstand temperature changes.

Some ideas may result from new realization of old products or materials supported by new needs or desires. For example, colored cotton has been around since 2700 BC in Indo-Pakistan, Egypt and Peru. At that time, it was common to grow cotton in a variety of natural colors such as mocha, tan, gray and red-brown. However, those cottons were not long enough or strong enough to be processed on normal machinery. As a result, only white cotton has been used for hundreds of years. In the early 1980s, Sally Fox, an expert in pest management was introduced to colored cotton while working for a cotton breeder, whose focus was on developing pest-resistant strains of cotton. A small amount of brown cotton seeds sparked a rediscovery of naturally colored cottons with characteristics suitable for processing in today's high production machinery. This rediscovery was also supported by a concern for the environment resulting from extensive misuse of pesticides. Obviously, Sally Fox did not invent colored cotton as it has always existed in nature. Her true contribution was in creating naturally colored cotton that could be spun on a machine and woven into fabric. This was made possible through her ten years of breeding efforts and careful selection from generation to generation to coax a usable variety of cotton from an initial cross-bred seed.

As one can see from the above historical lessons, there are no limitations or boundaries to creative work or innovative ideas. It is, therefore, worthwhile for engineers working in product development continuously to attempt to enhance and cultivate their creative abilities. This can be achieved by continuous learning, not only in the field of study but in other related or unrelated areas. In addition, understanding the surrounding factors associated with product development (economy, environment, regulations, etc.) can indeed enhance creativity and can result in robust solutions.

In this chapter, the basic elements of design conceptualization have been discussed. These are creativity, brainstorming, decision making and concept optimization. The importance of this discussion is to increase the awareness of many engineers that design conceptualization is fundamentally different from design analysis. This is despite the fact that they represent two sides of the same coin and they both aim at the development of innovative product model. Unfortunately, design conceptualization is not taught to a degree of depth equivalent to its importance in many engineering programs or in the training process of many organizations.

5.5   References

1. EL MOGAHZY Y E and CHEWNING C, *Fiber To Yarn Manufacturing Technology*, Cotton Incorporated, Cary, NC, USA, 2001.
2. ERNEST J, *Human Factors in Engineering and Design,* 4th edition, McGraw-Hill, New York, 1976.
3. DIETER G, *Engineering Design: A Material and Processing Approach*, McGraw-Hill Series in Mechanical Engineering, New York, 1983.
4. PAUL R and ELDER L, 'Critical thinking: the nature of critical and creative thought', *Journal of Developmental Education*, 2006, **34**, 26–34.
5. SPERRY R W, 'Mind–brain interaction: mentalism, yes; dualism, no'. *Neuroscience*, 1980, **5**, 195–206.
6. TAMA C, 'Critical thinking has a place in every classroom', *Journal of Reading*, 1989, **33**, 64–5.
7. VAN FRANGE E, *Professional Creativity*, Prentice-Hall, Englewood Cliffs, NJ, 1959.
8. ALGER J R M and HAYS C V, *Creative Synthesis in Design*, Prentice-Hall, Englewood Cliffs, NJ, 1964.
9. KIVENSON G, *The Art and Science of Inventing*, Van Nostrand Reinhold, New York, 1977.
10. DE BONO E, *Six Thinking Hats*, Little, Brown and Co, New York, 1985.
11. DE BONO E, *Serious Creativity: Using the Power of Lateral Thinking to Create New Ideas*, Harper Business, New York, 1992.
12. SHAFER G and PEARL J (eds), *Readings in Uncertain Reasoning*, Morgan Kaufmann, San Mateo, CA, 1990.
13. RAIFFA H, *Decision Analysis: Introductory Readings on Choices Under Uncertainty*, McGraw Hill, New York, 1997.

14. DE GROOT M, *Optimal Statistical Decisions*, Wiley Classics Library, New York, 2004.
15. BERGER J O, *Statistical Decision Theory and Bayesian Analysis*, 2nd edition, Springer Series in Statistics, New York, 1980.
16. HARTIGAN J A, *Bayes Theory*, Springer-Verlag, New York, 1983.
17. HOWSON C, 'Some recent objections to the Bayesian theory of support', *British Journal for the Philosophy of Science*, 1985, **36**, 305–9.
18. EL MOGAHZY Y E, *Statistics and Quality Control for Engineers and Manufacturers: from Basic to Advanced Topics*, 2nd edition, Quality Press, Atlanta, USA, 2002.
19. TRIBUS M, *Rational Descriptions, Decisions and Designs*, Pergamon Press, NY, 1969.

Abstract: Design analysis is essentially a decision-making process in which analytical tools derived from basic sciences, mathematics, statistics and engineering fundamentals are utilized to develop a product model that can be converted into an actual product. The type of analysis required will depend on the product concept established, the specifications of the actual product intended and the application(s) in question. Common analytical tools used in design analysis are examined. The discussion is restricted to the principle of the analytical tool and potential applications in fiber-to-fabric engineering. Many examples of fibrous products are used to illustrate the principles of these analytical tools.

Key words: transparent system; gray-box system; black-box system; mathematical modeling; system behavior; empirical modeling; regression model; artificial neural networks; optimization analysis; linear programming; genetic algorithms; simulated annealing; fuzzy logic; membership function; finite-element analysis; failure analysis; reliability analysis; survival analysis.

6.1   The purpose of design analysis

The task of design conceptualization discussed in Chapter 5 should ultimately result in a product concept or a product image represented by thoughts and stated features and perhaps supported by a rough sketch of the product idea. Before this product concept can be converted into an actual product, it should be first converted into a product model that largely resembles the product idea and reflects the attributes of the intended product. This task is called design analysis and it is the subject of this chapter.

Design analysis is essentially a decision-making process in which analytical tools derived from basic sciences, mathematics, statistics and engineering fundamentals are utilized for the purpose of developing a product model that is convertible into an actual product. The type of analysis required will depend on the product concept established, the specifications of the actual product intended and the application(s) in question. In general, a product model should meet four basic criteria:[1,2] (1) resemblance of a design concept, (2) visibility and predictability, (3) optimum functional performance and (4) manufacturability. Figure 6.1 illustrates the analyses associated with these criteria.

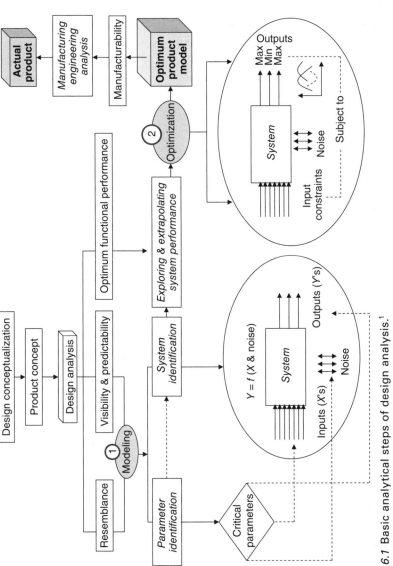

6.1 Basic analytical steps of design analysis.[1]

The first criterion implies that the product model should reflect and resemble a well-defined design concept that is stated clearly in the problem definition and supported by creative thoughts and sketches or images of the anticipated product. This criterion requires two types of design analysis: parameter identification and system identification. Parameter identification is a simple task in which each parameter associated with the product idea is defined, evaluated, weighted with respect to its anticipated relative contribution to the product function, and ranked by the order of its importance. This task typically begins in the design conceptualization phase, using largely qualitative descriptions of the different parameters and continues in the design analysis phase with more in-depth analysis in which only critical parameters are revealed and carefully evaluated for the purpose of determining their impact on the product function. System identification is a more complex task in which the product concept is simulated by a system with well-defined input and output parameters, as well as possible noise or uncontrollable parameters.

The second criterion for a product model implies that it should be largely visible and highly predictable. This criterion can only be met through exploratory and extrapolation analysis. In order to carry out this analysis, reliable models relating output parameters to input parameters should be developed; this task represents the foundation of system identification. Indeed, a system that is fully identified is the one that exhibits high visibility or transparency and high predictability within and outside its performance range. Using the models developed, exploratory analysis can be carried out by evaluating the effects of input parameters on the performance of output parameters. Exploratory analysis also involves evaluation of the interactive nature of different parameters. In this regard, statistical techniques such as design of experiment and analysis of variance[1] can be very useful. Predictability is another key criterion as it not only means anticipating particular outcomes with some degree of certainty but, more importantly, providing opportunities to revise and control the product system so that key design principles such as simplicity, support, familiarity, encouragement and safety can be fulfilled. These principles were discussed in Chapter 5. Similar to exploration, predictability is also a result of modeling analysis. However, the models that are most suitable for prediction are those of the empirical type. In this regard, multiple regression analysis and neural network techniques can be very useful. These tools are discussed later in this chapter.

The third criterion for a product model is that it should yield an optimum functional performance at optimum utilization of natural resources. This requires optimization analysis in which the values of the critical input parameters are carefully selected and controlled so that optimum functional performance is achieved with careful use of natural resources. In general, an optimization analysis aims at maximizing or minimizing the values of some system parameters subject to some constraints associated with pre-specified input factors or other external parameters. The success of the optimization analysis will largely depend on how well the system was identified in the modeling phase of analysis. This is a direct result of the simple fact that transparency and predictability can increase the options and the alternatives associated with optimization techniques. One of the common engineering tools of optimization is linear programming. The principles underlying this tool are discussed later in this chapter.

Finally, the product model should be manufacturable into an assembly in which design specifications are coordinated with manufacturing parameters to yield a final product with optimum performance. Although this criterion represents a bridging aspect between design analysis and manufacturing, it should be highly considered in the design phase so that a smooth transition from a product model to an actual product can be achieved. For instance, modeling analysis should account for noise or external parameters that may not be observed at the prototype scale but are anticipated in actual operating conditions. Similarly, optimization analysis should account for constraints anticipated during manufacturing.

In light of the above discussion, the two key analytical tasks of design analysis are modeling and optimization. This point is illustrated in Fig. 6.1. Modeling techniques can be used throughout the process of product development, beginning with developing a product concept and ending with developing a product model. In some situations, a user or an implementer's model may be developed to resemble the usability of a product.[1] Optimization is an essential task of any design analysis without which practical and compromising solutions to design problems cannot be obtained. In the following sections, common analytical tools used for modeling and optimization are discussed. The focus of the discussion will be on three key aspects: objectives, principle and potential applications. Many examples will also be provided to illustrate some of the analyses presented. It should be pointed out that a great deal of mathematical analysis will be covered in this chapter. Readers who are not involved directly in performing analytical work can benefit from this chapter by reading the objectives and the principles associated with each type of analysis, without getting into the mathematical details. On the other hand, readers who are interested in more mathematical details of these analyses can begin with this chapter and then move to more specialized references related to their area of interest.

6.2   Textile modeling techniques

Webster's Dictionary defines a model as 'a mathematical representation of facts, factors, and inferences of an entity or situation'. In the context of design, a model may be defined as a simulation of an entity or a situation using reliable mathematical techniques for the purpose of exploring, interpreting, predicting or controlling a system.[1] From an engineering viewpoint, the term 'system' implies an entity that involves clearly defined inputs, highly realized outputs and perhaps some noise or system uncertainties. This is different from the more general term 'process', which typically encompasses one or more of five basic elements: material, machine, methodology, people and environment. A design process may involve some or all of these elements. In order to simplify a design project, a product should be treated as a system or a set of subsystems, with the final design outcome being a linear or non-linear integration of these subsystems.

6.2.1 Product system classification

In the classic sense, one may define modeling analysis as a method of identifying a system, or conveying an understanding of the components that make up a system, by exploring all interactive relationships governing its performance. In this regard, engineering systems may be divided into three categories:[1] a transparent system, a gray-box system and a black-box system. A transparent system is typically self-identifiable and self-explored. This means that all basic inputs and outputs of the system are well identified, their interactive relationships are well established and they are likely to exhibit static or consistent patterns over time and under most circumstances. A product resembled by a transparent system is highly visible and highly predictable. In the real world, such a system hardly exists as it requires extreme simplicity, or high clarity in all system aspects such as design, technology, material, human handling, monitoring and environment.

A more realistic system is the gray-box system. This system is typically characterized by reasonable visibility and realization of some of its basic input–output interrelationships. However, the nature and the specific order of some of the equations governing the system may not be fully known. This makes the system only partially predictable. A black-box system represents the most difficult system category. This is often a dynamic system that is associated with limited knowledge of basic properties such as system linearity, system memory and the true interactive modes. As indicated earlier, one of the key goals of engineering design is to explore and transpire systems associated with the product being developed. The extent of meeting this goal will depend on the system category, reflected by its complexity and dynamic nature.

The above classification of systems can serve as a guideline for characterizing the logical evolution of a product concept during the design process. In the beginning of a design project, the problem idea is typically ill-defined and possible solutions are not clearly identified. This reflects a black-box system model. As more information is gathered and more brain-storming is consumed, design problems become more clearly defined and feasible solutions become possible, reflecting a gray or a transparent box model. However, an optimum solution to a design problem may not necessarily result in complete transparency of a system, as this can be time-consuming and a resources drain; this is where long-term scientific research can play a significant role. As indicated in Chapter 3, a great deal of scientific research is consumed in exploring the performance character-istics of products that have already been introduced into the market in order to gain a better understanding of how the product works. Other research activities are devoted to rationalizing the cause of failure of some existing products.

In the fiber-to-fabric system, one can find many examples of systems that belong to the different categories discussed above. On the machine side, the drafting system of staple fiber strands can be modeled by a gray-box system with respect to the relationships between the average values of the characteristics of input and output strands. However, when the dynamics of fiber flow through the drafting system are considered, a great deal of complexity is encountered, leading to a black box model. Indeed, the true dynamics of fiber flow through a drafting or a carding system have not been fully understood despite the extensive research activities in this area. On the end product side, one can list numerous products associated with systems that have been explored over the years for the purpose of making them more transparent. For example, a bulletproof vest system by virtue of its functional nature and the difficulty associated with its testing has often been treated as partially gray, partially black box system model.[1] The transparency of this system has been limited by the risky tests required to test the actual performance of this product and the lack of reliable analyti-cal tools fully to explore the system. In recent years, efficient computa-tional models that can simulate complex and high-energy impacts have been developed, producing results that are remarkably close to the results of real-life tests. This development has paved the way for entertaining multiple-component and high-speed deformable vest models.[3]

6.2.2 Model classification

In the engineering world, models may be classified in many different ways depending on the model structure and its intended purpose. Common clas-sifications include[2] static versus dynamic and deterministic versus probabi-listic. These represent general categories that mainly describe the behavior of model outputs or response variables. A static model is one whose prop-erties do not change with time. A dynamic model describes a phenomenon that is likely to change with time. A deterministic model is a model that describes system outcomes that occur with a high degree of certainty. A probabilistic model deals with events and system outcomes that are likely to vary in a random fashion. In addition, engineers may use other types of models such as iconic and analog models. The former is primarily used to describe entities rather than phenomena (e.g. maps, photographs, engineering drawings, wood or paper models), and the latter can be used to simulate real systems (e.g. common graphs, and process flow charts).

Regardless of the way models are classified, the common purpose of a model is either to explore a system or to predict its performance over time or under different levels of input values. In the design process, exploring a system can greatly assist in understanding the critical properties of its components and their interrelationships. When service life and potential failure of a product are of concern, prediction becomes a critical aspect of system identification. As indicated earlier, exploratory models are typically represented by mathematical formulae that can be derived from known physical laws (mathematical models). Predictive models often require statistical analysis to provide the capability to predict random and patterned outcomes (empirical models). In many applications, a combination of these two types of model may be required as random outcomes always represent a possibility of system performance.

6.2.3 Mathematical modeling

Mathematical modeling is a theoretical approach which typically aims to idealize a system through simplifying its components and making some assumptions which allow exploration of system behavior or prediction of patterned system changes.[1,2] Almost all design applications involve some form of mathematical modeling. A universal procedure for performing mathematical modeling does not exist. However, engineers should typically begin with a generalized conceptual model of the effects of critical system inputs and the anticipated responses of the outputs under consideration. In this regard, it is often useful to begin the modeling analysis by sketching all input parameters that can possibly influence a system and list all important output parameters. This procedure provides a great insight into the critical parameters influencing the overall performance of a product and the functional forms that may be needed to relate input and output parameters.

Figure 6.2 shows a generalized flow chart of a product represented by a system in which multiple inputs characterizing the raw material and multiple outputs characterizing the performance of end product are listed. Input parameters may include dimensional or geometrical parameters, mechanical parameters, thermal parameters and mix parameters (when a mixture of different materials is used). The performance of the end product

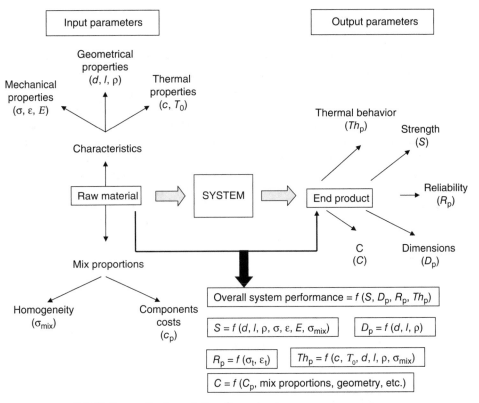

6.2 Mathematical modeling flow chart.

may be characterized by multiple output parameters characterizing product performance (e.g. thermal behavior, strength, reliability, dimension stability, cost, etc).

When complex products are considered, it will be useful to breakdown a system into subsystems in which fewer parameters can be analyzed at once. In addition, some assumptions can be made to simplify the initial modeling analysis. These assumptions often aim to reduce the task of system identification down to parameter identification (parameter-by-parameter analysis) so that numerical solutions can readily be obtained. In addition, some parameters may be neglected or maintained at constant levels as a result of their small influence on system performance or because of their predictable behavior. A good illustration of this simplification is modeling the body-clothing system, which has been performed using various mathematical modeling techniques. This model is discussed below to demonstrate the principles and the merits of mathematical modeling.

Mathematical model example: heat transfer through a clothing system

A body-clothing system involves many mechanisms such as heat transfer through the human body and clothing systems, moisture transport, pumping effects, breathability, ventilation and interaction with the wearer's activity.[4-8] A simple way to begin modeling this system is to assume dry clothing and utilize basic equations of heat transfer. In this case, the focus of modeling is on the heat transfer mechanism as shown in Fig. 6.3. This requires consideration of many parameters including[4,5] the metabolic heat produced by the body, heat transfer to the skin, changes in skin temperature and sweat and the intrinsic clothing insulation. The latter parameter may be modeled by another set of inputs including[6] the body surface area, skin–clothing temperature gradient and clothing thermal conductivity. The modeling analysis should also account for different forms of heat transfer in the body-clothing system. These include conduction, convection and radiation.

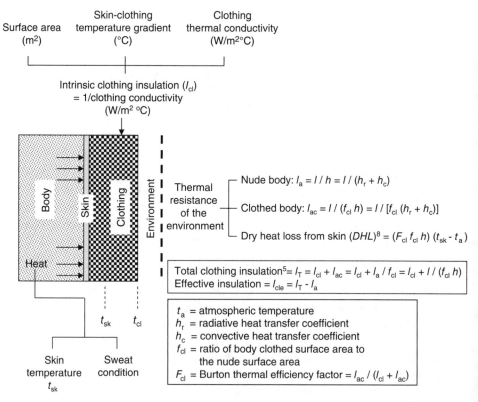

6.3 Mathematical modeling of heat transfer through clothing.[4-8]

The simple model of body-clothing system shown in Fig. 6.3 accounts for basic heat transfer equations including:[5]

- thermal resistance of the environment for a nude body (l_a)
- thermal resistance of the environment for a clothed body (l_{ac})
- dry heat loss from skin (*DHL*)

Since the surface area for heat transfer of a clothed body is increased by the thickness of the clothing layer, it is important to consider the parameter f_{cl}, which is the ratio of the clothed surface area of the body to the nude surface area of the body.[7] This is a difficult parameter to determine, particularly for an active body. As a result, it is typically estimated or obtained from photographic techniques, anthropometric scanners, or copper manikin measures. The total insulation associated with the simple body-clothing system model is a function of the intrinsic clothing insulation, l_{cl}, and the thermal insulation of the environment for a clothed body, l_{ac}. This parameter has to be adjusted for the effect of thermal insulation of the environment on a nude body, l_a, leading to the more convenient term, l_{cle}, which is the effective clothing insulation.

The above example represents a simplified model under which the body-clothing system is simulated under stationary conditions. This is a classic case of a combination of parameter identification. As a result, it provides only an approximation to more complicated dynamic situations or system identification. A design engineer may use this model as a guideline that must be followed by testing the thermal insulation of clothing under various environmental and physical activity conditions[8] in order to reach a complete simulation of the body-clothing under consideration. Alternatively, more mathematical modeling may be performed to accommodate other parameters. In this regard, a two-parameter model in which both wet and dry conditions are considered should be the next logical analytical step so that both heat and moisture transfer can be considered independently and interactively as moisture can indeed transfer heat between the body and the environment, particularly when the skin sweats. More detail of this sort of analysis will be presented in Chapter 12 in the context of thermo-physiological comfort and in Chapter 15 in the context of comfort-protection modeling.

The role of system behavior in mathematical modeling

One of the critical tasks in mathematical modeling is realization of the behavior of system variables or the way that one system variable behaves when another is varied. This realization can be the result of prior knowledge of the physical phenomena associated with the product or the system simulating the product, or of examination of engineering data. It is also

useful to breakdown the system into a number of input and output parameters that exhibit some form of relationship. Typically, the analysis of
parameter behavior begins with evaluating simple relationships each relating an independent variable, x, and a dependent variable, y. In this regard,
many questions can be addressed. These include:[1]

- What is the general behavior of y over the entire range of values of the
  variable x?
- What happens when the x value is small?
- What happens when the x value is large?
- Are there any x values for which the y value has local maximum or
  minimum?
- Are there any x values for which $y = 0$?
- Is there a trend in the x–y relationship?
- Is there a clear periodic pattern in the x–y relationship?
- Is there an asymptotic behavior in the x–y relationship?

In addition to exploring system relationships, the above questions also
represent important guidelines for optimizing product performance.
Examples of x–y relationships that can be observed in many engineering
systems are shown in Figs 6.4 and 6.5. Familiarity with these general relationships is a key requirement in mathematical modeling. It should be

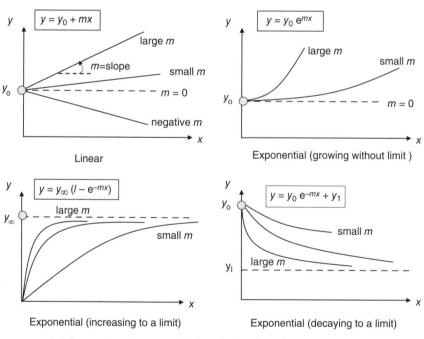

6.4 Examples of two-variable relationships.[1]

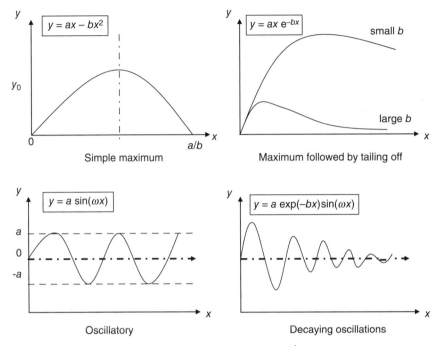

6.5 Examples of two-variable relationships.[1]

pointed out, however, that many mathematical relationships are based on assumptions that should be validated prior to using them. For example, the mechanical behavior of a fibrous product may be assumed to be purely elastic under small levels of external stresses for the sake of simplicity. This will yield simple stress–strain relationships or a simple model such as a spring system with known elastic constant. As mathematical modeling progresses, consideration of higher levels of external stresses may be necessary and the influence of time and temperature should be considered. This may lead to more in-depth modeling analysis to characterize the visco-elastic nature of fibrous materials. Similarly, fluid may be assumed to be Newtonian viscous, leading to simple models such as dashpots, particularly when deviation from this assumption will have a little influence on the system or product performance.

6.2.4 Empirical modeling

As indicated in the previous section, mathematical modeling is a theoreti-cal approach, which aims to idealize a system in order to explore or predict system behavior. Unfortunately, not all physical entities can be fully modeled using purely theoretical means. Hindering factors such as inadequate knowledge of all the relevant physical laws and complex computation often prohibit complete mathematical modeling. In addition, any physical phenomenon will be associated with some inherent variability or stochastic behavior. In these situations, theoretical models may be adequate to explore the average behavior of a system, yet inappropriate to predict the effects of possible noise or variability sources. Design analysis should ultimately lead to a realistic model that is fully supported by outcomes and verifiable by data under various conditions. As a result, the final model should be both exploratory and predictive in the full sense of these two criteria with greater emphasis on the model predictive power of both patterned behavior and possible random outcomes. These objectives can largely be met through empirical modeling.

Empirical modeling is an applied mathematical approach in which actual data related to the system plays a key role in both the modeling analysis and the verification process. The term 'empirical' implies practicality and possibility of application in the real world. This implication often leads to the common misconception that empirical modeling is a scientifically diluted approach. Indeed, it is quite the opposite. Reliable empirical modeling should make use of mathematical, statistical and physical concepts to complement the database used for developing the models.[1] The strength of an empirical model lies in the fact that despite the complex analysis involved, the model itself is represented by a simple input/output relationship that can be used by a wide range of personnel regardless of their qualifications. In this context, a reliable empirical model is a true representation of a successful marriage between theory and practice. Most empirical models are developed using well-established statistical techniques.[9-11] When many non-parametric variables are involved in the analysis and in situations where a great deal of subjectivity (owing to the lack of solid knowledge) is presented, other techniques such as neural network analysis or fuzzy logic techniques may be used to develop empirical models.[12,13]

Before performing empirical modeling, engineers should be fully aware of three basic differences between pure mathematical models and empirical models:[1]

- Empirical models are typically unique to the type of data and the specific process they are developed for. In other words, they often fail to generalize or hold for other systems or applications.
- The extent of revealing causes and effects by empirical models is limited as it often depends on the analyst's knowledge of the nature of the model and the underlying assumptions associated with the system.
- While mathematical models are bounded by balanced dimensions and units on both sides of the developed relationship, empirical models are

dimension-independent. For example, an empirical model may be used to relate the strength of a spun yarn, expressed in N/tex, to the speed of a machine expressed in m/min. This provides greater flexibility in exploring and predicting a system.

There are many methods that can be used to perform empirical modeling. The reliability and the success of any method will depend on several factors including[1] the database used for modeling, the extent of meeting the assumptions associated with the analysis, the criteria associated with the method used and, most importantly, the extent of understanding of the nature of data and the purpose of modeling. Empirical models may generally be divided into three main categories: (1) descriptive (or exploratory) models, (2) predictive models and (3) descriptive/predictive models. The first category aims to improve our understanding of possible interrelationships or interactions between the different variables used in the model. The second category aims to use the model to predict future values of the variables. The third category satisfies both the exploratory and the predictive functions.[1]

Regression models

The most common technique used in empirical modeling is correlation and regression analysis. Correlation is a way to determine the strength of the relationship between two variables, x and y. Statistics provides ways to determine how well a linear or a non-linear equation describes or explains a relationship between variables. Regression is a statistical procedure for estimating the parameters necessary in order to use a certain relationship in prediction applications.[9] The simplest form of relationship is the one that can be described by a linear function; that is when all data points in an x–y plot seem to lie near a line or follow a linear trend. This relationship is represented by the following regression model:

$$y = \beta_o + \beta_1 x + \varepsilon \qquad (6.1)$$

where β_0 and β_1 are called parameter estimates and ε is an error term. Regression analysis aims to estimate the parameters β_0 and β_1 by the corresponding estimates b_0 and b_1 leading to the simple regression equation:

$$y = b_o + b_1 x \qquad (6.2)$$

This equation is commonly called the best-fitting straight line. The estimates b_0 and b_1 are determined on the ground that the best-fitting line is the line which minimizes the sum of squares of the distances from the observed data points on the scatter diagram to the fitted line (see Fig. 6.6). This is the underlying concept of the familiar least-squares method,[9–11]

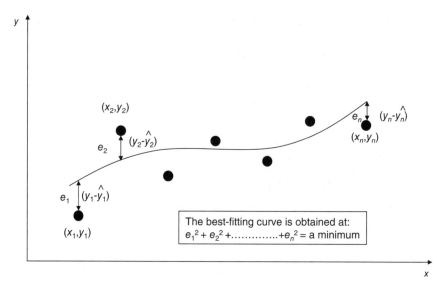

6.6 The concept of least squares.[1]

which represents the foundation of regression analysis. According to this method, the smaller the deviations of observed values from this line, the smaller the sum of squares of these deviations of the observed values from the line and, consequently, the closer the best-fitting line will be to the data. The least squares method solution aims to find the values of b_0 and b_1 (the estimates of β_0 and β_1) for which the sum of squares of the deviations is a minimum.

The strength of a linear relationship can be visualized by simply examining the trend and the cluster of points around the straight line. In this regard, there are three distinguishing situations that may be observed (Fig. 6.7): strong positive linear correlation, strong negative linear correlation and a weak relationship. The strength of the relationship can be determined statistically using the so-called coefficient of correlation; r. This is a dimensionless quantity, which varies from −1 to +1. The closer the value of r to −1 or +1 the more perfect the linear association between x and y. If r is close to zero, we conclude that there is a very weak linear association between x and y. This could mean no association at all or a highly non-linear relationship.

Since data is a key aspect in empirical modeling, it is important to use data in a plausible range unless the relationship is expected to be highly linear. This is because of the fact that different ranges of data may be associated with different forms of the x–y relationship. For example, consider the x–y relationship in Fig. 6.8. Over the entire range of data, the relationship exhibits a clear non-linear shape with an initial increase in y

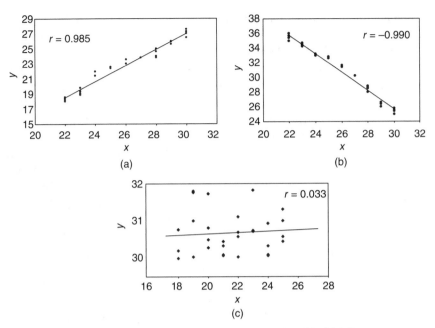

6.7 Different forms of a simple linear relationship.[1] (a) Strong positive linear association, (b) strong negative linear association, (c) weak linear relationship: no association.

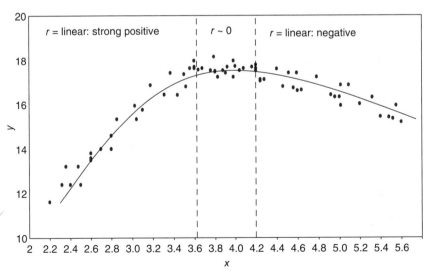

6.8 Importance of considering the entire plausible data range in developing a simple linear relationship.[1]

following an increase in x, followed by a leveling off of the y value with an increase in x, and finally a decrease in the y value with an increase in x. If only the initial range of data (i.e. from 2.2 to 3.6) is used to develop this relationship, a strong positive linear relationship will be obtained. If only the middle range is used (i.e. from 3.6 to 4.2), the relationship revealed will be a constant value of y over that range. If only the final range of data (i.e. from 4.2 to 5.6) is considered to develop this relationship, a negative linear relationship will be observed.

The key feature of regression modeling is the ability to predict system behavior. In this regard, the best prediction comes from a linear relationship or a well-defined non-linear relationship. In the absence of knowledge of the physical laws governing a system, data for input and response variables represent the only source of information. Therefore, it is important to examine the data reliability and the patterns generated by the data. It is also important to realize that a poor linear relationship does not mean a poor overall relationship between variables as this could mean the existence of a strong non-linear relationship. For example, consider the relationship between the rate of machine failure and production speed shown in Fig. 6.9. As shown in this figure, a linear relationship between the two variables yields a weaker coefficient of correlation than a non-linear relationship. This means that a non-linear model will be more suitable to represent the machine failure rate as a function of production speed.

When many independent variables are considered, multiple regression analysis should be performed to develop input/output relationships. Computations involved in multiple regression analyses are quite lengthy and

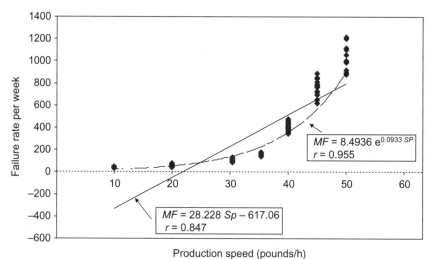

6.9 Machine failure versus production speed relationship.[1]

often complicated.[9-11] However, the availability of computers and powerful statistical software programs has made it possible to carry out these computations accurately and more efficiently. A multiple regression model may take one of the following forms:

$$Y = \beta_o + \beta_1 x + \beta_2 x^2 + \ldots + \beta_k x^k + \varepsilon$$

or   (6.3)

$$Y = \beta_o + \beta_1 x_1 + \beta_2 x_2 + \ldots + \beta_k x_k + \varepsilon$$

where x^i or x_j are the independent variables, β_0, β_1, β_2, ... β_k are the regression coefficients and ε is a random error. Again, the purpose of multiple regression analysis is to estimate the coefficients β_0, β_1, β_2, ... β_k at the minimum error using the familiar least-squares method.[9] In this regard, the random error ε is assumed to be normally distributed with mean of 0 and a variance of σ^2 (the variance of the response variable).

Earlier in this section, it was pointed out that the strength of a regression relationship is determined by the coefficient of correlation, r. In case of multiple regression, the square of this term, r^2 or R^2 is more useful for evaluating the strength of the relationship. In this regard, R^2 is called the coefficient of determination and it indicates the percent of information explained by the multiple-regression model in view of the independent variables considered. For example, a value of R^2 of 0.95 simply means that about 95% of the information about the response variable y can be explained by the independent variables x_i considered in the model and 5% is unexplained.

In the area of fiber-to-fabric engineering, regression analysis has been considered to be one of the most effective modeling approaches for fibrous structures. This is largely due to the many sources of variability associated with fibrous structures and their discrete nature. Examples of applications in which regression models were used include[14] correlations between different fiber properties of fibers, fiber-to-yarn models, cost-related models of fiber blending and yarn-to-fabric models. Many of these models were based on actual mill data supported by the findings of exploratory mathematical models of fiber-to-fabric relationships.[15]

6.3   Artificial neural networks

Artificial neural networks (ANN) are algorithms for cognitive tasks, such as learning and optimization that have gained their popularity from their resemblance to the human brain in terms of acquiring knowledge through a learning process and using interneuron connection strengths (known as synaptic weights) to store the knowledge. This resemblance is reflected in the following features:[16]

- Information processing occurs at many simple elements called neurons, units, cells or nodes.
- Individual neurons are gathered together into groups called slabs.
- Slabs can receive input (input slabs), provide output (output slabs) or be inaccessible to both input and output, with connections only to other slabs (internal slabs).
- Each connection link has an associated weight, which, in a typical neural network, multiplies the signal transmitted.
- Each neuron applies an activation function, usually non-linear, to its net input (or the sum of weighted input signals) to determine its output signal.

In light of the above features, a neural network typically consists of three basic elements:[17] architecture, activation function and learning methodology. The architecture represents a pattern of connections between the neurons. This is the particular way the slabs are interconnected to receive and handle input and output information. In this regard, a neural network typically consists of a large number of simple processing elements called neurons, units, cells or nodes. Each neuron is connected to other neurons by means of directed communication links, each with an associated weight. The weights represent information being used by the net to solve a problem. In addition, each neuron has an internal state, called its activation or activity level, which is a function of the inputs it has received. In principle, an activation function describes the output of a neuron given its input. Typically, a neuron sends its activation as a signal to several other neurons. Figure 6.10 a shows a neuron Y, which receives inputs from neurons X_1, X_2, and X_3. The activations (output signals) of these neurons are x_1, x_2 and x_3, respectively. The weights on the connections from X_1, X_2, and X_3 to

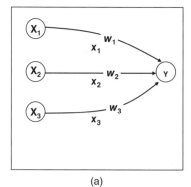 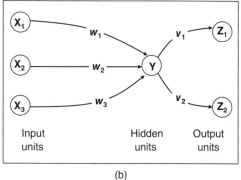

(a)   (b)

6.10 Principle of a neural network (modified from references 16 and 17). (a) Simple artificial neuron, (b) simple artificial neural network.

neuron Y are w_1, w_2 and w_3, respectively. The net input, y_{in} to neuron Y is the sum of the weighted signals from neurons X_1, X_2 and X_3. Mathematically, this is represented by the following expression:

$$y_{in} = w_1 x_1 + w_2 x_2 + w_3 x_3 \qquad (6.4)$$

The activation y of neuron Y is given by some function of its input, $f(y_{in})$. For example, a logistic sigmoid function (an S-shaped curve) is expressed as follows:[18,19]

$$f(x) = \frac{1}{1 + e^{-x}} \qquad (6.5)$$

Neuron Y may also be connected to other neurons Z_1 and Z_2, with weights v_1 and v_2, as shown in Fig. 6.10b. This is a case of a very simple neural network in which we have input units, output units and one hidden unit (a unit that is neither an input unit nor an output unit). Typically, the values received by neurons Z_1 and Z_2, will be different, because each signal is scaled by the appropriate weight, v_1 and v_2. In addition, the activations z_1 and z_2 of neurons Z_1 and Z_2 would depend on inputs from many neurons, not just the one shown in this simple neural network.

Similar to the human brain, the neural network can learn by example or training data. Commonly, the examples are pairs of input and output information. They are inputs for which the user knows the output or at least has some experience-based expectation of the outcomes. Training data can be obtained from historical problem data in which the outcomes are known, or by creating sample problems and solutions with the help of experts. Learning consists of determining a weight factor so that for every input, the network generates the corresponding output. This is called supervised learning. The weights are found by successively better approximations. First, a random guess is made. With these values of the weights, the output is calculated for a particular input. This output may not be totally correct, because the weights were guessed. However, this initial guess can then be adapted to minimize an error term describing the system behavior. This procedure can be used to update the weight in proportion to the error term. In this case, a learning rate is normally used to determine the stage of the learning process. This process is normally used in the commonly known back-propagation algorithm.[16–19]

A typical neural net will have three layers of neurons (see Fig. 6.11), each of which is connected to the neurons in the next layer. Each connection has a weight associated with it. Input values in the first layer are weighted and passed on to the hidden layer. Neurons in the hidden layer produce outputs by applying an activation function to the sum of the weighted input values. These outputs are then weighted by the connections between the hidden and output layer. The output layer produces the desired

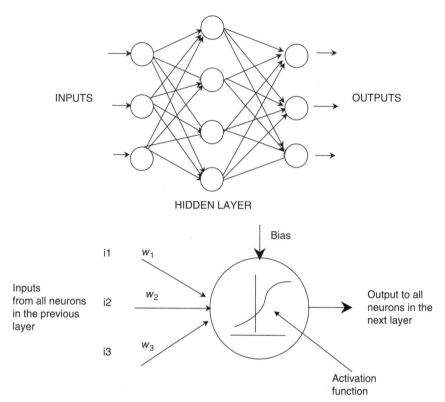

INPUTS

OUTPUTS

HIDDEN LAYER

Bias

i1 w_1

Inputs
from all neurons
in the previous
layer

i2 w_2

Output to all
neurons in the
next layer

i3 w_3

Activation
function

6.11 Elements of neural network modeling (modified from references
16 and 17).

results. The net learns by adjusting its interconnection weights repeatedly
so that the output neurons produce results close to the correct outputs in
the training data. Eventually, if the problem is learned, the weights become
stable.

Similar to regression modeling, neural networks are used for specific
applications and one application may not be universally applied to another
without significant adjustments and extensive learning. In this regard, it
should be pointed out that new inputs derived from other applications may
not follow the same input/output patterns as the ones used in the original
analysis. This necessitates a continuation of the learning process. Models
are often generalized for the sake of simplicity. In doing so, many gross
assumptions that involve a great deal of bias may be encountered.

In the context of design, neural network modeling is often used for
control purposes. Traditionally, it has been used in many applications such
as developing input–output relationships in order to predict non-linear
outcomes, noise suppressing on electronic devices, pattern recognition,

artificial diagnostic and treatments, speech recognition, business decisions, production control and logistics.[16]

6.4   Optimization analysis: linear programming

As indicated earlier, the design process is essentially an iterative analysis in order to reach the best or the most compromising solution to a design problem. This simple definition makes optimization an essential task of engineering design. In principle, any optimization analysis will aim to maximize the value of a desired parameter or minimize the value of an undesired parameter under some boundary conditions or subject to well-defined constraints.

One of the classic techniques of optimization is linear programming. This is a mathematical approach for maximizing or minimizing a linear objective function that specifies the benefit or cost associated with each decision variable. When many decision variables and many constraints are involved in the optimization process, the so-called 'simplex method' can be used to perform optimization analysis.[20,21] This is an iterative procedure that progressively approaches and ultimately reaches an optimal solution to linear programming problems. Computerized linear programming (LP) routines automatically arrange the inputs of the problem (objective functions and constraints), perform the iterative analysis and produce an output of the LP program. The general mathematical forms of the problem solved in the simplex method are discussed below.

Suppose that the goal is to minimize the cost of a mixture of raw material, say cotton and polyester that will be used in the design of a certain fabric structure. This is essentially a blending problem in which the decision variable is the proportion of a particular fiber type to be used in the blend or the fiber mix so that the cost of raw material is minimal. In this case, the objective function may be represented by a linear function of fiber cost, which can be represented by the following expression:[22,23]

$$\text{Minimize } Z(COST) = \sum_{i=1}^{i=k} a_i c_i = a_c c_c + a_p c_p \tag{6.6}$$

where Z is the total cost, a_i is the proportion of the ith fiber type in the blend and c_i is the cost of the ith fiber type.

Associated with this function, there may be a set of constraints that perhaps are dictated by some threshold values of fiber properties, some design criteria of the fabric to be made from these fibers, or even by supply and inventory conditions. Mathematically, these constraints are expressed by equations or inequalities, depending on the nature of the constraint. In this example, an obvious constraint is that the sum of proportions is unity (or a 100%):

$$a_c + a_p = 1 \qquad (6.7)$$

Constraints based on fiber characteristics can be expressed by formula of the following linear forms:

$$a_c L_c + a_p L_p \geq L_m$$
$$a_c F_c + a_p F_p \leq F_m \qquad (6.8)$$
$$a_c S_c + a_p S_p \geq S_m$$

where L_i, F_i and S_i are, say, fiber length, fiber fineness and fiber strength of the fiber types used, respectively, L_m, F_m and S_m are the corresponding threshold values desired for these characteristics and subscript c refers to cotton and p refers to polyester. In establishing these constraints, two basic assumptions are made: linearity and additivity. These assumptions may be valid from physical knowledge or they may be verified using modeling analysis.

When constraints associated with the design criteria of the end product (the fabric in this case) are required, fiber-to-fabric relationships should be available so that the design constraints can be driven directly from the fiber characteristics.[22,23] For the example above, suppose that fabric strength is a key design parameter that should be considered in the optimization analysis. In this case, two options may be entertained to account for this constraint. The first option is to consider the fiber strength constraint listed above as an indirect constraint on fabric strength. The choice of this option may be based on the assumption that for a given fabric structure, high fiber strength should result in high fabric strength. The second option is to use a pre-developed relationship in which fabric strength is a function of fiber properties. An empirical linear relationship may take the following form:

$$FS = b_0 + b_1 S + b_2 L + b_3 F \qquad (6.9)$$

where FS is fabric strength and L, F and S are fiber length, fiber fineness and fiber strength, respectively.

Using the above relationship, fabric strength can be estimated at different levels of fiber properties for each of the fiber types used in the optimization analysis:

$$FS_c = b_0 + b_1 S_c + b_2 L_c + b_3 F_c \qquad (6.10)$$
$$FS_p = k_0 + k_1 S_p + k_2 L_p + k_3 F_p \qquad (6.11)$$

Using these estimated values, a constraint of threshold fabric strength, FS_m can be established as follows:

$$a_c FS_c + a_p FS_p \geq FS_m \qquad (6.12)$$

It is important to point out that in using this option, the relationships developed for fabric strength should be highly reliable. In addition, one has to carefully examine whether the equation of fabric strength constraint actually meets the assumptions of linearity and additivity.[24]

The outcome of linear programming is represented by optimum values of the decision variables. For the example above, this will be the proportion of cotton and the proportion of polyester at which the cost of the blend is minimal and the different constraints are met.

6.5   Problem solving tools: genetic algorithms and simulated annealing

A genetic algorithm (GA) is a mathematical technique in which computer capabilities are used to mimic biological evolution as a problem solving methodology.[24-27] In principle, GA operates on the basis of establishing an input for a given specific problem in the form of a set of potential solutions to the problem, encoded in some fashion, and a metric called 'fitness function' that allows each solution to be evaluated quantitatively. Selected solutions may typically represent ones that are already known to work and others may be randomly generated. The candidate solutions are evaluated according to the fitness function. Since candidates are generated randomly, many may not work at all, but some may actually hold promise by pure chance. These are the ones that show effectiveness or 'activity' towards solving the problem.

Invalid solutions are deleted from the population of candidate solutions and promising candidates are kept and allowed to reproduce through making multiple copies of them. Since these copies may not be perfect, random changes are introduced during the copying process resulting in a new pool of candidate solutions that are also subject to a second round of fitness evaluation. Obviously, some random changes may result in either more inferior solutions that should be deleted or improved solutions, by pure chance, that should be selected and copied over into the next generation with random changes, and the process repeats. The expected outcome of this genetic algorithm is that the average fitness of the population will increase each round and hundreds or thousands of repeats can eventually lead to the discovery of very good solutions to the problem. In connection with neural network modeling, one can easily see the immense benefits of using genetic algorithm for training neural networks.

Key aspects associated with genetic algorithms are the method of representation of potential solutions, the method of selecting candidate solutions and the method of inducing random changes. The method of representation of potential solutions implies the technique used for encoding solutions in a form that can be understood by computers. The common encoding approach is to use binary strings: sequences of ones (1) and zeros (0), where the digit at each position represents the value of some aspect of the solution. Another method is to encode solutions as arrays of integers or decimal numbers, with each position representing some particular aspect of the solution. The latter approach allows for greater precision and higher resolution.[24] A third method is to represent individuals as strings of letters, where each letter stands for a specific aspect of the solution.[25] The idea underlying all these methods is to make it easy to establish operators that can cause the random changes in the selected candidates. These are called genetic operators.

The method of selecting candidate solutions, or the individuals to be copied over into the next generation, is a critical component of genetic algorithm. In this regard, there are many methods, some of which can be used in combination and others which are mutually exclusive. An obvious approach is to guarantee the selection of the best fit members of each generation. This is called the 'elitist selection'. This method may be modified so that the single best or a few of the best individuals from each generation are copied into the next generation.[26] It is also possible to implement a probabilistic approach in which more fit individuals are more likely, but not certain, to be selected. This method is called 'fitness-proportionate selection'. Using a probabilistic approach opens the door to many creative methods of selection.[27] One example is when a selection is made in which the chance of an individual's being selected is proportional to the amount by which its fitness is greater or less than its competitors' fitness. This simulates a game of roulette in which each individual gets a slice of the wheel, but more fit ones get larger slices than less fit ones. The wheel is then spun and whichever individual 'owns' the section on which it lands each time is chosen.

When all individuals have relatively high fitness or only small differences in fitness can be detected, 'scaling selection' may be used.[25,26] It is also possible to divide the population into subgroups of individuals. Members of each subgroup compete against each other so that ultimately only one individual from each subgroup is chosen to reproduce. This method is called 'tournament selection'. Another approach is to assign a numerical rank to each individual in the population based on its fitness. This yields a selection based on ranking rather than absolute differences in fitness so that dominance of very fit individuals can be prevented early at the expense of less fit ones. This method is commonly called 'rank selection' and it is intended to maintain the integrity of the population's genetic diversity.

It is also possible that the offspring of the individuals selected from each generation become the entire next generation.[27] In other words, no individuals are retained between generations. This is called 'generational selection'. Alternatively, the offspring of the individuals selected from each generation go back into the pre-existing gene pool, replacing some of the less fit members of the previous generation. In this way, some individuals are retained between generations. This is called 'steady-state selection'.

Inducing random changes follows the selection of fit individuals. The purpose of random changes or alteration is to improve the fitness of individuals for the next generation. The two basic methods used for random changes are 'mutation' and 'crossover'. Mutation is the simpler of the two and it simulates mutation in living things as it changes one gene to another.[25] Accordingly, mutation in a genetic algorithm causes small alterations at single points in an individual's code. Crossover simulates the analogous process of recombination that occurs to chromosomes during sexual reproduction. Accordingly, it implies choosing two individuals to swap segments of their code, producing artificial 'offspring' that are combinations of their parents.

In practical applications, random changes may include flipping a 0 to a 1 or vice versa, adding or subtracting from a certain value in a random fashion, or replacing one letter by another. A more sophisticated method of random changes is called 'genetic programming'. This method is based on using branching data structures called trees.[25,26] Using these structures, random changes can be made by changing the operator (see Fig. 6.12), altering the value at a given node in the tree, or replacing one sub-tree with another.

In the context of design, the merits of the genetic algorithm have been numerous and apply in various applications. A few applications are listed below with their corresponding references:

- task assignment and dynamic scheduling of manufacturing jobs[28,29]
- control systems design[30]
- optimization of multi objectives[31,32]
- design of conductive polymers[33,45]
- applications in molecular design[34]
- telecommunications and network routing.[35,36]

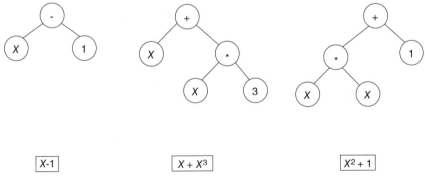

$$X-1 \qquad X + X^3 \qquad X^2 + 1$$

6.12 Genetic programming: examples of simple program trees (modified from references 24–27).

Another advanced problem solving technique is so-called simulated annealing. This is essentially an optimization analysis that derives from the familiar annealing process in which a material is heated to above a critical point to soften it, then gradually cooled in order to erase defects in its crystalline structure, producing a more stable and regular lattice arrangement of atoms.[37] Similar to genetic algorithms, simulated annealing uses a fitness function that defines solution suitability. However, unlike genetic algorithms in which a population of candidates is dealt with, simulated annealing only deals with the fitness of one candidate solution. In addition, and by virtue of its simulation, it uses the concept of 'temperature' as a global numerical quantity, which gradually decreases over time.

At each step of the analysis, the solution mutates (similar to moving to an adjacent point in the fitness landscape). The fitness of the new solution is then compared to the fitness of the previous solution; if it is higher, the new solution is kept. Otherwise, the algorithm makes a decision whether to keep or discard it based on temperature. If the temperature is high, even changes that cause significant decreases in fitness may be kept and used as the basis for the next round of the algorithm, but as temperature decreases, the algorithm becomes more and more inclined only to accept fitness-increasing changes. Ultimately, the temperature may reach zero, at which point the system 'freezes' leading to the optimum solution at this point.

6.6   Modeling human judgment: fuzzy logic

In situations where a great deal of human judgment must be made about a product or a system, most traditional models fail to characterize fully the nature of human judgment. In this case, a method of rule-based decision making should be used in conjunction with perhaps artificial intelligence systems and process control. The objective of this would be to emulate the rule-of-thumb thought process used by human beings. Such a method is now commonly known as 'fuzzy logic', a revolutionary approach to system identification and control pioneered by Lotfi Zadeh in 1965. The principles underlying this approach are discussed below using a simple example. Detailed discussion of the mathematical concepts underlying fuzzy logic is obviously outside the scope of this book. Interested readers can refer to the numerous publications on this subject including some that are cited in the references of this chapter.[38-41]

6.6.1 Modeling classer cotton grading using fuzzy logic membership functions

Suppose that the goal is to establish objective evaluation of the cotton grade. Traditionally, cotton is graded in the marketplace by subjective classing in which a classer observes the appearance and the color of fibers and touches the fibers to determine their smoothness or roughness to the hand. Quantitatively, the color appearance is measured using color reflectance, Rd, and fiber roughness can be evaluated through testing a combination of parameters including fiber friction, bulk resiliency under cyclic pressing and un-pressing of the fibers, and the energy consumed to open the fiber bulk.[14] This simple case illustrates the superiority of expert human judgment of certain phenomena to laboratory testing, as it takes many quantitative parameters and a significant amount of time to make a judgment of cotton grade that a classer can make in a few seconds. When a cotton classer observes, touches and handles the cotton sample, all these quantitative parameters come into play in a very complex interactive fashion that only the human brain supported by experience can comprehend. In recent years, efforts were made to convert subjective grading of cotton into instrumental grading. These efforts involved comparative analysis between subjective measures and instrumental measures, with expert classer opinion being one of the main references for calibrating the objective measures. This is a case where fuzzy logic analysis can be very useful as it can result in characterizing human judgment on the basis of a combination of expert opinion supported by instrumental grading.

To demonstrate the above point, let us take, for the sake of simplicity, one of the quantitative parameters, say color reflectance, Rd, and assume that it is the only quantitative parameter required to make a judgment on the cotton grade. The traditional approach to relating the quantitative value (Rd values) to human judgment has been based on the discrete classification of the parameter of interest. For example, experts in the field may come together and establish the following criteria:

- poor grade $<75\ Rd$
- middle grade $75\text{--}80\ Rd$
- good grade $>80\ Rd$

These criteria may be represented by the simple graph shown in Fig. 6.13. The number zero is given for both poor and good grades, and the number 1.0 is given for middle grade cotton. This binary judgment is represented in Fig. 6.13 by a crisp set, or a set that only involves discrete judgment (e.g. middle grade or not, in this case). This approach is very common in many human judgments owing to its simplicity. Indeed, people often use terms such as 'good or bad', 'true or false' and 'stop or go' to characterize their judgments. The problem with this approach, however, is that it often masks a great deal of information and judgment resolution. A cotton classer being a human may easily judge extreme conditions such as extremely dark or extremely light color. As colors deviates from these

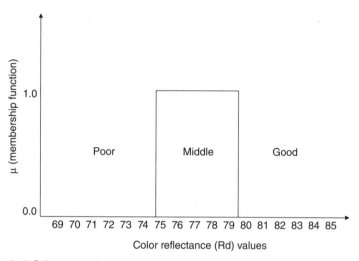

6.13 Crisp set–middle cotton grade.

extremes, human judgment may become progressively fuzzy. In this case, binary scoring may not be accurate.

If the grading criteria of *Rd* listed above are accepted by experts, we will find that the classer's judgment of very poor and very good grades represent a clear cut. As we approach the middle grade from the low or the high value, different classers may have different opinions about whether the cotton is 'poor to middle', 'definitely middle' or 'middle to good'. This fuzziness can be resolved by the so-called fuzzy logic membership function. This is developed based on asking a panel of classing experts to vote on the matter. The fraction of the panel that regards cotton with a given *Rd* value as 'middle grade' will give a number from 0 and 1, which will indicate the strength of their judgment. An expert who examines a cotton sample of, say, 77 *Rd* value, may determine that it is too close to the poor category, but not close enough to be judged poor. In this case, he/she may give a number close to zero (say 0.3). This number implies the uncertainty of his/her judgment about how to consider this cotton sample, a 'middle' or 'poor grade'. Similarly, an expert who examines a cotton sample of, say, 79 *Rd* value, may determine that it is too close to the good category. As a result, he/she may again give a number close to zero; say 0.2 which implies a judgment uncertainty of middle grade. Finally, an expert who examines a cotton sample of, say, 78 *Rd* value, may determine that it is definitely middle-grade cotton and gives a number 1 or close to 1 (say, 0.8) to imply judgment certainty.

The above example illustrates one way in which fuzzy logic handles the extent of certainty of human judgment; it is through a continuous pattern, not a discrete pattern, of decision-making. The fuzzy set arrived at in this way is illustrated in Fig. 6.14. It indicates a more resolute judgment, particularly in the representation of 'middle cotton grade'. Fuzzy sets corresponding to poor or good grades might also be defined in the same way. This is illustrated in Fig. 6.15. As shown in this figure, the three sets developed overlap. This reflects the fact that a certain value of color reflectance may lead to a two-way judgment. The figure also indicates that there are no sharp changes in *Rd* groups. As *Rd* increases, membership of the 'good grade' gradually increases to 1.0; or as *Rd* decreases, membership of the 'poor grade' gradually declines to 0. Thus, each of the fuzzy sets described in Fig. 6.15 can be regarded as the definition of a corresponding linguistic value, in this case 'poor grade', 'good grade' and 'middle grade'. This point leads us to two different variables related to grade:

1. Color in *Rd*: a numerical variable with integer numerical values
2. Color group: a linguistic variable taking the linguistic values 'poor grade', 'good grade' and 'middle grade'.

In the context of determining cotton grade, color *Rd*, though numerical, represents the underlying measurement, which in this case drives everything else, but it is too fine a level of detail to be easily interpreted. Color group, on the other hand, ranges over a more limited set of values. This makes it easier to comprehend and easier to use.

The simple example discussed above illustrates the concept of membership function. In practice, there may be many parameters and many ways to characterize each parameter. In addition, parameters may indeed interact with one another leading to many decisions regarding the subjective

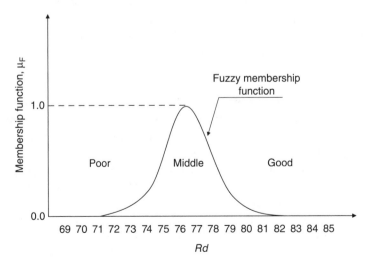

6.14 Fuzzy set–middle cotton grade.

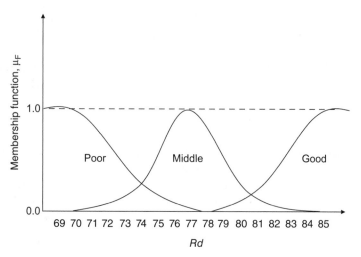

6.15 Fuzzy sets for cotton grade categories.

measure. As we indicated earlier, cotton grade reflects many parameters that are interacting in a complex manner. Accordingly, the above approach should be expanded to accommodate all these parameters, individually and combined. This gives rise to a combination matrix of inputs, in response to which a fuzzy logic controller or a decision scheme can be developed.

Mathematically, the concept of membership function can be understood by defining some basic terms used in fuzzy logic.[38,39] The first one is the set in a universe of discourse. There are two types of set: an ordinary 'crisp' set and a 'fuzzy' set. A crisp set is a well-defined set. A characteristic function X_A can be used to characterize a crisp set A. Each item in the universe is given a corresponding value of either 1 or 0 depending on whether the item is either in the set (true) or not (false). The mathematical notation in this case is the familiar set notation:

$$A = [x | P(x)] \qquad (6.13)$$

where the set A consists of those items x for which the property P is true. In this case, $P(x)$ is true, if and only if $X_A(x) = 1$. Accordingly, the membership value assigned to an item is restricted to just two possibilities, 0 or 1 (this point was illustrated in Fig. 6.13).

Fuzzy sets are based on detailing the range of the characteristic function so that it covers the real numbers in the interval. In this case, the membership value will have limiting levels of 0 and 1, but it can take on any value in-between. This is basically like saying that true or false can be relative according to people judgment. By analogy with the characteristic function of the crisp set, the membership function will exhibit two extreme values of 0 and 1, and an infinity of possible values in-between. To illustrate this point, suppose U is a collection of objects denoted by $\{u\}$. U is called the universe of discourse and u represents a generic element of U. A fuzzy set F in a universe of discourse U is characterized by a membership function μ_F, which takes values in the interval $[0,1]$.

A fuzzy set F in U is usually represented as a set of ordered pairs of elements u and grade of membership value:

$$F = [(u, \mu_F(u)|u \varepsilon U]$$ (6.14)

If U is continuous, a fuzzy set F could be written concisely as:

$$F = \int U \mu_F(u)/u$$ (6.15)

If U is discrete, a fuzzy set F could be written concisely as:

$$F = \sum \mu_F(u_i)/u_i$$ (6.16)

where the $\int$ or Σ refer to set union rather than to arithmetic summation, and the solidus '/' is simply used to connect an element and its membership value and has no connection with arithmetic division. This point was illustrated in Fig. 6.14 and 6.15.

The concept of membership function is often confused with the probability concept. Obviously, many statisticians believe they have solutions to all the problems in the world. Yes, there is a similarity in that both membership values and probability values exhibit a range from 0 to 1. However, the interpretation of this range of values is substantially different in the two cases. Fuzzy membership provides a measure of judgment, whereas the probability indicates the proportion of times the result is true or false in the long run.

It should be pointed out that the above example was only presented as a simple demonstration of some of the critical elements of fuzzy logic analysis. In recent years, fuzzy logic analysis has been used in numerous applications in which human judgment was accurately represented or simulated for control or modeling.[40,41] In the area of developing fibrous products, many subjective phenomena such as appearance, hand and comfort can be characterized using fuzzy logic analysis.

6.7   Finite element analysis

The subject of design analysis cannot be fully completed without discussion of one of the most commonly used analyses in engineering design, that is finite element analysis. This analysis represents a combination of simulation, modeling and optimization analysis that is based on considering that a solid or a fluid is built up from numerous tiny connected elements. The fact that those tiny elements can exhibit numerous arrangements results in the possibility of modeling even the most complex shapes by analyzing the stress–strain behavior of the different elements. In the context of design, finite element analysis represents one of the most powerful analytical approaches for reaching optimum designs without the need for formulating an idealized model that must be later evaluated to examine its deviation from the real product. Using finite element analysis, design engineers can indeed verify a proposed design and examine client's specifications prior to manufacturing or construction. In addition, it can be used effectively in modifying an existing product or structure for the purpose of improving its performance. Examples of design applications in which finite element analysis can be used are listed in Table 6.1.

In principle, a loaded structure is modeled with a mesh of separate elements that may take different shapes as shown in Fig. 6.16. The elements are connected to one another using connecting points called nodes. Accordingly, a mesh grid of many nodes is formed in which the material and structural properties are defined to describe the reactions of the structure to certain loading conditions. A key aspect of the analysis is the density of nodes throughout the structure.[42,43] This should be assigned in accordance with the anticipated stress levels in particular regions. Areas that are expected to exhibit high or complex stress distributions should have a higher node density than those which exhibit little or no stress.

Table 6.1 Examples of applications of finite element analysis[3,38,39]

Application	Description
Structural analysis	• Stress–strain analysis of various structures • Linear models using simple parameters and assuming that the material is not plastically deformed • Non-linear models via stressing the material past its elastic capabilities
Fatigue analysis	• Prediction of the life of a material or structure via analysis of the effects of cyclic loading • Analysis can show the areas where crack propagation is most likely to occur. • Failure due to fatigue may also show the damage tolerance of the material.
Vibration analysis	• Testing a material against random vibrations, shock and impact • Each of these incidences may act under the natural vibration frequency of the material which, in turn, may cause resonance and subsequent failure.
Heat transfer analysis	• Modeling the conductivity or thermal fluid dynamics of the material or structure • This may consist of a steady-state or transient transfer. • Steady-state transfer refers to constant thermoproperties in the material that yield linear heat diffusion.

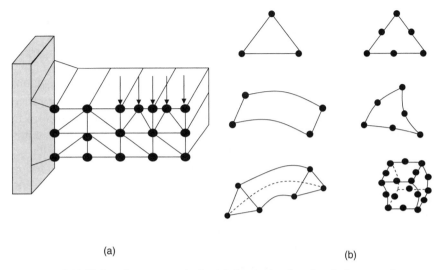

(a)   (b)

6.16 Finite element analysis. (a) Analysis of a simple beam using finite element analysis, (b) common elements used in finite element analysis.

Finite element analysis utilizes basic stress–strain equations to determine the deflection in each element by the system of forces transmitted from adjacent elements through the nodes. The stress is determined from the strain, which is in turn determined from the deflection of the nodes. There are many ways to conceptualize finite element analysis. One may consider the mesh as a spider web in which from each node, there extends a mesh element to each of the adjacent nodes. This web of vectors is what carries the material properties to the object, creating many elements. One can also think of the elements of a mesh grid as a group of springs each of which deflects in accordance with the external loads applied until all forces are in equilibrium. It is important therefore to use matrix algebra to deal with this complex system of simultaneous equations. It is also important to consider the stiffness matrix for each element. By analogy with a spring system, the deflection of nodes under a system of applied forces can be described by the matrix notation, $\{f\} = [k]\{\delta\}$, where $\{f\}$ is the column matrix (vector) of the forces acting on the element, $[k]$ is the stiffness matrix for the element and $\{\delta\}$ is the column matrix of the deflections. The stiffness matrix is constructed from the coordinate locations of the nodes and the matrix of elastic constants of the materials.

When all the elements of the systems are assembled, the basic matrix equation is $\{F\} = [k]\{\delta\}$, where $\{F\}$ is the matrix of external forces at each node, $[K]$ is the master stiffness matrix, assembled from the $[k]$ for all the elements and $\{\delta\}$ is the column matrix of the displacements at each node.

The force matrix is determined from the numerical values of loads and reactions computed prior to the start of the finite element analysis. The unknown displacements are determined by transposing the stiffness matrix.

With today's computing power, the mathematical details of finite element analysis are automated so that the analysis can be used for a wide range of applications. In practice, there are generally two types of analysis: 2D modeling and 3D modeling. The former is obviously simpler and it tends to yield less accurate results. The latter produces more accurate results but requires high computing capabilities. Within each of these modeling schemes, numerous algorithms (functions) can be utilized, some yielding linear system behavior and others yielding non-linear behavior. Obviously, linear systems are less complex but they generally do not take into account plastic deformation. Non-linear systems do account for plastic deformation and many also are capable of testing a material to the fracture point.

Many of the available computer programs used for finite element analysis also include a wide range of objective functions (variables within the system) for minimization or maximization of various parameters such as mass, volume, temperature, energy, force, displacement, velocity and acceleration. In addition, many finite elements analysis programs have readily available sample elements that can be used for various applications. Examples of these elements include: rod elements, beam elements, plate/shell/ composite elements, shear panel, solid elements, spring elements, mass elements, rigid elements and viscous damping elements. Some programs are also equipped with the capability to use multiple materials within the structure such as isotropic (or identical throughout), orthotropic (identical at 90 degrees) and general anisotropic (different throughout).

6.8   Failure analysis

Failure analysis is commonly performed by design engineers to examine potential failures in some materials or components of a product assembly or to diagnose the cause of failure after a failure has occurred. Normally, analysis performed after the occurrence of failure can be time consuming and often very costly, as the failure incident must be recreated (e.g. reassembling an aircraft after a crash or examining a bulletproof vest after a bullet penetration). Commonly, failures occur owing to design errors which are often attributed to structural failures resulting from inappropriate material selection or deficiencies in material properties. The case of the failure of the supersonic Concorde airplane mentioned in Chapter 4, which led to a major disaster in 2000, was primarily material related (tire failure). Another example of a disastrous failure was the crash of a German high-speed passenger train in 1998. This was attributed to a damaged wheel caused by metal fatigue. Numerous other failure incidents can be mentioned in which inappropriate material was the primary cause. The purpose of failure analysis is to determine the primary causes leading to failure, develop ways to prevent future failure occurrences and modify design and manufacturing procedures in order to develop a product or system that is robust against potential failures.

Causes of failure may be divided into two main categories:[2] design deficiencies and inappropriate material selection. Examples of design deficiencies include inadequate consideration of stress concentration points, lack of knowledge of service loads and environment, and difficulty of stress analysis in complex parts and loadings. Material-related deficiencies include poor match between service conditions and selection criteria, inadequate data on material, too much emphasis given to cost and not enough to quality, imperfections in material caused during manufacturing, inadequate maintenance and repair, and environmental factors.

One of the key aspects of failure analysis is testing and evaluation of material before and after failure. This requires two main types of testing: mechanical testing and microscopic examination. Mechanical testing of material should be performed under conditions simulating structural failure. It aims to test material strength, toughness and stiffness under loading conditions equivalent to (or extrapolatable to) the anticipated stresses that are likely to be applied on the material in a product form. Failure over time should be tested using fatigue tests such as cyclic stress, abrasion, friction and wear tests. When external effects such as temperature or moisture are anticipated, tests should be performed under different levels of these effects. Microscopic testing aims to evaluate structural points at which failure may potentially occur, or examine material after failure (e.g. stress concentration points and crack propagation).

In performing failure analysis, it is important to define the meaning of failure as it may imply different things for different products. In the above discussion, failure was primarily a structural failure under stress. For an automobile airbag, failure could mean a malfunction in the deployment mechanism or a gas leak. For a bulletproof vest, failure may imply unexpected bullet penetration at low speed. For commercial carpets, failure could mean wear out of fibers under abrasion stresses (fatigue) or failure to resist some forms of stains. The need to identify the meaning of failure stems from the strong relationship between the failure mechanism and the specific causes of failure.

In general, failure analysis involves three basic tasks: cause and effect analysis, techniques of failure evaluation and corrective or preventive actions. The first task is primarily a quality control and maintenance task. Cause and effect analysis can be performed using many techniques.[1] The most common technique is the so-called 'cause-and-effect diagram', which is typically developed through brainstorming of various causes of an effect. This requires generating and stimulating the maximum number of ideas by a team consisting of all personnel familiar with the effect and its associated causes, including manufacturers, design engineers and quality control personnel. In general, the specific failure is identified (the effect), then the main potential causes of the problem effect are suggested by different personnel. Once the main potential causes are determined, a closer look at each main cause should be made to determine how it contributes to the failure problem. In this process, many specific sub-causes under the main cause category may be listed. The final stage of cause-and-effect analysis is to narrow down the causes to a few causes that are believed to have a direct contribution to the failure incident. Figure 6.17 shows an example of a cause-and-effect diagram for the failure of automobile tires. This is only a generalized example to demonstrate the concept of a cause-and-effect diagram. As shown in this diagram, the main causes of tire failure assigned by the experts for this case were[1] manufacturing flaws, service conditions, design-related causes and material-related causes.

In situations where multiple effects associated with multiple modes of failure are presented, failure analysis can be performed using the so-called failure-experience matrix.[2] The matrix represents a three-dimensional assemblage of information cells. As shown in Fig. 6.18, one dimension describes the functions of the product that have potential impacts leading to failure, the second describes the various potential failure modes and the third describes possible corrective/preventive action. The example of the failure-experience matrix shown in Fig. 6.18 demonstrates the analysis of

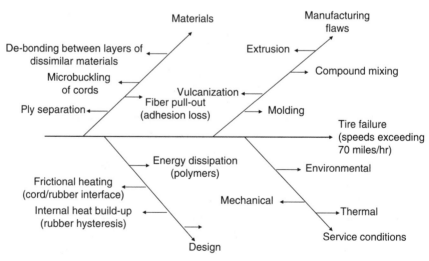

6.17 Cause-and-effect analysis of tire failure.

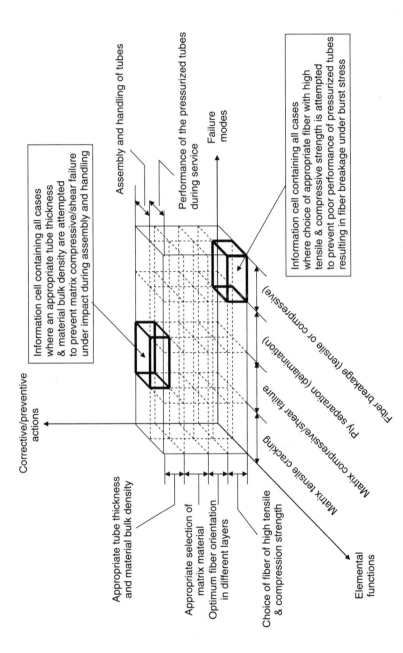

Corrective/preventive actions

Information cell containing all cases where an appropriate tube thickness & material bulk density are attempted to prevent matrix compressive/shear failure under impact during assembly and handling

Assembly and handling of tubes

Performance of the pressurized tubes during service

Failure modes

Information cell containing all cases where choice of appropriate fiber with high tensile & compressive strength is attempted to prevent poor performance of pressurized tubes resulting in fiber breakage under burst stress

Fiber breakage (tensile or compressive)

Ply separation (delamination)

Matrix compressive/shear failure

Matrix tensile cracking

Appropriate tube thickness and material bulk density

Appropriate selection of matrix material

Optimum fiber orientation in different layers

Choice of fiber of high tensile & compression strength

Elemental functions

6.18 Failure against experience matrix of thin-walled composite tubes under low energy impact (developed from a series of work in this area).[1,2,46]

filament-wound composite thin tubes that may be made from fiberglass or carbon composites for various applications such as pressure vessels or automobile components. The key concerns are failure caused by accidental impact by drop weights during assembly or handling, and potential failure caused by tube bursting under pressure. These are common problems with this type of product that often result in progressive failure imposed by cracks and crack propagation. Note that the failure-experience matrix reveals information cells that contain actions needed to prevent particular failure modes corresponding to some elemental functions, as shown in the two selected cells in Fig. 6.18.

Both cause-and-effect and failure-experience matrix analyses aim to gather efforts to focus on the true causes of failure and identify the few that are most probable. These analyses must be supported by more technical failure analysis in which structural and environmental effects are considered. This type of analysis will obviously vary depending on the nature of failure and whether it is a potential failure or a failure that had already occurred. For potential failure, the modeling techniques and the finite element analysis discussed earlier can be very useful. Post-failure analysis is one task that engineers typically are not pleased to do. It is typically performed under the influence many surrounding factors particularly when failure was associated with loss of life or a pressure to recall products. As indicated earlier, post-failure analysis requires re-creation of the failure incident, which may take time and effort. In addition, some form of backward projection and modeling analysis may be necessary to determine the root causes of failure and provide guidelines or regulations to prevent future incidents.

6.9   Reliability and survival analysis

One of the key aspects of design is to assure that a product will last its expected service life without a failure. The common measure of this assurance is called reliability or survival of a product over time. A major contributor to product or system reliability is material properties. A material of low strength and low resistance to temperature changes is likely to degrade over time leading to an unreliable product. Similarly, mixing or assembling two materials that have incompatible properties, particularly at their interface, will also result in an unreliable product or system.

In theory, reliability $R(t)$ is the probability that a product (or system) will perform without failure for a specified period of time under specified operating conditions. This definition yields the following expression:[46]

$$R(t) = P(t > t_0) \tag{6.17}$$

where t is a continuous random variable denoting time to failure. The parameter, t_0, indicates some specified time and it should be treated as a product characteristic value.

The definition given above can only be verified through evaluation of the probability of the time-to-failure of a certain material or a product under realistic operating conditions via a simulating test or actual use. Obviously, different materials will exhibit different levels of physical reliability depending on their durability characteristic such as strength, ductility, stiffness, temperature resistance, corrosion and other physical properties. Typically, the reliability of metals can be determined with a greater degree of certainty than that of non-metal materials. This is due to the low to moderate variability in metal characteristics particularly at moderate temperatures.[44] At elevated temperatures, all materials including metals will be associated with greater uncertainty in their fatigue or failure behavior.

The chance of failure, or unreliability, $F(t)$, represents the complementary probability of reliability with respect to time. Thus:

$$R(t) + F(t) = 1 \tag{6.18}$$

Suppose that there are N_t components in a system that were tested for failure at some point in time, t. If the number of surviving components is $N_s(t)$ and the number of failing components is $N_f(t)$, then:

$$N_t = N_s(t) + N_f(t) \tag{6.19}$$

This equation leads to one of the common expressions of reliability, $R(t)$:

$$R(t) = \frac{N_s(t)}{N_t} = 1 - \frac{N_f(t)}{N_t} \tag{6.20}$$

Reliability analysis can also be performed after a component has survived for some time, t_1. This leads to another parameter directly related to reliability, which is called hazard rate or failure rate, $h(t)$. This is defined as the chance that a certain component selected randomly will fail in the time period between t_1 and $t_1 + dt_1$ when it already has survived t_1. When many random system components are tested, a frequency distribution and a cumulative frequency distribution of time-to-failure can be obtained as shown in Fig. 6.19a and 6.19b, respectively. From these distributions, a hazard rate function can be obtained as follows:[2]

$$h(t) = \frac{f(t)}{1 - F(t)} = \frac{f(t)}{R(t)} = P(t_1 \leq t \leq t_1 + dt) \quad t \geq t_1 \tag{6.21}$$

Hazard rate also reflects the instantaneous failure rate, or the number of failures during a unit time per the number of items examined in the same time period. This is classically defined by:

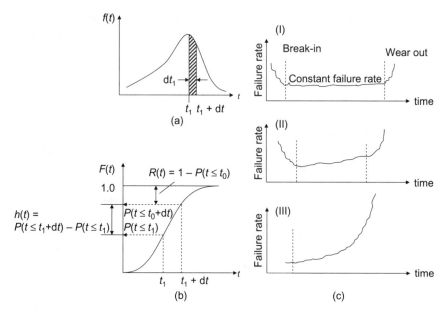

6.19 Reliability functions.[1,2] (a) Time-to-failure frequency distribution, (b) time-to-failure cumulative frequency distribution, (c) examples of failure–time profiles.

$$h(t) = \frac{dN_f(t)}{dt} \frac{1}{N_s(t)} \qquad (6.22)$$

In light of Equations (6.21) and (6.22), the value of the hazard rate is represented by a frequency value (or percent of components) determined over a certain period of time before failure occurs. Commonly, this is expressed as percent per hour unit (e.g. 1% per 1000 h, or 10^{-5} per hour). Components in the range of failure rates of 10^{-5} to 10^{-7} per hour exhibit a good commercial level of reliability.[2]

Failure in product or system components over time may take different shapes depending on many factors such as[1] product type, the nature of use, maintenance, quality control and other unpredictable factors. Figure 6.19c illustrates some examples of failure patterns over time. Case I in this figure is commonly known as the 'bathtub curve'. It indicates that for some products, the rate of failure is high at the initiation stage of the product life-cycle. This is commonly called 'infant mortality' of products, which typically arises from design errors or manufacturing defects that have not been detected, perhaps due to time-to-market intense pressure. These early failures are normally expected and they can be overcome through extensive initial testing of product performance and strict quality control. After the initiation period, product performance typically stabilizes and, except for few random occurrences, the rate of failure becomes more or less constant. Finally, and after the product reaches its expected service life, failure rate will increase as a result of wear out. Most electronic equipments follow this failure model. Case II in Fig. 16.9c represents a failure profile in which a constant failure rate does not exist; instead after the initial break-in period, the failure rate continues to increase until the end of wear out period. Most production machines follow this profile. Case III in Fig. 16.9c represents no initial break-in period, instead a steady rise in failure rate reaching maximum levels at the wear out phase. This case represents products that are intended for short-period use such as dust-air filters.

6.9.1 Constant failure rate analysis

In theory, a constant failure rate may be expressed by the condition, $h(t) = \lambda$, where λ is the number of failures per unit time. Using the classic characteristics of the frequency distributions:

$$f(t) = \frac{\mathrm{d}F(t)}{\mathrm{d}t} = \frac{\mathrm{d}[1 - R(t)]}{\mathrm{d}t} = \frac{-\mathrm{d}R(t)}{\mathrm{d}t} \tag{6.23}$$

From Equation (6.21):

$$h(t) = \frac{f(t)}{R(t)} = \frac{-\mathrm{d}R(t)}{\mathrm{d}t} \frac{1}{R(t)} \tag{6.24}$$

Thus,

$$\frac{-\mathrm{d}R(t)}{\mathrm{d}t} = h(t)\,\mathrm{d}t$$

or

$$\ln R(t) = -\int_0^t h(t)\,\mathrm{d}t$$

Accordingly:

$$R(t) = \exp\left[-\int_0^t h(t)\,\mathrm{d}t\right] \tag{6.25}$$

Thus, at a constant rate of failure, $h(t) = \lambda$.

$$R(t) = \exp\left[-\int_0^t \lambda\,\mathrm{d}t\right] = e^{-\lambda t} \tag{6.26}$$

The above equation indicates that the reliability $R(t)$ of a product under a constant rate of failure, λ, is an exponential function of time in which product reliability decreases exponentially with the passing of time. The parameter λ is related to the mean time between failures, T, via $T = 1/\lambda$. This yields the following expression:

$$R(t) = e^{-\lambda t} = e^{-t/T} \tag{6.27}$$

The failure rate, λ, or the mean time between failures (MTBF), can be determined from past history of the performance of a product or system, or through testing systems over specified periods of time during which system failures are expected. These should be considered as characteristic values of systems or products. Knowledge of these values can assist in determining the reliability for a given time period, t. For example, if a certain component is known to have a failure rate of 0.2 failures per one million hours (0.2×10^{-6}), the reliability of this device over a period of say 100 000 hours will be:

$$R(100\,000) = E^{-\lambda t} = e^{-(0.2 \times 10^{-6}) \cdot 100\,000} = E^{-0.02} = 0.9803$$

This means that the component will have a chance of survival over 100 000 hours of about 98%. Suppose that the producer of this component has distributed 5000 units to different users, then $N_s = N_t . R(t) = 5000 \times 0.98 = 4901$. Accordingly, about 98 components are likely to fail during the 100 000 hours. Obviously, this is a significant failure record that must be overcome by selecting more appropriate material, modifying product design, or implementing strict quality control.

6.9.2 Variable failure rate

In the above example, the failure rate was assumed to be constant ($\lambda = 0.2 \times 10^{-6}$). Thus, the mean time between failures will also be constant at a value of $1/\lambda$ or 500 000 hours. In some situations, the failure rate may not be constant but rather variable. As discussed earlier, the bathtub curve indicates three patterns of failure rate, two of which are variable. These variable failure rates can be modeled using the familiar Weibull distribution:[1,2,46,47]

$$f(t) = \frac{\beta}{\eta}\left(\frac{t-\gamma}{\eta}\right)^{\beta-1} e^{-\left(\frac{t-\gamma}{\eta}\right)^{\beta}} \tag{6.28}$$

$$f(t) \geq 0, t \geq 0, \beta > 0, \eta > 0$$

where β is called shape parameter, also known as the Weibull slope, η is called scale parameter, and γ is called location parameter.

In most applications, the location parameter, γ, is not used, or set at zero. This yields the following function:

$$f(t) = \frac{\beta}{\eta}\left(\frac{t}{\eta}\right)^{\beta-1} e^{-\left(\frac{t}{\eta}\right)^{\beta}} \tag{6.29}$$

The above equation represents a two-parameter Weibull distribution. When the shape factor, β, is known beforehand, the Weibull distribution becomes a one-parameter distribution. In this case, only the scale parameter, η, will need to be estimated, allowing for analysis of small data sets. One has to be very careful in this case as a very good and justifiable estimate for β should be used before using the one-parameter Weibull distribution for analysis.

The key parameter of the Weibull distribution is the shape parameter, β. The value of β is equal to the slope of the line in a Weibull probability plot.[1] Different values of β result in different shapes of the Weibull distribution, as shown in Fig. 6.20a. Looking at the same information on a Weibull probability plot, one can easily understand why the Weibull shape parameter is also known as the slope. Figure 6.20b shows how the slope of the Weibull probability plot changes with β. Note that the models represented by the three lines all have the same value of η.

The most important aspect of the effect of β is that on the failure rate or hazard function shown in Fig. 6.21a. Weibull distributions with $\beta < 1$ have a failure rate that decreases with time. This simulates the infant mortality zone presented in Fig. 6.19c, case I and II. When β is close to 1, the failure rate is fairly constant. When $\beta > 1$, the failure rate increases with time, simulating the wear out situation shown in Fig. 6.19. Accordingly, a mixed Weibull distribution with one subpopulation with $\beta < 1$, one subpopulation with $\beta = 1$, and one subpopulation with $\beta > 1$ would have a failure rate resembling the entire bathtub failure curve.

The parameter η of the Weibull distribution, known as the scale parameter or the characteristic value. It also has important features related to reliability. As shown in Fig. 6.21b, the value of η at a constant β will simply stretch out the probability density function (pdf) of the Weibull distribution. Since the area under the pdf curve is always 1, this will result in a decrease in the peak of the function.[1]

For a three-parameter Weibull distribution, reliability is defined by the following equation:

$$R(t) = e^{-\left(\frac{t-\gamma}{\eta}\right)^{\beta}} \tag{6.30}$$

Note that for a two-parameter Weibull distribution, γ is zero, and for the one-parameter Weibull distribution, β is constant.

The Weibull failure rate function is:

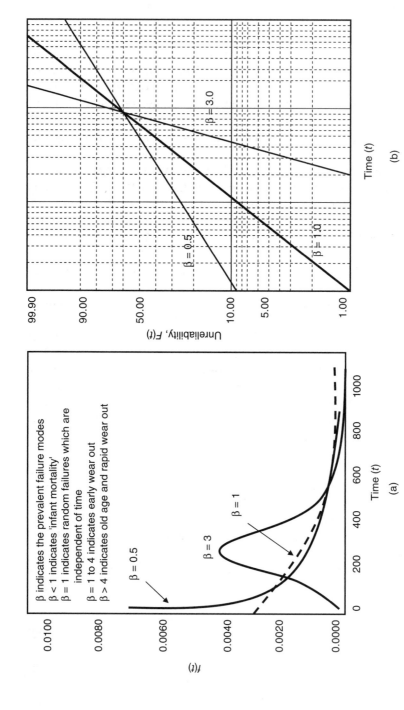

β indicates the prevalent failure modes
β < 1 indicates 'infant mortality'
β = 1 indicates random failures which are
        independent of time
β = 1 to 4 indicates early wear out
β > 4 indicates old age and rapid wear out

6.20 The Weibull distribution.[1,2,46,47]

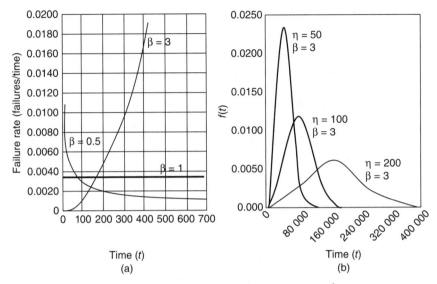

6.21 Effects of Weibull distribution b and h values.[1]

$$h(t) = \frac{f(t)}{R(t)} = \frac{\beta}{\eta}\left(\frac{t-\gamma}{\eta}\right)^{\beta-1}$$ (6.31)

The Weibull mean life or MTF, is given by:

$$\bar{t} = \gamma + \eta \cdot \Gamma\left(\frac{1}{\beta}+1\right)$$ (6.32)

where Γ is the gamma function defined by:

$$\Gamma(n) = \int_0^\infty e^{-x} x^{n-1} dx$$ (6.33)

Modeling reliability and failure rate using the Weibull distribution has become an easy task using today's software programs. The key, however, is to gather reliable data for time-to-failure and to interpret the analysis outcome properly. The concepts presented above can be useful in this regard.

In closing this section, it will be important to point out that when different material types are combined or blended to form a certain product, the reliability of the combined assembly may not be linearly related to the reliability of the individual components. In practice, we are often interested in the overall reliability of a system or a product. This will depend on how the failure of one component will influence the other components forming the product. In this regard, system/product failure is divided into two categories:[47]

1. Series system – in which the individual components are so arranged that a failure of any component will cause the whole system to fail. For this system, reliability is expressed by $R_{system} = R_A . R_B . R_C \ldots R_n$
2. Parallel system – this is a better arrangement because it requires that all components in the system fail in order for the whole system to fail. For this system, reliability is expressed by $R_{system} = 1 - (1 - R_A).(1 - R_B).(1 - R_C) \ldots (1 - R_n)$.

6.9.3 Safety factor

The uncertainty associated with a material capability of meeting some product performance requirements calls for the establishment of a failure preventive measure or index. Indeed, no matter how appropriate material selection can be, there will always be some degree of uncertainty that must be accounted for in the final design analysis. This uncertainty is primarily associated with loading and performance conditions with respect to an allowable application value. When this uncertainty exists, a factor of safety, also known as the safety factor, is established. This factor is used to provide a design margin over the theoretical design capacity to allow for uncertainty in the design process. The value of the safety factor is associated with the extent of confidence in the design process. The simplest expression of the factor of safety is:[2]

$$S_f = \text{nominal strength/allowable strength} = S/S_a$$

The nominal strength represents the strength of material as obtained from the characteristic database or through testing and the allowable strength is the expected load applied to the material. For example, if a material needs to withstand a load of 100 N, then a factor of safety of three means that the material strength should be 300 N.

According to the above expression, an increased factor of safety can result from a stronger material, heavier (or denser) part of the material, or improved design to enhance product safety and reliability. As a result, the choice of an appropriate factor of safety for a design application represents a careful compromise between the cost added, the dimensions added (weight, thickness and density) and the benefits associated with increasing safety and/or reliability.

In addition to strength, safety factors should also be established for other parameters including toughness, creep and stiffness. A factor of safety is always greater than 1; it can be as low as 1.2 and as high as 20. Low factors of safety are typically assigned in situations where material properties are known in detail, operating conditions are highly predictable, loads and corresponding stresses and strains are fully realized and environmental conditions are highly anticipated. These conditions should be supported by material test certificates, proof loading, regular inspection and maintenance. When any of these conditions is violated, higher factors of safety are assigned. The magnitude of the factor of safety will also depend on the material category under consideration. For example, brittle materials are likely to have a greater variation in their properties than ductile materials and hence require a larger factor of safety. The anticipated mode of loading is also a key condition; typically, fatigue conditions (cyclic stress) are associated with higher factors of safety than steady conditions.

6.10   References

1. EL MOGAHZY Y, *Statistics and Quality Control for Engineers and Manufacturers: from Basic to Advanced Topics*, 2nd edition, Quality Press, 2002.
2. DIETER G, *Engineering Design: a Material and Processing Approach*, McGraw-Hill Series in Mechanical Engineering, 1983.
3. CARBAJAL L A and DIEHL T, 'Design news on nonlinear mechanics', *DuPont Engineering Technology*, 2005, May 16.
4. PARSONS K C, 'Heat transfer through human body and clothing systems', *Protective Clothing Systems and Materials*, Mastura Racheel (ed.), Marcel Dekker, NY, 1994, Chapter 6.
5. PARSONS K C, 'Protective clothing: heat exchange and physiological objectives', *Ergonomics*, 1988, **31**(7), 991–1007.
6. EL MOGAHZY Y, 'Understanding fabric comfort', *Textile Science 93, International Conference Proceedings*, Vol. 1, Technical University of Liberec, Czech Republic, 1993.
7. FATMA S K, *A Study of the Nature of Fabric Comfort: Design-oriented Fabric Comfort Model*, PhD Thesis, Auburn University, AL, USA, 2004.
8. MCCULLOUGH E A and JONES B W, *A Comprehensive Database for Estimating Clothing Insulation*, IER Tech Report 84-01, Institute for Environmental Research, Kansas State University, 1984.
9. BELSLEY D A, KUH E and WELSCH R E, *Regression Diagnostics*, John Wiley and Sons, New York, 1980.
10. DRAPER N R and SMITH H, *Applied Regression Analysis*, 2nd edition, Wiley, New York, 1981.
11. KLEINBAUM D G, KUPPER L L and MULLER K E, *Applied Regression Analysis and Other Multivariable Methods*, PWS-KENT Publishing, Boston, 1988.
12. ANDERSON D Z, *Neural Information Processing Systems*, American Institute of Physics, New York, 1988.
13. YAN J, RYAN M and POWER P, *Using Fuzzy Logic Towards Intelligent Systems*, Prentice Hall, Englewood Cliffs, NJ, 1994.
14. EL MOGAHZY Y E and CHEWNING C, *Fiber To Yarn Manufacturing Technology*, Cotton Incorporated, Cary, NC, USA, 2001.
15. HEARLE J W S, GROSBERG P and BACKER S, *The Structural Mechanics of Fibers, Yarns, and Fabrics*, Wiley-Interscience, NY, 1969.

16. MULLER B and REINHARDT J, *Neural Networks–An Introduction*, Springer-Verlag, Berlin, 1990.
17. HOPFIELD J J, 'Neural networks and physical systems with emergent collective computational abilities', *Proceedings of the National Academy of Scientists*, 1982, **79**, 2554–8.
18. CLARK J W, 'Statistical mechanics of neural networks', *Physics Reports*, 1988, **158**(2), 91–157.
19. CLARK J W, *Introduction to Neural Networks in Nonlinear Phenomena in Complex Systems*, A. N. Proto, Elsevier, Amsterdam, 1989.
20. MURTAGH B A, *Advanced Linear Programming, Computation and Practice*. McGraw-Hill, NY, 1981.
21. OZAN T M, *Applied Mathematical Programming for Production and Engineering Management*, Prentice-Hall, Englewood Cliffs, NJ, 1986.
22. EL MOGAHZY Y, 'Optimizing cotton blend cost with respect to quality using HVI fiber properties and linear programming, Part I: fundamental and advanced techniques of linear programming', *Textile Research Journal*, 1992, **62**(1), 1–8.
23. EL MOGAHZY Y, 'Optimizing cotton blend cost with respect to quality using HVI fiber properties and linear programming, Part II: combined effects of fiber properties and variability constraints' *Textile Research Journal*, 1992, **62**(2), 108–114.
24. FLEMING P and PURSHOUSE R C, 'Evolutionary algorithms in control systems engineering: a survey', *Control Engineering Practice*, 2002, **10**, 1223–41.
25. MITCHELL M, *An Introduction to Genetic Algorithms*, MIT Press, Boston, MA, 1996.
26. KOZA J, KEANE M, STREETER M, MYDLOWEC W, YU J and LANZA G, *Genetic Programming IV: Routine Human-Competitive Machine Intelligence*, Kluwer Academic Publishers, The Netherlands, 2003.
27. KOZA J, KEANE M and STREETER M, 'Evolving inventions', *Scientific American*, 2003, February, 52–9.
28. CHRYSSOLOURIS G and SUBRAMANIAM V, 'Dynamic scheduling of manufacturing job shops using genetic algorithms', *Journal of Intelligent Manufacturing*, 2001, **12**(3), 281–93.
29. JENSEN M, 'Generating robust and flexible job shop schedules using genetic algorithms', *IEEE Transactions on Evolutionary Computation*, 2003, **7**(3), 275–88.
30. FLEMING P and PURSHOUSE R C, 'Evolutionary algorithms in control systems engineering: a survey', *Control Engineering Practice*, 2002, **10**, 1223–41.
31. FONSECA C and FLEMING P, 'An overview of evolutionary algorithms in multi-objective optimization', *Evolutionary Computation*, 1995, **3**(1), 1–16.
32. HANNE T, 'Global multiobjective optimization using evolutionary algorithms', *Journal of Heuristics*, 2000, **6**(3), 347–60.
33. GIRO R, CYRILLO M and GALVÃO D S, 'Designing conducting polymers using genetic algorithms', *Chemical Physics Letters*, 2002, **366**(1–2), 170–5.
34. GLEN R C and PAYNE A W R, 'A genetic algorithm for the automated generation of molecules within constraints', *Journal of Computer-Aided Molecular Design*, 1995, **9**, 181–202.

35. HE L and MORT N, 'Hybrid genetic algorithms for telecommunications network back-up routeing', *BT Technology Journal*, 2000, **18**(4), 42–50.
36. HAUPT R and HAUPT S E, *Practical Genetic Algorithms*, John Wiley & Sons, New York, 1998.
37. KIRKPATRICK S, GELATT C D and VECCHI M P, 'Optimization by simulated annealing', *Science*, 1983, **220**, 671–8.
38. ZADEH L A, 'Fuzzy sets', *Information & Control*, 1965, **8**, 338–53.
39. ZADEH L A, 'The concept of a linguistic variable and its application to approximate reasoning, I, II, III', *Information Science*, 1975, **8**, 199–251, 301–57 and **9**, 43–80.
40. ZADEH L A, 'The role of fuzzy logic in the management of uncertainty in expert system', *Fuzzy Sets and Systems*, 1983, **11**, 199–227.
41. YAN J, RYAN M and POWER P, *Using Fuzzy Logic Towards Intelligent Systems*, Prentice Hall, Englewood Cliffs, NJ, 1994.
42. COOK R D, *Concepts and Applications of Finite Element Analysis*, John Wiley & Sons, New York, 1974.
43. VAN PAEPEGEM W and DEGRIECK J, 'Modelling damage and permanent strain in fibre-reinforced composites under in-plane fatigue loading', *Composites Science and Technology*, 2003, **63**(5), 677–94.
44. DAVIS J R, *Metals Handbook*, American Society for Metals, Metals Park, Ohio, 1961, **1**, 185–7.
45. SEYMOUR R B, *Engineering Polymer Sourcebook*, McGraw-Hill, NY, 1990.
46. HERTZBERG R W and MANSON J A, *Fatigue of Engineering Plastics*, Academic Press, NY, 1980.
47. BOMPAS-SMITH J H, *Mechanical Survival*, McGraw-Hill Book Company, London, 1973.

Part II

Material selection

7

Material selection for textile product design

Abstract: Material selection is an essential aspect of product design and development. An appropriate material should meet many basic criteria including efficient manufacturability, performance, reliability non-degradability and recyclability. Basic steps in material selection suitable for fibrous products are discussed. These include consideration of the design-problem statement, the conditions under which the material is likely to be processed, the service and environmental conditions under which the material is likely to perform and material performance-related criteria. The key tasks for evaluating material candidates for a particular fibrous product are also discussed. Many examples of fibrous products, traditional and technical are used to demonstrate the different concepts presented here. In addition, key aspects of material selection such as understanding the technology and the difference between design-direct and value-impact performance characteristics will be discussed, using two examples of fibrous products.

Key words: cost-performance-value relationships; cost-performance equivalence; design-direct performance; value–impact performance; metals; metal alloys; ceramics; polymers; composites.

7.1   Introduction

As indicated in Chapter 4, material selection is an essential phase of the product design cycle. The critical importance of this phase stems from the fact that inappropriate choice of material for a certain product not only will result in a failure to reach the optimum solution to the design problem, but also may lead to increase in the cost of manufacturing and product handling. In selecting a material for a particular product, the two key factors that should be taken into consideration are material properties and material processing. The former directly influences product performance and the latter determines product manufacturability. For some products, the interaction between these two factors can provide many possible combinations of materials for engineers to select from.[1,2]

In general, an appropriate material should meet the following basic criteria:

- *Efficient manufacturability*: Material should be formable to the desired shape at the lowest cost possible.

- *Performance related*: Material properties should be directly related to the performance of the end product.
- *Reliability*: Material should exhibit minimum changes in its properties over time.
- *Non-degradable*: Material should not be adversely influenced by external effects or environmental conditions.
- *Recyclable*: Material is preferably recyclable or can be reclaimed after use.

When a new product is to be developed, or when several materials are to be combined together to form a product, the decision about what material to select can be complicated. For a new product, one has to select the most appropriate material from the enormous material database available today,[1,2] amounting to over 45000 different metallic alloys, 15000 different polymers and hundreds of other materials that fall into the categories of wood, ceramics, soils and semiconductors. On the other hand, this enormity of material sources can provide ample opportunities for flexible design and innovative alternatives. When many materials are to be combined together to form a product, issues such as material compatibility, characteristic equivalency and the linearity of properties of material mixture should be considered in the design analysis. In a previous study by the present author,[3] concepts and analytical procedures required to blend different fibers were addressed in great detail. These concepts can be generalized to different materials including non-fibrous materials.

In this chapter, the basic steps of material selection will be discussed. These are critical steps that can assist in making decisions regarding the selection of any material. Other important material-related issues discussed in this chapter will include cost–performance–value relationships and cost–performance equivalence. In addition, key aspects of material selection such as understanding the technology and the difference between design-direct and value-impact performance characteristics will be discussed, using two examples of fibrous products.

7.2   Basic steps of material selection

Material selection, as a key element of the design cycle, should primarily be driven by the design problem statement;[1,2] this is the starting point of any material selection process, as shown in Fig. 7.1. In this regard, the problem statement should address two key questions:

- Does material type represent one of the core issues in the problem statement?
- Does the material type anticipated represent a common one or a new one with respect to the intended product?

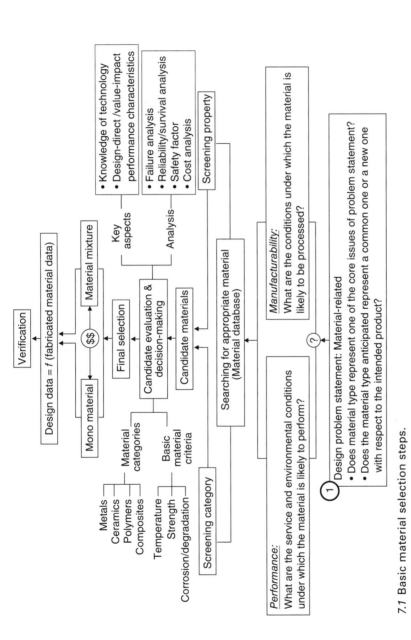

7.1 Basic material selection steps.

In some situations, the core issue of the design problem is the type of material needed to achieve better performance or prevent product failure. This situation is common with existing products that do not perform well under some circumstances owing to some inherent limitation in material characteristics. For example, if the design problem is poor appearance retention and excessive pilling (or fuzzing) of a commercial carpet, then a selection of a suitable fiber type will represent the core task of carpet design. This is a direct result of the strong association between fiber attributes and carpet performance characteristics.[4-6] Indeed, when carpet is the product in question, fiber type will represent the single most critical design factor. This point will be illustrated further at the end of this chapter.

In other situations, material selection may represent a straightforward process by virtue of the limited options available of appropriate materials or the well-established performance of some materials in the particular product in question. For example, the most commonly used material for denim jeans, which has been used since the 19th century, is cotton fiber. This is a direct result of the fact that denim fabric derives its familiar texture and desired appearance from the nature of cotton fibers.[3] Obviously, this does not exclude attempts to use alternative fibers, or add other fiber types to cotton for specific fashion effects or performance applications. Indeed, denim fabrics have been made from blends of cotton with other fiber types such as flax, polyester, rayon and spandex as will be discussed later in Chapter 12. Similarly, nylon fiber is the dominant material used for safety airbags owing to its superiority in key performance-related aspects, as will be discussed in Chapter 14.

In view of the design problem statement, material selection may proceed using the basic steps shown in Fig. 7.1. As indicated above, the two key factors that should be taken into consideration are material properties and material processing. The former reflects product performance and the latter reflects manufacturability. In this regard, two key questions should be addressed:

1. What are the conditions under which the material is likely to be processed?
2. What are the service and environmental conditions under which the material is likely to perform?

The first question deals directly with the manufacturing-related characteristics of the material. Obviously, a material that is not easily formable into a useful product will be of very limited use or will add significantly to the cost of manufacturing. Therefore, manufacturing criteria should be addressed in the early stage of material selection. This calls for a joint effort by both design engineers and manufacturing engineers[7] to address material manufacturability. Common manufacturing-related material characteristics include[1,2] dimensional characteristics, strength, flexibility (formability), brittleness and heat resistance. In addition to these characteristics, other properties that are unique to fibrous materials include resilience (bulk recovery upon loading and unloading), moisture regain, inter-fiber friction, fiber-to-metal friction and bulk integrity.[3,4]

The second question deals directly with material performance criteria. Obviously, the performance characteristics of a material and their levels will vary depending on the particular product in question. For most traditional fibrous products, performance characteristics include[8] aesthetic, appearance, durability, comfort, safety, care, biological resistance and environmental resistance. In Chapter 12, detailed discussions of some of these characteristics will be presented. For function-focus fibrous products, key criteria include[9] strength, stiffness, fiber density, cross-sectional shape, surface morphology, strength-to-weight ratio, durability under mechanical and environmental conditions and time-dependent behavior. In Chapters 13 through 15, detailed discussions of these characteristics will be presented using different examples of function-focus fibrous products.

In most applications, good performance criteria of materials also result in good material manufacturability. In other words, meeting the performance criteria of a product may not come at the expense of meeting the requirements for good manufacturing performance. For example, a strong and flexible fiber, which typically results in a strong yarn or fabric, will also perform well during manufacturing, particularly in resisting the rigorous mechanical stresses applied on the fibers during opening, cleaning and carding.[3] In other situations, however, a fibrous material that is desirable from a performance viewpoint can indeed create manufacturability problems. A classic example of this situation is that of microdenier fibers discussed below.

Microdenier fibers are synthetic fibers (polyester, nylon, acrylic, glass fibers, etc) with very fine diameters; they can be one hundred times finer than human hair and some have a diameter that is less than half the diameter of the finest silk.[10,11] They are commonly defined in the category of fibers of less than 1.0 denier. Some fibers with a denier as low as 0.001 were designed by Toray of Japan.[12,13] Microdenier fiber is the primary choice for many products, traditional and function-focus. For traditional apparels, they simulate natural fibers such as silk, cotton and wool fibers. The high aspect ratio (length/diameter ratio) of these fibers provides great flexibility and a high packing density of fiber strands. These are the important attributes required for key performance criteria such as comfort, fittability, porosity and air or water permeability. For function-focus fibrous products, the same characteristics make microdenier fibers an excellent choice for many applications such as filtration, wipe cleaning, thermal applications and sound insulation.[13]

The desirable attributes of microdenier fibers also introduce many challenges during fiber production and during manufacturing. The required superfine diameter of this type of fiber makes them very fragile in production directly from extrusion units. As a result, most microdenier fibers are generated from bicomponent polymers using precise spinning technologies.[12,13] Some are made using bicomponent fibers that split apart mechanically (e.g. hydro-entangling). Others are made by making bicomponent fibers from which one material is dissolved, leaving the microdenier fibers behind. During manufacturing of microdenier fibers into yarns or fiber webs, other challenges are introduced. The high aspect ratio of these fibers makes them extremely flexible and this increases the tendency of fibers to curl and entangle during manufacturing, forming a great deal of fiber neps.[10] In addition, the high density of microfiber strands can result in excessive fiber breakage by the mechanical wires of opening and carding units.[11] Accordingly, when microdenier fiber is the material of choice, the manufacturing process must be optimized through consideration of critical parameters, such as the number of stages of processing, the wire area density of the opening units, draft settings, carding speed and production rates.[10,11]

The above example demonstrates a unique case in which the same material attributes that yield desirable performance criteria also create processing difficulties. In other situations, materials may be selected despite prior knowledge of anticipated manufacturing difficulties provided that some external treatments can be applied to overcome these difficulties. For example, all staple-fiber yarns cannot be woven as spun owing to poor surface integrity which makes them weak against the high abrasive actions imposed by the weaving process. These yarns generally exhibit low abrasion resistance and a high level of hairiness.[3] These problems are largely caused by the discrete nature of staple fibers and the limited ability of the current spinning systems to consolidate fibers into compacted structures.[14] Since staple-fiber yarns represent an inevitable material choice for numerous traditional fibrous products, they must be woven into fabrics despite their poor surface integrity. This is achieved using a chemical surface treatment called 'sizing' or 'slashing', applied to the yarn surface prior to weaving. This treatment reduces yarn hairiness significantly and improves abrasion resistance. Chemicals used in the sizing process are later removed during fabric finishing in a process called 'desizing'. This issue represents an excellent opportunity for design engineers to overcome a historical problem by developing fibers that can be spun into yarns of good surface integrity so that the cost of chemical treatment, which can be as high as 80% of the cost of sizing, can be saved.

In light of the above discussion, both the manufacturing and performance criteria of material should represent the foundation for any material selection process. When a conflict between these two criteria exists, efforts should be made to minimize or eliminate this conflict through close cooperation between design engineers and manufacturing engineers. The lack of such cooperation can ultimately result in many adverse effects including failure of product functions and excessive manufacturing cost. Indeed, most reoccurring processing or performance problems can be rooted back to a selection of inappropriate materials or inherent limitations in material characteristics.

The next basic step in the material selection process (see Fig. 7.1) is searching for material candidates suitable for the intended product. This search should be made in view of the problem statement established and both the manufacturing and the performance criteria. This step aims to screen different material categories and material properties to find a set of material candidates suitable for the product in question. In this regard, two of the important terms that are commonly used among engineers are[1,2,7] screening category and screening property. Screening category is the best type of material (or material blend) that is suitable for the product in question. Screening property is any material property for which an absolute lower (or upper) limit can be established for the application.

In Chapter 4, the task of gathering information was mentioned as a key element in design conceptualization (see Fig. 4.2). A critical piece of information is the database of potential materials. Without such a database, the material selection process can be lengthy and costly. It is important, therefore, to establish a material information database capable of achieving reliable and efficient screening. This database should include both traditional and non-traditional materials as this provides the design engineer with a more global view of all material possibilities and alternatives.

The outcome of the search process should be a set of material candidates that have great potential to meet the desired product performance and manufacturing criteria. The next step is to evaluate the candidate materials more closely so that a decision can be made about the best material category, material type and material properties for the design application in hand. In this regard, the decision-making analysis discussed in Chapter 5 can be very useful as a guiding tool in selecting the best material and the best levels of characteristics required for the application. In situations where a new product is being developed or when limited information is available about past material performance, more in-depth analysis should be made, supported by extensive testing of relevant material characteristics.

Evaluation of candidate materials is a highly variable practice that largely depends on many factors including the complexity of the product, the extent of prior knowledge of material performance and the clarity of material contribution to the product. In general, traditional products are typically associated with familiar materials and high predictability of their performance. Function-focus products will generally be more complex than traditional products by virtue of the high specificity associated with their performance. As a result, they may require continuous development and an ongoing search for more advanced materials to satisfy and improve their specific functions. The clarity of material contribution to the product is a critical aspect of material evaluation which requires specific tasks including understanding the relative merits of using one material over another and evaluating basic material criteria such as temperature resistance, strength and corrosion or degradation behavior. In addition, some analyses associated with determining the appropriateness of a material for a certain product may be necessary. These include failure analysis, reliability and survival analysis and safety factors. These analyses were discussed in Chapter 6. In addition, material cost is often a major factor in selecting the appropriate material for a certain product. This issue will be discussed later in this chapter.

The outcome of the material selection process may be represented by a single material with associated characteristics, or two or more materials that can be mixed together to meet the desired application. In either case, it will be important to evaluate design data properties. These are the properties of the selected material in its fabricated state.[1,2,7] The need for knowledge of these properties stems from the fact that a material in isolation could have a substantially different performance from that exhibited by the same material in a product assembly. This point is particularly true for fibrous products. Indeed, it is well established that as fibers are converted into different structures (yarns or fabrics), some of their characteristics can be altered in their magnitude and in their relative contribution to end product performance.[3,15] Knowledge of material design properties can be based on long experience with existing products, a reliable database, or extensive testing that aims at simulating material performance in fabricated forms.

In light of the above discussion, the key tasks in evaluating material candidates for a particular product are:

- knowledge of the common material categories
- understanding the basic material criteria
- determining the optimum cost of material with respect to its performance and its contribution to the value of the end product
- understanding the effects of technology on material selection
- understanding the differences between design-direct and value-impact performance characteristics.

These tasks are discussed in the following sections.

7.3   Material categorization

The basic component of all fibrous products is fiber. This is essentially a polymeric material with characteristics directly determined by the properties of the polymeric substance, the orientation of molecules and the molecular arrangement. Conventional fibers may be from natural sources (e.g. cotton, wool, and silk) or from synthetic sources (e.g. rayon, polyester and nylon). All natural fibers, except natural silk, are in the short staples or the discrete form.[3,4] Synthetic fibers are produced in continuous filament form and they can be further processed in this form, or cut into discrete lengths or staples.[5,6] Some specialty fibers may be derived from non-polymeric substance such as metals, ceramics or rocks. In Chapter 8, a review of different fiber types and their applications will be presented.

Qualified fibers may be converted into different structural forms or fibrous assemblies to meet the desired performance characteristics of end products. For traditional fibrous products, yarns and unfinished fabrics are treated as the building blocks of the end product (e.g. apparels and furnishings). For function-focus fibrous products, fibers may be represented in different forms including staple fibers, continuous filaments, non-woven fiber webs or mats, yarns and fabrics. Discussions of different types of fiber assemblies and their characteristics are presented in Chapters 9 through 11.

In addition to the conventional forms above, fibers may be combined with other non-fibrous materials such as metals or ceramics. They may also be incorporated into composite structures or fiber-reinforced structures in which other polymeric materials are used. Therefore, it is important that design engineers of fibrous products be familiar with different material categories and their properties. In this regard, sources of information may include material data listed in books and the scientific literature, specialty websites and technical reports published by material producers. In the discussion below, a brief overview of major material categories is presented. In Chapter 13, many product applications utilizing these materials, some in conjunction with fibrous materials, will be discussed in the context of the development of function-focus fibrous products.

Over 95% of available materials consist of four main categories: metals or metal alloys, ceramics, polymers and composites. The key factor in this categorization is material structure. In other words, each one of these four categories of material has structural features that make it uniquely distinguished from the other categories in both properties and applications. In general, material structure can be divided into four levels: atomic structure, atomic arrangement, microstructure and macrostructure.[2] The role of conventional design is typically to understand and manipulate both the microstructure and the macrostructure of material. However, the atomic and crystal structures of materials must be first understood in order to allow such manipulation.

The importance of understanding the atomic structure of materials stems from the fact that materials are essentially made up from atoms.

Although, there are only about 100 kinds of atoms in the Universe,[16] how they are put together is what makes a certain material form trees or tires, ashes or animals, and water or air. They are all made from atoms, many of which are used several times. In general, the atomic structure influences how the atoms are bonded together, which in turn provides specific ways of categorizing materials as metals, ceramics and polymers and permits exploratory analysis of the mechanical properties and the physical behavior of these different material categories. Fundamentally, the atom consists of a nucleus containing neutrons and protons around which electrons orbit in more or less confined radii.[2,16] The smaller the diameter of the orbit, the greater the attractive force between the electron and the nucleus and the greater the absolute binding energy. By convention, binding energies are considered to be negative. This means that the innermost orbit will exhibit a large negative binding energy and the electrons in the outer orbits are bound less tightly. Indeed, the outermost electrons may be considered to be loosely bound and not necessarily residing in well-defined orbits. These are the so-called valence electrons and are the ones that are involved in the bonding together of the atoms and hence strongly affect all material properties, physical, mechanical and chemical. The nature of this bonding is what determines whether the substance is a metal (bonds between like atoms), or a ceramic or a polymer material (bonds between dissimilar atoms).

7.4   Common material categories

In the following sections, the main categories of material are briefly reviewed. The discussion will be restricted to examples of materials within each category, their important structural features and general applications that are suitable with respect to materials capabilities. Obviously, each material category will require many books in order to be fully covered. However, the objective of reviewing these categories of material is to provide polymer and fiber engineers with generalized concepts of material categories in the hope of attracting them to read more about these materials in the highly specialized literature.

As indicated above, metals, ceramics, polymers and composites represent the main categories of materials. In addition, other types of materials may also be identified on the basis of their unique characteristics. These include wood, foam, porous ceramics and rubber. As an aid to the discussion given below, the reader should refer to Fig. 7.2 to Fig. 7.6 as they illustrate some of the main differences between the major categories of materials. These figures represent simplified versions of Cambridge material charts,[17] verified by other sources of material information.[1,2,16] Note that in Fig. 7.2 to Fig. 7.5, yield strength under tension is considered

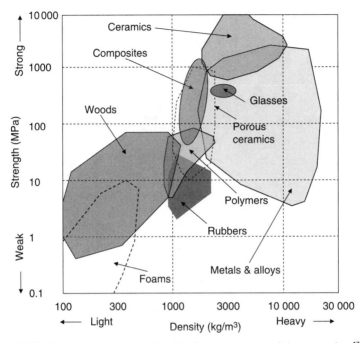

7.2 Yield strength versus density for major material categories.[17]

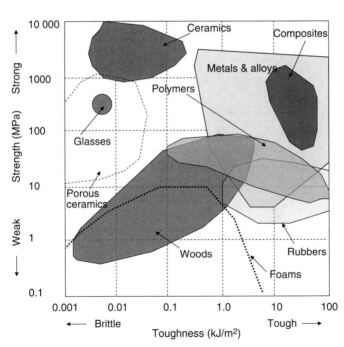

7.3 Yield strength versus toughness for major material categories.[17]

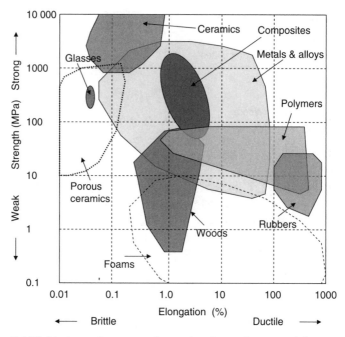

7.4 Yield strength versus elongation for major material categories.[17]

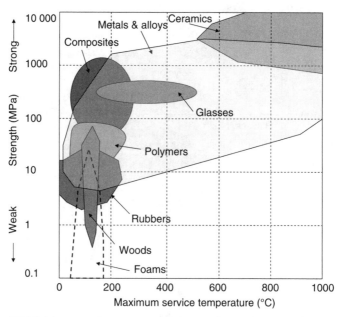

7.5 Yield strength versus temperature for major material categories.[17]

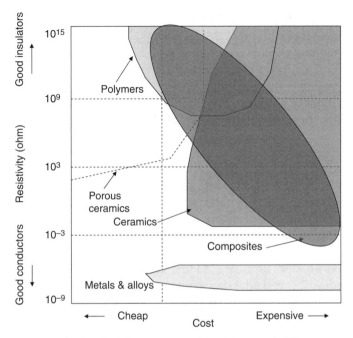

7.6 Electrical resistivity versus estimated cost of different materials.[17]

for all materials, except for ceramics for which yield strength under compression is considered. The reason for using this parameter is that ceramic tensile strength is typically about 10% of its compressive strength.

7.4.1   Metals and metal alloys

Among the four main categories of material, metals are the oldest materials as they have been used for thousands of years in numerous products including nowadays appliances and automobile bodies (low-carbon sheet steels), cutlery and utensils (stainless steels), aircraft frames and surfaces (aluminum alloys), and for electrical wiring and water pipes (unalloyed copper). Metallic materials are mainly inorganic substances composed of one or more metallic elements, but they may also contain non-metallic elements. An important classification of metals is by the ferrous versus non-ferrous category.[16,18] For ferrous metals, the primary metallic element is iron (e.g. steels and cast irons). Non-ferrous metals, on the other hand, may contain elements other than iron (e.g. copper, aluminum, nickel, titanium and zinc).

A metallic structure is primarily a crystalline structure consisting of closely packed atoms arranged in an orderly fashion.[1,2] This provides metals with good light reflectance and high density in comparison with other materials, as shown in Fig. 7.2, and with their familiar high strength and exceptional toughness in comparison with most other materials, as shown in Fig. 7.2 to Fig. 7.4. The high strength of most metals is largely maintained at elevated temperatures, as shown in Fig. 7.5. Other mechanical properties of metals, such as hardness, fatigue strength, ductility and malleability, are largely influenced by the presence of defects or imperfections in their crystal structure. For example, the absence of a layer of atoms in its densely packed structure enables a metal to deform plastically and prevents it from being brittle.[18]

Metals are commonly alloyed together in the liquid state so that, upon solidification, new solid metallic structures with different properties can be produced. In addition, metals are typically manufactured in their nearly final shape (e.g. sheet ingots or extrusion billets) through a casting process in which metals and alloys are cast into desirable geometries.[18]

Products in this form are commonly called castings and they can be subsequently worked using common processes such as rolling and extrusion into fashioned or wrought products (e.g. sheets, plates and extrusions).

Metals are also good electrical conductors, as shown in Fig. 7.6. They are also good thermal conductors, as will be seen in Chapter 8. The high electrical and thermal conductivities of simple metals are best explained by reference to the free-electron theory,[1,2] according to which the individual atoms in such metals have lost their valence electrons to the entire solid and these free electrons, which give rise to conductivity, move as a group throughout the solid. In the case of complex metals, conductivity is better explained by the band theory, which takes into account not only the presence of free electrons but also their interaction with the so-called *d* electrons.[1,2,18]

7.4.2   Ceramics

Ceramic material is essentially a combination of one or more metals and non-metallic substance chemically bonded together.[1,2] For this reason, they are often defined as inorganic non-metallic materials and classified according to a non-metallic element such as oxides, carbides, nitrides and hydrides depending on whether the metal is combined with oxygen, carbon, nitrogen or hydrogen. In contrast with most metals, ceramics can be crystalline, non-crystalline or mixtures of both. They exhibit high hardness and high strength at elevated temperatures (see Fig. 7.5). However, the biggest flaw that hinders their widespread use in various applications is brittleness and a propensity to crack propagation.[16] They also have medium to high electrical insulation, as shown in Fig. 7.6. Furthermore, their melting points are significantly higher than those of metals and they are much more resistant to chemical attack. These features make ceramics useful for electrical and thermal products with high insulation.

Ceramic materials can generally be categorized into three main categories:[1,2,16] (a) conventional ceramics, (b) advanced or technical ceramics and (c) glasses. Conventional or traditional ceramics consist of three basic components: clay, silica and feldspar. Clay is a form of ceramic before hardening by a firing process and it makes up the major body of material. It consists mainly of hydrated aluminum silicates ($Al_2O_3 \cdot SiO_2 \cdot H_2O$) with smaller amounts of other oxide impurities. Silica (SiO_2) has a high melting temperature and provides the refractory component of traditional ceramics. Feldspar ($K_2O \cdot Al_2O_3 \cdot 6H_2O$) has a low melting temperature and produces a glass when the ceramic mix is fired; it bonds the refractory components together. Advanced or technical ceramics are typically pure or nearly pure ceramic components or mix of components. They are normally a higher price because of the necessary control required to produce them. Examples of advanced ceramics include aluminum oxide (Al_2O_3), zirconia (ZrO_2), silicon carbide (SiC), silicon nitride (Si_3N_4) and barium titanate ($BaTiO_3$). Common applications for advanced ceramics include alumina for auto spark-plug insulators and substrates for electronic circuitry, dielectric materials for capacitors, tool bits for machining and high-performance ball bearings. The third category of ceramics is glasses, which are different from all other ceramics in that their constituents are heated to fusion and then cooled to a rigid state without crystallization.[2] The solid form of silica is a glass with SiO_2 being the main glass ingredient, but it is still a non-crystalline ceramic material.

7.4.3 Polymers

Polymers are mainly carbon-containing long molecular chains or networks.[19] Interestingly, they are known to the public as plastics, a term that typically defines materials that can be easily molded. Although most polymeric materials are non-crystalline or partially crystalline, some can be made of highly crystalline structures. As a result, some polymeric materials can be found that demonstrate a wide range of strength and toughness which overlaps with some metals, as shown in Fig. 7.3 to Fig. 7.5. In addition, some are highly ductile and many are moisture absorbent. As shown in Fig. 7.2, most polymers have lower densities than metals and ceramics. They also have relatively low softening or decomposition temperatures and many are good thermal and electrical insulators as, shown in Fig. 7.6.

In general, polymeric materials can be classified into three classes:[1,2,19] (a) thermoplastics, (b) thermosets and (c) elastomers. Thermoplastic polymers consist of very long chains of carbon atoms strongly (covalently) bonded together, sometimes with other atoms, such as nitrogen, oxygen and sulfur, also covalently bonded in the molecular chains. In addition, weaker secondary bonds bind the chains together into a solid mass. It is these weak bonds that make thermoplastic polymers soften under heat, forming a viscous state that allows geometrical and shape manipulation of this material.

Solidifying or setting these conditions via cooling into rigid solid states will result in them retaining their shapes and geometries. Examples of thermoplastic polymers include[19] polyethylenes, polyvinyl chlorides and polyamides or nylons. Products generated from these polymers include plastic containers, electrical insulation, automotive interior parts, appliance housings and, of course, fibrous materials.

Thermosets are polymeric materials that do not have long-chain molecules. Instead a network of mainly carbon atoms covalently bonded together to form a rigid solid. Again, nitrogen, oxygen or other atoms can be covalently bonded into the network. Thermosets are typically manipulated to form shapes and geometries, then cured or set using chemical processes that involve heat and pressure. Once they are cured or set, they cannot be re-melted or reshaped into other forms.[1,19] The common product application for thermosets is as a matrix substance for fiber-reinforced plastics (e.g. epoxy).

Elastomers are long, carbon-containing molecular chains with periodic strong bond links between the chains. As the name implies, elastomers (commonly known as rubbers) can easily deform elastically when subjected to a force and can recover perfectly to their original shapes upon removing the force. They include both natural and synthetic rubbers, which are used for auto tires, electrical insulation and industrial hoses or belts.

7.4.4 Composites

Composite material is not inherently a single category of material as it is made from a mixture of two or more materials that differ in form, structure and chemical composition, but are essentially insoluble in each other.[1,2] Most composites are produced by combining different types of fiber with different matrices in order to meet specific performance criteria such as strength, toughness, light weight and thermal stability. The basic idea is simple; while the structural value of a bundle of fibers is low, the strength of individual fibers can be boosted if they are embedded in a matrix that acts as an adhesive that binds the fibers and lends solidity to the overall structure. The matrix also plays a key role; it protects the fibers from environmental effects and physical damage, which can initiate cracks. Both the fibers and the matrix combined act together to prevent structural fracture.[16]

In contrast to composites, a monolithic (or single) material suffers fractures that can easily propagate until the material fails.

Materials considered for composite matrices include[16,19] polymers, metals and ceramics. The most common type is the polymeric-matrix composites. These can be found in applications at temperatures not exceeding 200–400°F (100–200°C). One of the earliest types of polymeric-matrix composites is glass-fiber composite in which short glass fibers are embedded in a polyester plastic matrix to form a lightweight structure suitable for many applications including appliances, boats and car bodies. In addition to light weight, this type of composite also exhibits a number of important features such as ease of fabrication into different shapes, corrosion resistance and moderate cost.[16] Other polymeric-matrix composites include carbon and aramid fibers embedded in heat-resistant thermoset polymeric matrices.[1,2] These have been used in many applications such as aircraft surface material and structural members. Metal-matrix composites are fabricated by embedding fibers such as silicon carbide and aluminum oxide into aluminum, magnesium and other metal alloy matrices. The role of fibers here is to strengthen the metal alloys and increase heat resistance. These types of composite are commonly used for automotive pistons and missile guidance systems. Ceramic-matrix composites include the reinforcement of alumina with silicon carbide whiskers. These are used to enhance the fracture toughness of ceramics.

The structure and shape of fiber composites vary greatly depending on the application and the direct purpose for which the material is used.[1,2] In general, the most effective approach to form composite structures is by using long fiber strands employed in the form of woven structures, non-woven structures, or even layers of unidirectional fibers stacked upon one another until a desired laminate thickness is reached. The resin may be applied to the fibrous assembly before laying up to form the so-called prepregs, or it may be added later to the assembly. In either case, the structure must undergo curing of the net assembly under pressure to form the fiber composite. The way different components in a composite structure contribute to its properties is simply realized using the familiar rule of mixtures, which expresses the composite stress, σ_c, as follows:[2]

$$\sigma_c = \sigma_f V_f + \sigma_m V_m \tag{7.1}$$

where σ_f is the strength of the fiber component, V_f is the volume fraction of fibers, σ_m is the strength of the matrix component and V_m is the volume fraction of the matrix.

The rule of mixtures is analytically associated with a number of key criteria. The first criterion is that fibers must be securely bonded to the matrix in the sense that atoms of each component react and bond together. Obviously, different components also interact and bond together. In this case, the rule of mixture will still hold as long as the potential failure of the composite is unlikely to occur at the interface between the different components.[2] This assumption can be gross in some situations in which the interface represents the weakest region of the fiber composite. This is obviously a design issue that must be handled through selection of appropriate materials and optimum fiber orientation. The second criterion is that the fibers must be either continuous or overlap extensively along their respective lengths. The third criterion is that there must be a critical fiber volume, $V_{\text{f-critical}}$ for fiber strengthening of the composite to be effective. The fourth criterion is that there must be a critical aspect ratio or a fiber length/diameter ratio for reinforcing to occur. Detailed analysis of the rule of mixtures is outside the scope of this chapter but they can be found to a great extent in the literature.[1,2]

From a design viewpoint, engineers should realize that for optimum composite performance not only is the choice of appropriate fibers important but also the way fibers are oriented and incorporated in the matrix. The latter is directly cost associated. One way to avoid expensive hand lay-up operations is to use nonwoven structures or chopped fibers that are arranged in mat form, or use loose fibers that may be either blown into a mold or injected into a mold along with the resin. In choosing matrix materials, other challenges, or design problems, can be faced. These include[1] the choice of appropriate material (e.g. epoxies, polyimides, polyurethanes and polyesters), processing cost, processing temperature (curing temperature if using a thermoset polymer and melting temperature if using a thermoplastic), flow properties in the molding operation, sag resistance during paint bake out, moisture resistance and shelf life. These multiple factors require full collaboration between fiber and polymer engineers, and engineers of various fields in which composite structures are utilized.

7.5   Basic criteria for the material

In order to make an appropriate decision about which material to select for a specific application, engineers should understand the basic criteria for the material as well as the specific criteria required for the intended product. For existing products that require modification, this may be a simple task. For newly developed products, this task can be more complex. As G. T. Murray describes,[1] 'Often, more questions are raised than are answered and the engineer is not always aware of all the parameters required in the material selection process'.

As indicated in the basic steps discussed earlier, engineers should begin by searching for a set of appropriate materials in view of the problem statement established and both manufacturing and performance criteria. Material candidates should be selected on the basis of a careful trade-off between many factors. Obviously, cost is a critical factor in selecting the appropriate material. However, this factor should not be considered as the first one in selecting the best material required for a specific design application. Only after a list of candidate materials is selected on the basis of their technical criteria should cost be considered. For most design applications, these criteria include:[1,2,7,16,19] temperature, strength, and corrosion or degradation resistance. Key points associated with these criteria are discussed below.

7.5.1 Temperature

For applications where changes in temperature are inevitable, temperature will represent the most critical criterion in determining which material to select. For most materials, strength may deteriorate with the increase in temperature. In addition, oxidation and corrosion of the material are likely to increase with temperature. Some categories of material are eliminated in applications involving elevated-temperature usage (above 500°C). These may include most polymers and low-melting point metals.[16] On the other hand, design applications under room temperature or lower (e.g. −10 to +50°C) can be associated with thousands of potentially useful materials including polymeric materials.[1,2]

The importance of temperature in selecting a certain material has resulted in some materials being categorized as high-temperature materials. These are materials that serve above about 1000°F (540°C). Figure 7.5 illustrates plausible ranges of strength and maximum-service temperature for the major categories of material. As can be seen in this figure, some metals and ceramic materials dominate in high-temperature applications. Specific materials known for their high-temperature applications include[18] stainless steel, austenitic superalloys, refractory metals, ceramics or ceramic composites, metal-matrix composites, and carbon or graphitic composites. Even among these materials, the first three are well-established in industrial applications and the other materials are still under extensive research to determine whether they can be used as substitutes or extensions to the capabilities of austenitic superalloys in high-temperature applications such as aircraft jet engines, industrial gas turbines and nuclear reactors.

When fiber is the material of choice, most conventional fibers decompose at temperatures below 300°C. The only inherently temperature-resistant fiber is asbestos (naturally occurring mineral fibers). This fiber does not completely degrade at high temperature. However, its extreme fineness makes it a health hazard as it can be breathed into the human lungs. Glass fibers have been used as a substitute for asbestos fibers because of their high heat resistance (up to 450°C). Unfortunately, these fibers have poor aesthetic characteristics, high densities and can be difficult to process.[20,21]

Another fiber type that can be used for high temperature applications is aramid fiber (e.g. DuPont Nomex and Teijin Conex). These are highly thermally resistant fibers as they char above 400°C and may survive short exposures at temperatures up to 700°C. The key characteristic of this type of fiber is that they can resist temperatures of up to 250°C for 1000 hours with only 35% deterioration in breaking strength of that before exposure.[20,21] This makes them good candidates for fire protection or flame retardant applications. Another fiber type that is basically a high-performance aromatic fiber made from linear polymers is the so-called PBO, or poly(p-phenylene benzobisoxazole). This type of fiber exhibits very high flame resistance and has exceptionally high thermal stability, with the onset of thermal degradation reaching 600–700°C. They also have very good resistance to creep, chemicals and abrasion. However, they exhibit poor compressive strength (they kink under compression), which restricts their use in composite structures.

7.5.2   Strength

Material strength is perhaps the most critical criterion in all design applications, as most products must have minimum acceptable durability to be able to perform properly and to prevent structural failure. In this regard, it is important to identify the specific type of applied stress that a product will encounter during use and the level of this stress. In most design applications, yield strength, stiffness and toughness represent the key material strength parameters determining the product performance. In addition, the mode of stress should be realized including tension, compression, torsion and bending. Figure 7.2 to Fig. 7.5 clearly demonstrate the differences in strength parameters between different material categories. In Chapter 8, more discussion on the mechanical behavior of these material categories in comparison with fibers will be presented.

Typical questions related to strength in a design project include: (a) What is the maximum applied stress that is likely to be encountered during the manufacturing or the use of a product? (b) Can some plastic deformation or permanent set be tolerated? (c) Is the applied stress static or dynamic? and (d) What is the pattern of dynamic stress (random or periodic)? In addressing these questions, engineers should have good knowledge of the strength properties of the specific material used in the design application. In addition, they should have a good grasp of the basic concepts of material strength. Table 7.1 provides definitions and criteria of some of the key strength parameters of material.

When fiber is the material of choice, the importance of strength stems directly from the type of application or the intended end product.[20,21] Most fibers are essentially polymeric-based materials and they largely share

Table 7.1 Some strength-related parameters considered in material selection[1,2,16,18,19]

Strength parameter	General criteria
Yield strength	• The most critical strength parameter in design applications. • It is often more critical than the ultimate strength as a result of the permanent deflection that the material encounters when it is loaded with stresses exceeding its yield strength. • Typically, yield strength under torsion is only about one-half of that under axial tension. • Cyclic stresses are associated with fatigue behavior that must be carefully examined in design applications. • Variability in material strength can be the most dominant factor influencing structural failure, as failure often occurs at the weakest point of material structure.
Flexibility or stiffness	• Flexibility is the ease of deforming or changing in dimension under applied stresses. • Elastic modulus (Young's modulus) is used to measure the stiffness under tension. The higher the elastic modulus, the less flexible or more stiff the material. • Stiffness under bending is measured by the flexural rigidity of the material. • Stiffness under torsion is measured by the torsion rigidity of the material. • All stiffness measurements are largely constant for a certain material. Material inherent modulus cannot be altered without altering material composition or using special mechanical treatments (e.g. mechanical conditioning). • Flexural rigidity and the torsion rigidity are highly sensitive to material dimensions; they are proportional to d^4, where d is the diameter of circular material (e.g. rounded fiber or yarn). • For fibrous materials, specific modulus is typically expressed in terms of modulus/linear density ratio (e.g. g/denier or N/tex). • For metals, the modulus of a soft metal is identical to that of the same strain-hardened metal. • For polymers, the moduli are relatively very small and not well defined. • Traditional fibers exhibit elastic modulus values in the range from 10 to 30 g/denier, some industrial fibers can be in the range from 30 to 100 g/denier. High performance fibers can be much greater than that (300 to over 1000 g/denier or 50 to 600 GPa). • Ceramics have the highest moduli under compression because of their strong covalent–ionic bonds. • Composite materials can be made to a wide range of desired moduli.

Table 7.1 Continued

Material system	General criteria
Ductility	• Ductility is the property of material which allows it to be formed into various shapes (e.g. wires or filaments). • In general, the higher the strength and the lower the elongation, the lower the ductility of the material. • When both strength and ductility are design requirements, some compromise must be made. • For metals, the strength can be increased significantly by reducing the grain size without a significant reduction in ductility. • For composite materials, volume fraction and component orientation can be adjusted to provide a wide range of desired ductility with minimum reduction in strength. • Polymers exhibit wide variation in their ductility and glass temperature. As a result, more compromises must be made because of their relative weakness in comparison with other material categories. • Ductile ceramic hardly exists. • Brittle materials (e.g. cast iron, ceramics and graphite) are generally much stronger in compression than in tension as tensile stress can cause crack propagation and compressive stress tends to close cracks.
Toughness	• Toughness is the resistance to fracture of a material when stressed. • It is typically defined by the energy (ft-lb) that a material can absorb before rupturing. • Toughness can be a temperature/time sensitive parameter, particularly at elevated temperatures. • It should be highly considered in applications involving shock or impact loading. • Since this parameter is highly application oriented, no standard distinction between different material categories can be easily specified and extensive testing should be made to realize the effectiveness of a particular material under the application in question.

many of the inherent characteristics of polymers. However, the unique structure of fiber being a long molecular chain provides many additional strength advantages that exceed those of conventional polymeric structures. This point will be clearly demonstrated in Chapter 8.

7.5.3 Corrosion and degradation

Corrosion resistance is the parameter that describes the deterioration of intrinsic properties of a material caused by reaction with surrounding environments. Corrosion occurs with the help of corrosive chemicals, solids, liquids or gases that are capable of harming living tissues or damaging surfaces with which they are in contact. Corrosive chemicals include[1] acids, bases (caustics or alkalis), dehydrating agents (e.g. phosphorous pentoxide and calcium oxide), halogens or halogen salts (e.g. bromine, iodine, zinc chloride and sodium hypochlorite), organic halides or organic acid halides (acetyl chloride and benzyl chloroformate), acid anhydrides and some organic materials (e.g. phenol or carbolic acid). In the case of metals, corrosion is determined by the extent of oxidation of metals reacting with water or oxygen. For example, iron can be weakened by oxidation of iron atoms, a phenomenon well known as electrochemical corrosion,[18] or more commonly known as 'iron rust'. Many structural alloys corrode merely from exposure to moisture in the air.

Most ceramic materials are almost entirely immune to corrosion owing to the strong ionic and/or covalent bonds that hold them together, which leave very little free chemical energy in the structure.[16] However, in some ceramics, corrosion can be realized by the dissolution of material reflected by obvious discoloration. In other words, and unlike metals, when corrosion occurs in ceramics it is almost always a simple dissolution of the material or chemical reaction, rather than an electrochemical process.

In polymers and fibers, corrosion and degradation may result from a wide array of complex physiochemical processes that are often poorly understood. The complexity of these processes stems from the realization that owing to the large molecular weight of polymers, very little entropy can be gained by mixing a given polymer mass with another substance, making them generally difficult to dissolve. In general, polymer corrosion or degradation can be observed in various forms including[2] swelling or volume change, and reduction in polymer chain length by ionizing radiation (e.g. ultraviolet light), free radicals, oxidizers (oxygen, ozone) and chlorine. Treatments such as UV-absorbing pigment (e.g. titanium dioxide or carbon black) can slow polymer degradation. One particular example of a fiber known for its easy degradation is rubber or spandex. This type of filament, being elastomeric can degrade by exposure to many chemicals or ultraviolet light. One way to avoid this degradation is to use this filament as a core in a yarn wrapped in another fiber for protection, as will be discussed in Chapter 9.

In general, corrosion or material degradation is a difficult parameter to evaluate owing to the non-standard way of reporting its values for various materials. In addition, the environment causing corrosion is not always known. Indeed, corrosion or degradation can also be caused or promoted by microorganisms that attack both metals and non-metallic materials. This is generally known as microbial corrosion or bacterial degradation.[16]

7.6   Material cost

In many design applications, the cost of material is considered to be the most critical criterion in determining which material to select. Different material types will have different costs depending on a number of factors including[7,22] (a) supply and demand rules, (b) property uniqueness, (c) availability, as determined by the costs of extracting, fabricating, or modifying material structure, (d) technology cost, as determined by machining, forming and heat treatments, (e) ease of material handling, as determined by the cost of assembly, number of components to be manufactured, storage, retrieval, packaging and transporting, (f) the energy required to process or handle the material, (g) reliability and (h) material by-products (derivatives, waste, etc.).

The factors above result in distinct cost differences between different categories of materials. For example, materials such as composites, ceramics and metals could have comparable market prices, particularly if they are competing for the same market. On the other hand, polymers and rubbers are generally less expensive than these three categories of material, followed by wood and porous ceramics. Within the same category of material, one can find a substantial price difference as a result of the cost of modifying a material to meet certain performance characteristics. For example, a galvanized steel sheet is expected to be more expensive than billets, blooms or steel slabs. In addition, high-modulus pitch-based carbon fibers are substantially more expensive than low-modulus, non-graphitized mesophase-pitch-based fibers.

Another important cost factor that can greatly assist in the decision-making process of material selection is the relative cost contribution of material with respect to the total cost of manufacturing a product. This relative cost can range from approximately 20% to 95% depending on the type of material and the product in question. For example, products such as expensive jewelry and dental alloy in which precious metals are used (e.g. gold, platinum and palladium) can be associated with material cost of up to 85% or 90% of the cost of the final product. Similarly, some electronic products use metals that are sold as bonding wires, evaporation wire and slugs, and sputtering targets for the deposition of thin films. This often results in setting prices largely in accordance with the metal price with a minor fabrication charge. The other extreme can be found in products such as precision non-gold watches in which the cost of metals provides a minor contribution to the final product cost, which is mainly fabrication. For fibrous products, one can also find a wide range of relative cost contributions of material, which typically falls between the two extremes mentioned above. For example, the cost of cotton fiber typically contributes by a range from 50% to 70% to the total final cost of spun yarn. For some function-focus fibrous products, the contribution of raw material cost can vary widely depending on the type of fiber used and the weight or volume fraction of fibers in the product composition.

In light of the above discussion, the key aspect in evaluating material cost is how this cost is translated into a product value, reflected in an optimum product performance, and whether this value is ultimately appreciated by the user of the product. The interrelationships between cost, value and performance can be visualized using a simple triangle as shown in Fig. 7.7. This type of demonstration was discussed in an earlier book by the present author.[22] A cost–value–performance triangle is determined by three basic dimensions, each is scaled from 0% to 100%. These may be considered as ordinal values used to rank materials according to the three criteria of the triangle, or as interval values used to rank materials in such a way that numerically equal distances on the scale represent equal distances in the criterion being measured. Simply, the scales of the three dimensions of the cost–value–performance triangle are determined by two extreme interrelationship scenarios:

1. High cost–high performance–high value (material A). In this case, the high cost of material is directly translated into high value of the product in the marketplace, reflected by superior product performance. In other words, a linear relationship exists between value, performance and cost. Most liability-oriented products (e.g. aircraft, jet engines, construction products, tires and airbags) will contain materials of this category.
2. Low cost–low value–poor performance (material B). This category of material hardly exists in the marketplace.

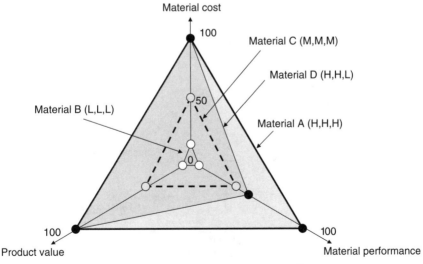

7.7 Material cost–value–performance triangle.[22]

Between these two scenarios, many materials can be found with a wide range of cost, value and performance (e.g. materials C and D). The cost–value relationship is often determined by a complex combination of market factors that must be considered collectively to establish a reliable relationship.[22] Cost–performance relationships on the other hand can be established during the design process and particularly in the material selection phase. This aspect is discussed below and examples will be presented later in this chapter to provide some guidelines on how performance and cost can be correlated in selecting raw materials for fibrous products.

7.6.1   Cost–performance relationship

Determining the relationship between material performance and cost is a common engineering practice. This type of relationship assumes that a certain performance parameter of material has a direct impact on the cost of fabricating a product or other costs. One common situation is the relationship between fiber strength and cost. For example, high-tenacity, high-modulus fibers are naturally more expensive than fibers with moderate values for strength or elastic modulus. This will result in an increase in the cost of converting these fibers into yarns or fabrics by virtue of their high price. This is illustrated in Fig. 7.8 by the increasing trend of the cost–performance relationship. In many situations, the use of high-strength fiber may result in cost reductions in other areas such as the number of fibers needed and the cost of finishing treatments. This may result in the descending curve in Fig. 7.8. The net result of these two curves is represented by a total cost–performance curve for which the optimum performance characteristic is determined at the minimum cost.

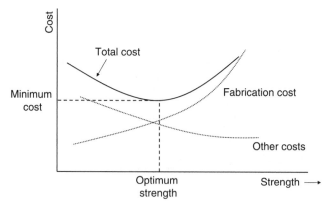

7.8 Material cost–value–performance relationship.

7.6.2 Cost–performance equivalence

Cost–performance analysis can also be performed using well-developed relationships between cost and some performance characteristics that allow a fair comparison between different material types.[1,2,7,22] One of the key parameters determining the cost of material is material weight or density. As a result, a cost–property comparison between two materials should be based on weight or structural equivalence. This is particularly important when two materials with different strengths or stiffnesses are compared. In this case, the relative weight of each material for equal strength or stiffness should be determined. To illustrate this point, suppose a fiber of cross-sectional area, A, is subjected to an axial tension, F. This will result in the following equation of working stress:

$$\sigma_w = \frac{\sigma_{yield}}{factor\ of\ safety} = \frac{F}{A} \tag{7.2}$$

When two fiber types M and N are compared under the same load, the condition of equal load-carrying ability in both fibers is given by:

$$A_M \sigma_{WM} = A_N \sigma_{WN}$$

$$\frac{\pi d_M^2}{4} \sigma_{WM} = \frac{\pi d_N^2}{4} \sigma_{WN}$$

or

$$\frac{d_N}{d_M} = \left(\frac{\sigma_{WM}}{\sigma_{WN}} \right)^{1/2} \tag{7.3}$$

where d_M and d_N are the diameters of fiber M and fiber N, respectively. The fiber weight is $W = \rho V = \rho A L = \rho \dfrac{\pi d^2}{4} L$

where ρ is the fiber density, V is the volume and L is fiber length. Thus,

$$\frac{W_M}{W_N} = \frac{\rho_M d_M^2 L_M}{\rho_N d_N^2 L_N} = \frac{\rho_M \sigma_{WN} L_M}{\rho_N \sigma_{WM} L_N} \tag{7.4}$$

For a constant length,

$$\frac{W_M}{W_N} = \frac{\rho_M d_M^2}{\rho_N d_N^2} = \frac{\rho_M \sigma_{WN}}{\rho_N \sigma_{WM}} \tag{7.5}$$

The above equation yields the weight ratio as a function of fiber density ratio and working-stress ratio. It indicates that the weight per unit strength is $W_s = \rho/\sigma$. It also allows for the inclusion of cost comparison provided that the cost per unit weight of each material (c_M and c_N) is known. For weights of fibers M and N of W_M and W_N, the total costs of fibers M and N are: $C_M = c_M \cdot W_M$ and $C_N = c_N \cdot W_N$. Accordingly,

$$\frac{C_M}{C_N} = \frac{c_M W_M}{c_N W_N} = \frac{c_M \rho_M \sigma_{WN}}{c_N \rho_N \sigma_{WM}} \tag{7.6}$$

This equation indicates that the cost per unit strength can be generally expressed by the following formula:

$$C_s = \frac{c\rho}{\sigma_W} \tag{7.7}$$

Similar expressions of cost per unit property can be derived for different cross sections and loading conditions.[7] The significance of the above equivalency analysis stems from the fact that it provides an objective comparison between different material types in relation to weight requirements and associated costs. To illustrate this point, suppose that materials with two fibers of the same length are being compared: steel (S) and Kevlar (K). If the length of the fiber is fixed and the working stresses of the two fibers are the same, the weight ratio will be:

$$\frac{W_K}{W_S} = \frac{\rho_K \sigma_{WS}}{\rho_S \sigma_{WK}} = \frac{\rho_K}{\rho_S} = \frac{1.44}{7.86} = 0.183 \tag{7.8}$$

This means that the weight of Kevlar required to meet this working stress is only a fraction (0.183) of the weight of the steel. If the cost of Kevlar is, say US$20 per pound and that of steel is US$5 per pound then the cost comparison of these two materials can be expressed by:

$$\frac{C_K}{C_S} = \frac{c_K \rho_K \sigma_{WS}}{c_S \rho_S \sigma_{WK}} = \frac{20 \times 1.44}{5 \times 7.86} = 0.733 \tag{7.9}$$

This means that the cost of Kevlar will also be a fraction of the cost of steel despite its higher cost per pound.

7.7   Effects of technology on material selection

Understanding the technology involved in making a product can have a significant effect on the choice of appropriate material for the intended product. In the area of fiber-to-fabric engineering, numerous examples can be listed in which the impact of material on the technology used or the effects of technology on material performance can be demonstrated. The example of microdenier fibers discussed earlier demonstrated how their unique characteristics can have adverse effects on their processing performance and how the technology should be adjusted to accommodate this type of fiber. The sizing process mentioned earlier demonstrates how the technology may compensate for a deficiency in raw material through chemical treatments that can enhance material processing performance. Indeed, independent segments of the industry such as finishing, coating and laminating are devoted to providing compensation for some of the deficiencies in raw material or to add features that cannot be provided by the raw material alone. This point will be illustrated further in Chapter 11 in the context of chemical finishing.

In some situations, the choice of raw material should be made in view of the technology involved, the capabilities offered by the technology and the cost of manufacturing. The simple example below demonstrates these situations.

7.7.1   Fiber selection for the design of spun yarns used for making strong and comfortable fabrics

The product in this example is a ring-spun yarn intended for making strong and comfortable fabrics that can be used in durable applications (e.g. military uniforms or working uniforms). For this product, fiber type and fiber properties are key aspects in the design analysis. Following the basic steps of material selection discussed earlier, the design problem established for this material may be stated as 'the need for a trade-off between strength and tactile comfort in the design of the ring-spun yarn'. The term trade-off is used here as a result of the fact that most efforts to achieve high strength normally results in a loss of flexibility and poor tactile features of yarns. Obviously, this may be considered as a sub-problem of a larger one that will involve other design aspects such as fabric construction and garment design. However, our focus in this example is strictly on the fibers required to make the spun yarn.

Following the problem statement, the next step is to establish the manufacturing and performance criteria of the product, as shown in Fig. 7.9. In this case, the manufacturing conditions under which the material is likely to be processed are those associated with the conventional ring spinning method.[3] One key criterion that can be mentioned in this regard is the need to spin the yarn at low twist to produce a soft yarn while maintaining the high efficiency of the spinning machinery. Another criterion is compatible blending when more than one fiber type is used to form the yarn. These are the areas where the effects of technology can be demonstrated. The performance criteria for this type of yarn are dictated by the desired combination of strength and tactile criteria required in the fabric made from the intended yarn. Accordingly, fibers that can provide high strength and good tactile behavior will be desirable for this application.

The next step in material selection is to search for the appropriate material. On the basis of screening category and based on experience with

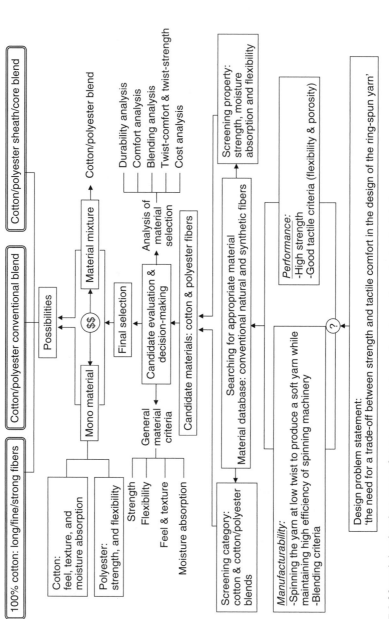

7.9 Material selection steps for spun yarns to make strong/comfortable fabrics.

existing products, two options may be considered: 100% cotton fibers or cotton/polyester blends. The use of 100% cotton fibers will provide many comfort-related advantages (e.g. natural feel and good absorption characteristics) and the use of cotton/polyester blend can provide a combination of comfort and strength characteristics owing to the relatively high strength of polyester fibers. On the basis of screening property, fiber properties such as strength, flexibility and moisture absorption of the candidate fibers are of critical importance to this application. Typical values of cotton fiber tenacity may range from 2.7 to 4.0 g/denier in dry conditions and 3.5 to 5.5 g/denier in wet conditions. Corresponding values for polyester fibers in both dry and wet conditions are 2.8 to 7.0 g/denier. Cotton stiffness, measured by the flexural rigidity, ranges from 60 to 70 g/denier. Corresponding values for polyester fibers are 12 to 17 g/denier. The moisture regain of cotton under standard conditions (e.g. relative humidity = 65 ± 2% and temperature = 70 ± 2°F or 21 ± 1°C) is about 8%. Polyester fiber, on the other hand hardly has any moisture regain (0.4%).

A good design process should aim to entertain all options, not only on the basis of performance but also on the basis of product value and manufacturing cost. For this type of product, the most economical option to use is 100% cotton, or cotton/polyester blends that can be processed using conventional spinning techniques. An alternative option would be to use a core/sheath yarn in which the core is made from 100% polyester fiber and the sheath is made from 100% cotton fiber. This option may be more costly as it often requires special spinning systems or additional spinning accessories and further analysis of the nature of the core/sheath interaction.[3] Accordingly, three specific options can be considered in this application: 100% cotton, conventional cotton/polyester blends and core-polyester/sheath-cotton blend.

With regard to the first option, the use of 100% cotton fiber for making strong/comfortable yarn can impose some challenges. The primary challenge is to meet the durability requirement of the yarn for this type of application since cotton fibers are not known for their high strength in comparison with polyester fibers. As a result, the choice of cotton will be limited to coarse or medium counts of spun yarns or to fine yarns that will be used in a high-dense fabric. In considering these options, the key design criterion will be the level of strength in the yarn. This should be high enough to provide durability with minimum impact on flexibility and fabric porosity. This brings up a critical issue that is directly related to both performance and cost, which is the choice of the appropriate level of twist. Although twist is primarily a technological issue, it has a direct impact on the choice of fibers and fiber properties. This point is explained below.

The way fibers are conventionally consolidated into a yarn is through twisting of the fibers together to a certain level of twist, or turns per unit length. This is a unique binding mechanism as it does not involve any use of external adhesives; as a result, fiber flexibility is highly preserved in the yarn. It is also a challenging engineering approach as discrete elements or fibers with very short lengths (e.g. 1.0–1.5 inch fibers or 25.4–38 mm) are twisted together to form a continuous yarn that can virtually have an unlimited length and exhibit acceptable strength. This challenge is met through innovative techniques of spinning.[3,14] The role of twisting is, therefore, to introduce strength to the yarn so that it can withstand the rigorous mechanical stresses during fabric forming and exhibit the necessary integrity.

The strength–twist relationship commonly takes the shape of curve A of Fig. 7.10. As can be seen in this figure, an increase in twist initially results in an increase in yarn strength. This trend continues to a certain point beyond which the strength decreases with the increase in twist. This point is commonly called the optimum twist point.[3,15] Twist also influences yarn stiffness and fiber compactness in the yarn, with the increase in twist resulting in an increase in stiffness and fiber compactness. As a result, excessive twist can lead to stiffer and low-porosity yarn, which can result in high discomfort to the wearer of the fabric made from this yarn.

In light of the above discussion, it is important that the twist level in the spun yarn be large enough to provide maximum yarn strength, yet be as small as possible to provide yarn flexibility and optimum fiber compactness, or good yarn porosity. As a result, the twist level should be optimized in view of two critical characteristics: strength and comfort. This point is illustrated in Fig. 7.10 in which the optimum twist level is determined by the intersection of two curves: curve A, the strength–twist curve, and curve B, the comfort–twist curve. For simplicity, in both curves the performance parameter is expressed by an index so that both parameters can be

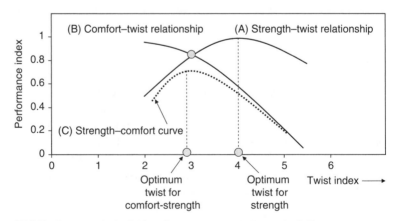

7.10 Performance–twist/performance–cost characteristic curves.

superimposed. In the case of the comfort–twist curve, the index ranges from 0 to 1, with 0 indicating high discomfort and 1.0 indicating the highest possible comfort level. In the case of the strength–twist curve, the strength index implies the ratio between the actual strength produced at a certain twist level and the maximum strength that can be obtained from the spinning system used and the fiber characteristics utilized. The net curve (C) representing these two performance parameters, is called strength–comfort characteristic curve.[3]

Twist is also a major cost factor as the increase in twist level is normally associated with a decrease in production rate and increase in energy consumption imposed by the increase in spindle speed.[3] Thus, the increase in twist will affect both the yarn performance and the manufacturing cost. One of the main approaches to reduce twist and, at the same time, maintain high strength and optimum comfort is to select an appropriate fiber material and certain values of fiber characteristics. If 100% cotton is the candidate material, then the choice of long, strong and fine cotton fibers can result in lower optimum twist levels (see Fig. 7.11). Since only expensive fibers can exhibit these levels of fiber characteristics, a trade-off should be achieved between the cost of material and the economical gains resulting from using lower twist levels.

If conventional cotton/polyester blends are considered, the addition of polyester can indeed provide more choices in the design analysis. As indicated earlier, polyester fibers are stronger than cotton fibers and they can be made in a wide range of fiber length and fiber fineness (including microdenier fibers). This allows the use of lower twist levels at acceptable strength values. Typically, the cost of polyester fibers used for these applications is compatible with that of cotton fibers. Accordingly, this may seem to be a more feasible option for this particular application. In addition to strength and flexibility, polyester fibers also provide key performance criteria such

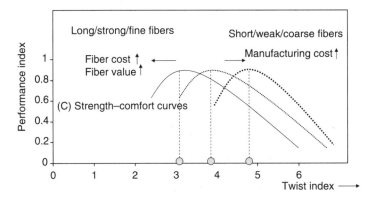

7.11 Effects of fiber properties on optimum twist.

as high wrinkle resistance and easy washing. The strength–comfort trade off can also be achieved by entertaining various blend ratios of the two fiber types.

The third option, which is the core/sheath yarn, may also be evaluated provided that the user of the product is willing to pay the additional cost for the higher value of product made by this option. Typically, non-conventional spinning techniques such as core spinning or friction spinning are used for this type of product.[3] The percent of core-to-sheath fiber ratio has to be precisely optimized to achieve the target strength at the most comfortable levels possible. When continuous filaments are used as the yarn core, sheath/core interface issues have to be resolved to avoid yarn peeling and poor maintenance.

The above example clearly explains how knowledge of the technology involved in making a product can have a significant impact on the choice of raw material. Indeed, the choice of raw material should be inseparable from the nature of the technology used to avoid unrealistic or very costly design options.

7.8   Design-direct versus value–impact performance characteristics

The choice of raw material should also be made in view of two key categories of performance characteristics: (a) design-direct performance characteristics and (b) value–impact performance characteristics. The first category implies the choice of a raw material that can directly yield the functional performance expected from the product. Typically, many materials may be suitable to meet the desired levels of this category of performance characteristics and the choice should be based on an optimum combination of material properties and cost-related factors. The second category implies the choice of raw material that can result in a significant added value to the product that can be felt and appreciated by the user of the product to such an extent that the user would not hesitate to pay a premium for the product. In this case, the cost of raw material may be relatively high but the added value can offset this high cost. This point is demonstrated by the example below.

7.8.1   Fiber selection for carpet piles

This example illustrates a common case where the choice of a certain material may seem to be economically inappropriate owing to its high market price in comparison with other competing materials, yet consideration of cost–performance relationships may indeed result in revealing other economical advantages that support the appropriateness of the expensive material. The product under consideration is carpet pile, the primary component of commercial carpets. The uniqueness of this product in relation to the subject of this chapter is that it can be made from many different materials, which makes material selection a key aspect. Most carpet piles commonly use four fiber types: wool, polyester, polypropylene and nylon fibers. Each one of these material candidates has strengths and weaknesses that should be fully recognized prior to selecting the appropriate fiber for this product.

Following the basic steps of material selection discussed earlier (see Fig. 7.12), suppose that the problem statement associated with this fibrous product is, 'What is the most appropriate fiber for developing carpet piles for heavy-traffic carpets that are likely to be subjected to high cyclic loading, potential stain, appearance-retention problems and possible flame?' Based on this statement, it is obvious that material type represents a core issue for this product. Moving forward to material criteria, one will find that all material candidates in this example can be manufactured to a high degree of efficiency as each one has been used in this type of product for many years and there is great experience of the manufacturing procedures and the criteria suitable for each type. In other words, the conditions under which any of these fiber types is likely to be processed are largely well-established. With regard to performance criteria, or the service and environmental conditions under which the material in this product is likely to perform, ten primary performance criteria can be listed for most applications. These are illustrated in Table 7.2.

Obviously, the ten criteria described in Table 7.2 cannot be met using one type of fiber material. This is evident by the general comparison between the different fiber types shown in Table 7.3. As a result, the design engineer should rank these criteria in the order of their importance with respect to the specific application or the target product. Alternatively, a weight factor can be assigned for each performance criterion depending on its value toward product performance. It is also useful to establish a scaled comparison of different materials to assist in making a decision about which materials to consider in the initial selection. Table 7.4 shows a rough scaled comparison on a ranking scale from 0 to 10, with 10 being excellent performance.

Upon establishing and prioritizing the key performance criteria, the next step in material selection will be to screen fiber category and fiber property. This will result in selecting a set of fiber types and fiber characteristics that have potential for use in this type of product. As indicated earlier, information gathering is a critical task in this basic step. The fiber categories in

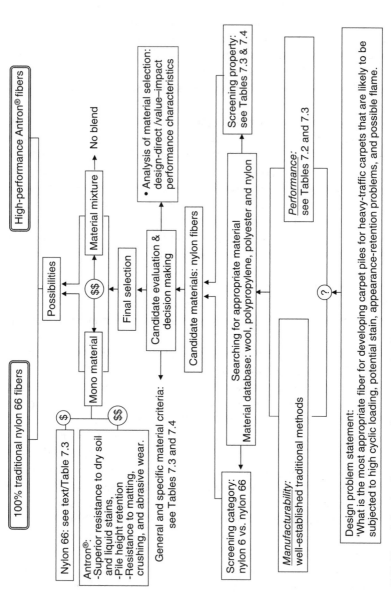

7.12 Material selection steps for carpet pile fibers.

Table 7.2 Basic performance criteria of carpet pile fibers[4,5,6,8,24]

Performance criterion	Contributing factors
Resiliency	Fiber structure, some enhancement modifications and pile density
Appearance retention	Fiber structure (cross-sectional shape and denier) and pile density
Abrasion resistance	Fiber type and the density of face fibers (more tightly packed yarns will result in high wear resistance)
Pilling and fuzzing	Fiber type, yarn structure and finish treatments
Soil and stain resistance	Fiber type, fiber surface treatment, color, texture, dyes and fiber structure
Sunlight resistance	Fiber bulk and surface structure and fiber treatments
Static	Fiber structure and treatments
Carpet feel or hand	Fiber type, surface density and cross-sectional shape
Resistance to mildew	Fiber structure and fiber treatments
Flammability	Fiber structure, modification, construction methods, dyes, padding and even carpet installation methods

Tables 7.3 and 7.4 represent one natural fiber (wool) and three synthetic fibers. Wool fibers are traditional carpet fibers which have been used for hundreds of years. They offer a deep, rich look and a special feel that are desired by the majority of carpet consumers. However, wool fiber has many limitations, some of which are performance related and others which are cost related. Wool is inherently staple fiber, making it sensitive to pilling or fuzzing; it tends to 'wear down' or wear away the piles. In some cases, bald spots may occur as a result of heavy traffic loads. Although it has high resilience, the fiber is relatively weaker than the synthetic fiber candidates. It is not easily treatable or modifiable in order to overcome soil and stain effects. However, it cleans especially well with appropriate cleaning methods. Since wool can hold ten times its weight in moisture, it is susceptible to shrinking and mold and mildew growth. When cost is an issue, the price of wool is relatively very high (almost double the price of nylon per yard) making it out of the reach of most consumers.

As a result of the above performance and cost constraints of wool, its choice may be restricted to specific applications such as indoor, infrequently used and luxurious areas. For most commercial carpets, only synthetic fibers are considered. In this regard, category and property screening can become very involved as the competition in this area is strong and developments in all synthetic fibers represent an ongoing effort to gain a larger market share in this highly profitable application.

Table 7.3 Descriptive comparison of performance characteristics of different carpet pile fibers[4,5,6,8,24]

Performance criterion	Wool	Polyester	Polypropylene	Nylon
Resiliency	Excellent	Good to fair	Good (avoid high piles)	Good to excellent
Appearance retention	Excellent	Fair	Fair	Excellent
Abrasion resistance	Good to excellent	Good to excellent	Excellent	Excellent
Pilling and fuzzing	Fair	Fair for staple fibers	Very good	Fair (for staple fibers) Excellent (for filament)
Soil and stain resistance	Good to excellent	Good to excellent (oily stains should be promptly treated)	Good if oily soils and stains are treated promptly	Good to excellent
Sunlight resistance	Poor (degrades under ultraviolet rays)	Good (weaken with prolonged exposure)	Poor (suffers strength losses and deteriorates unless chemically modified)	Good (some special dyes can be used to prevent sun damage)
Static	Builds up at low humidity (can be treated)	Builds up at low humidity (can be treated)	Builds up in low humidity but at a lower level than nylon or polyester	Builds up at low humidity (can be treated)
Carpet feel or hand	Warm, soft	Finer deniers are soft and silky	Waxy, soft	Varies from warm and soft to cold and coarse
Resistance to mildew	Poor if damp or soiled (can be treated)	Excellent	Excellent	Excellent
Flammability	Burns slowly	Burns slowly, melts; some are self-extinguishing; chemical odor	Melts at low temperatures (170°C); burns and emits heavy, sooty, waxy smoke; paraffin wax odor; pulling a heavy object across the carpet surface can cause enough friction to melt the carpet fibers.	Burns slowly, melts in direct flame; self-extinguishing. Structure may alter what occurs; celery-like odor

Table 7.4 Scaled-comparison of performance characteristics of different carpet pile fibers

Performance criterion	Wool	Polyester	Polypropylene	Nylon
Resiliency	10	5–6	5–6	6–9
Appearance retention	10	4–5	4–5	10
Abrasion resistance	6–9	6–9	9–10	9–10
Pilling and fuzzing	4–5	4–5 (staple) 8–10 (filament)	7–9	4–5 (staple) 8–10 (filament)
Soil and stain resistance	7–9	7–9	5–6	7–9
Sunlight resistance	3	5–6	3–4	5–6
Static	4	4	4	4
Carpet feel or hand	8	6	6	6
Resistance to mildew	5–6	8–10	8–10	8–10
Flammability	4	5	5	5

Excellent = 10, very good = 7–9, good = 5–6, fair = 4–5, poor = 0–3

With wool being eliminated from the list of potential candidates, the focus should then be shifted to synthetic fibers. Polyester fibers are made from terephthalic acid and ethylene glycol, offered primarily as a staple product for carpets. They are mainly used in residential and a few commercial applications. The fiber has good color clarity, colorfastness and resistance to water-soluble stains. It is also environmentally suitable as many staple polyester yarns can be produced by recycling plastic bottles. Indeed, some studies by polyester producers suggested that this 'food-grade' PET polyester fiber may exhibit better quality than 'carpet-grade' polyester fiber. Polypropylene fibers are known as olefin. The fiber-forming substance of olefin is any long-chain synthetic polymer composed of at least 85%, by weight, ethylene, propylene or other olefin units. Polypropylene fibers are offered primarily as continuous filaments with some staple product available. They are primarily sold as solution-dyed or pre-dyed fiber. Olefin fibers are famous for their resistance to fading, ability to generate low levels of static electricity, inherent resistance to stains and good resilience. They are also relatively cheaper than other synthetic fibers. Nylon fibers were first used for carpet in 1959. Now they are the most frequently used carpet fibers. It is a fiber-forming substance of any long-chain, synthetic polyamide that has recurring amide groups as an integral part of the polymer chain. It can be produced in continuous filament form or staple fiber form. Typically, nylon is produced as a solution-dyed fiber or white yarn to-be-dyed. Nylon fibers are highly desirable owing to their exceptional durability, versatility and reasonable pricing. They can be dyed in an endless variety of colors and made into numerous styles and textures for residential and commercial applications.

Suppose that resiliency and appearance retention is of primary performance concern. In this case, nylon fibers will represent the best candidate. Other fibers should be manufactured in high-density loop pile constructions to overcome crushing effects or pile flattening. If stain resistance is the primary concern, all fibers will perform equally unless the fiber is treated with special stain-resistant treatment. Similarly, flame resistance requires special treatment as no fiber is inherently immune to flame effects.

The general comparison above gives the design engineer some ideas about the differences between potential fiber candidates. Obviously, further comparative analysis should be made in which the basic steps of material selection may be repeated to narrow down the choices in view of the specific objectives and the performance criteria required. In this regard, more information about the candidate materials may be needed. For example, nylon fibers are available in two different types: nylon 66 and nylon 6. The former has a tighter molecular structure than the latter owing to a higher level of hydrogen bonding and maximum alignment between molecular chains. Nylon 6 does not have this level of internal bonding, resulting in a more open structure. In addition, nylon 66 has unsurpassed resilience properties. When subjected to 24 hours of compression at 100,000 psi, it demonstrates up to 100% recovery.[6,23,24] Accordingly, these two fiber types can present a considerable cost difference depending on performance, brand (producer) name and other supply and demand conditions. As a result, a nylon 66 carpet can have a substantial price difference of up to 30% over a comparable nylon 6 carpet.

Another type of nylon, which is a derivative of nylon 66, is the so-called Antron®, a four-hole hollow filament developed by DuPont.[24] This fiber enjoys superior performance in three key carpet criteria: resistance to dry soil and liquid stains, pile height retention and resistance to matting, crushing and abrasive wear. Soil resistance can also be enhanced by a treatment called DuraTech® soil resistant. Obviously, this fiber is more expensive than other nylon fibers.

The performance characteristics associated with the example above belong to the design-direct category mentioned earlier. These characteristics suggest that both conventional nylon 66 and Antron® represent good candidates for the raw material of carpet piles. The fact that Antron® is more expensive than conventional nylon 66 may represent a concern to the

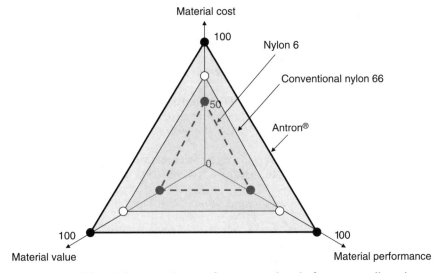

7.13 Material cost–value–performance triangle for carpet pile nylon fibers.

design engineer as the high cost must be translated into a true value that can be felt and appreciated by the user of the product. This is where value–impact performance characteristics are necessary to justify the higher cost. These may include longer lifecycle, minimum maintenance, lower cost of energy, easy installation, minimum water use, minimum air and land emission and easy cleaning.

According to DuPoint publications[24] based on a case study in which Antron® fibers were used in commercial carpets to cover the David L. Lawrence Convention Center in Pittsburgh, the use of Antron® resulted in significant savings in energy expenditure that are beneficial to the environment. The reported characteristics are different from the design-direct characteristics. With this value–impact information, a better comparison between nylon 6, nylon 66 and Antron® fibers can be established using an appropriate decision-making analysis as discussed in Chapter 5 which may yield a cost–value–performance triangle similar to that shown in Fig. 7.13.

The above example demonstrated how value–impact characteristics can be of critical importance in justifying the use of expensive and high performance fibers. Later in Chapters 12 to 15, many examples will be introduced to illustrate the differences between design-direct and value–impact performance characteristics of various fibrous products.

7.9   References

1. MURRAY G T, *Introduction to Engineering Materials: Behavior, Properties, and Selection*, Marcel Dekker, New York, 1993.
2. SHACKELFORD J F, *Introduction to Materials Science for Engineers*, 4th edition, P Imprenta, New Jersey, 1996.
3. EL MOGAHZY Y and CHEWNING C, *Fiber to Yarn Manufacturing Technology*, Chapters 1 to 5, Cotton Incorporated, Cary, NC, USA, 2001.
4. MORTON W E and HEARLE J W S, *Physical Properties of Textile Fibres*, The Textile Institute, Manchester, 1993.
5. BLOCK I, *Manufactured Fiber*, in AccessScience@McGraw-Hill, http://www. accessscience.com, DOI 10.1036/1097-8542.404050, last modified: May 6, 2002.
6. KLEIN W, *Man-made Fibers and their Processing.* The Textile Institute, Manchester, 1994.
7. DIETER G, *Engineering Design: A Material and Processing Approach*, McGraw-Hill Series in Mechanical Engineering, 1983.
8. HATCH K L, *Textile Science*, West Publishing Company, Minneapolis, NY, 1999.
9. HORROCKS A R and ANAND S C (editors), *Handbook of Technical Textiles*, Woodhead Publishing Limited, Cambridge, UK, 2000.
10. HWANG Y J, OXENHAM W and SEYAM A M, 'Formation of carded webs from microfibers', *International Nonwovens Journal*, 2001, **10** (1), 18–23.
11. EL MOGAHZY Y, 'Utilization of cotton/micro-denier polyester blends in air-jet spinning', paper presented at *Textile World Microdenier Fibers Conference*, Greenville, SC, Atlanta, USA, June 1–2, 1994.
12. OHMURA K, 'Japanese microfiber nonwovens', *Nonwoven Industry*, 1991, May, 16–18.
13. ROBERTS A M, 'What's the difference?', *International Fiber Journal*, 1992, **7** (2), 18–20.
14. STALDER H, *Compact Spinning – A New Generation of Ring-Spun Yarns*, Melliand International, Frankfurt, No. 3, E29–E31, 1995.
15. HEARLE J W S, GROSBERG P and BACKER S. *Structural Mechanics of Fibers, Yarns, and Fabrics*, Wiley-Interscience, New York, 1969.
16. SMITH W F, *Materials Science and Engineering*, in AccessScience@McGraw-Hill, http://www.accessscience.com, DOI 10.1036/1097-8542.409550, last modified: May 4, 2001.
17. CAMBRIDGE MATERIAL CHARTS, http://www-materials.eng.cam.ac.uk/mpsite/ interactive_charts, Department of Mechanics, Materials, Design, United Kingdom, 2002.
18. DAVIS J R, *Metals Handbook*, American Society for Metals, Metals Park, Ohio, 1961, Volume 1, 185–7.
19. CARRAHER C E, JR, *Polymer*, in AccessScience@McGraw-Hill, http://www. accessscience.com, DOI 10.1036/1097-8542.535100, last modified: March 12, 2004.
20. HEARLE J W S (ed), *High-performance Fibers*, Woodhead Publishing Limited, Cambridge, UK, 2001.
21. HONGU T and PHILLIPS G O, *New Fibers*, 2nd edition, Woodhead Publishing Limited, Cambridge, UK, 1997.

22. EL MOGAHZY Y, *Statistics and Quality Control for Engineers and Manufacturers: from Basics to Advanced Topics*, 3rd edition, Quality Technical Press, Columbus, GA, 2003.

23. MUKHOPADHYAY S K, 'High-performance fibers', *Textile Progress*, 1993, **25**, 1–85.

24. DUPONT REPORTS: *Carpets of DuPont Antron® Legacy nylon* http://pacificrest. com/products2/Legacy_Warranty.pdf, E. I. du Pont de Nemours and Company, 2001.

8

Structure, characteristics and types of fiber for textile product design

Abstract: This chapter is devoted to the discussion of fibers in the context of fiber-to-fabric engineering and product design. Basic differences between fibers and non-fibrous materials are discussed with respect to key aspects such as fiber structure, fiber deformation behavior, fiber strength parameters, fiber thermal conductivity, fiber aspect ratio and fiber surface characteristics. In addition, a review of different fiber types is presented, again from a design viewpoint. These include natural fibers, regenerated synthetic fibers, synthetic fibers and fibers for high performance applications.

Key words: degree of polymerization, degree of order, degree of orientation, elastic limit, cold drawing, viscoelastic behavior, Maxwell model, Voigt model, thermal conductivity, aspect ratio, surface morphology, gel spinning.

8.1 Introduction

Although fiber is the basic component of any fibrous product and this tiny component has altered the circumstances of human existence in many ways and made a huge contribution to the world in terms of its enormous number of applications, it is not a stand-alone component as it must be represented in a clustered form or in an assembly to play a significant role in the making or in the performance of a product. Fibers can be compacted and bonded into nonwoven structures, consolidated into a yarn, or converted into a woven or knit fabric via yarn structures. In a fiber-to-fabric engineering system, it is important to understand the potentials of these fiber assemblies in different applications and the effects of their characteristics on various fibrous products. These aspects will be discussed in the next four chapters. In this chapter, the focus will be on fibers.

8.2 Basic differences between fibers and other materials

Most fibers are essentially polymeric-based materials and they largely share many of the inherent characteristics of polymers. For example, the Cambridge charts[1] shown in Fig. 7.2 of Chapter 7 indicated that polymeric materials exhibit bulk density ranging from about 900 to 2800 kg m^{-3} (0.9 to 2.8 g cm^{-3}). Carbon-containing fibers exhibit bulk density falling within this range as shown in Fig. 8.1. However, fibers are uniquely different from other polymeric and non-polymeric materials by virtue of their molecular structures and the innovative techniques used to produce them. These aspects are discussed below.

8.2.1 Fiber structure

The internal structure of a polymeric fiber consists of small molecules or monomers joined together to form a long molecular chain. As shown in Fig. 8.2, this structure can be described by three basic parameters:[2,3] (1) the degree of polymerization, (2) the degree of order and (3) the degree of orientation. The length of the molecular chain depends on the number of molecules connected in a chain, which is generally known as the 'degree of polymerization'. This is typically determined by the ratio between the molecular weight at a certain point of time during polymerization and the molecular weight of one monomeric unit. Molecular chains may be arranged in a predominantly random order, commonly known as amorphous structure, or in a predominantly organized order, commonly known as crystalline structure. Most fibers exhibit structures that are partially crystalline

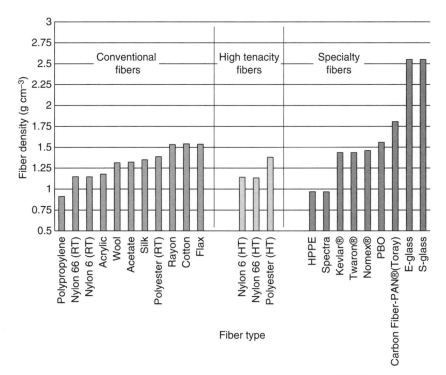

8.1 Typical values of density of different fiber types.[2,3,5–8]

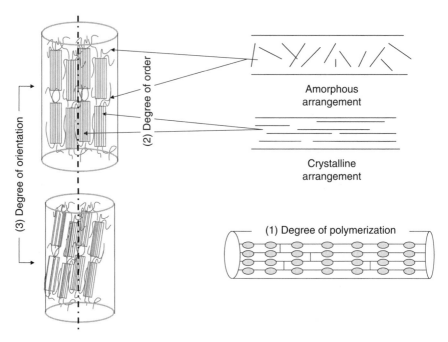

8.2 Basic structural features of fibers.

and partially amorphous. This is commonly known as the degree of order, with highly crystalline fibers having high molecular order. In addition, fibers can be made at different levels of molecular orientation by stretching the molecular chains to different levels of draw ratio; this is known as the degree of orientation.

8.2.2   Fiber deformation behavior

In order to understand the unique strength characteristics of fibers, it is important to provide a few comments on the nature of their mechanical behavior in comparison with other materials. In general, crystalline metals and ceramics will typically deform by a very small amount (often unnoticeable) at small levels of applied stress. Typically, small loads will result in small displacement of the metal atoms, less than 10% of their inter-atomic distances. As fibers are a polymeric material they will respond easily to external loading even at very low levels. Within elastic limits, polymer molecules recoil almost immediately from the load, but some polymers may exhibit a slight delay in elastic recoil.[4,5] When the applied stress exceeds the elastic limit, polymeric materials undergo the phenomenon of polymer yielding, which is the onset of a permanent or irreversible deformation. For non-crystalline (or amorphous) polymeric materials, polymer yielding results from molecular uncoiling, leading to a neck forming at the yield point, which is followed by an overall drop in stress. At the neck region, the folded chains become aligned. The microscopic thinning down in cross-section results in a local increase in the stress and any deformation occurs preferentially there. This helps the neck propagate crosswise under a steady load, in a process known as cold drawing. Any deformation produced beyond the yield point is not recoverable. In a crystalline polymer, the unfolding of chains begins in the amorphous regions between the lamellae of the crystals. This is followed by breaking-up and alignment of crystals.[4]

When time is a factor in rationalizing the mechanical behavior, metals and ceramics behave differently from fibrous materials. Typically, crystalline metals and ceramics deform plastically by dislocation motion, which is in step timewise with the applied stress. Fibers, after passing the elastic region, also deform plastically under stress when long-chain molecules slide past one another via the breaking of the weak secondary intermolecular bonds. The difference is that plastic deformation can increase with time without an increase in stress. Another difference is that the deformation is not purely plastic, particularly at the initial stage; it is a mix of plastic and elastic deformation. With further time, deformation can be purely plastic. This deformational behavior is uniquely described as a viscoelastic behavior, implying a combined behavior of fluid and solid.[2,3]

Theoretically, the viscoelastic behavior of fibers and polymeric materials has been simulated by conceptual models such as the so-called Maxwell and Voigt models which consist of a series of springs and dashpots.[2,3] In this regard, two common time-dependent phenomena should be realized: creep and stress relaxation. In simple terms, creep occurs when a constant force is continuously applied to a component, causing it to deform gradually with time. The result is a strain–time relationship with an increasing strain over time. The extent of creep will largely depend on the constant force applied. For thermoplastic polymers, creep will also depend on the temperature, with the increase in temperature leading to faster deformation. A classical model of the creep behavior of a material has three stages of strain versus time. The first stage consists of the initial elastic and plastic effects of loading and an initially high, but rapidly decaying creep rate. This is followed by steady state creep in stage two, where the rate is linear with time, and concludes with tertiary creep where the creep rate rapidly accelerates and ends in failure. Stress relaxation is almost exclusively a characteristic of polymeric materials under constant strain over time. It is typically manifested by a reduction in the force (stress) required to maintain a constant deformation. In classic polymer relaxation theory, the time-dependent behavior is governed by highly developed equations often based on Maxwell or Voigt models and governed by a spectrum of relaxation times for the material.

The effect of temperature on the deformational behavior of fibers is also well documented. Fibers may be subjected to high temperatures during processing and during chemical or thermal finishing. A polymeric fiber that melts, softens, or deteriorates under detergent treatment or hot water would make an unacceptable fabric. Fibers made from thermoplastic polymers can be manipulated during processing by raising their temperature above the softening or glass transition temperature (T_g), texturizing or deforming them and cooling them again below the T_g to set the intended textured structure.[6,7] Non-thermoplastic fibers can be treated using chemical cross-linking to reduce the loss of energy of deformation and provide better dimensional stability and shape recovery after deformation. For cotton fibers, this type of treatment is commonly known as permanent press finish and it functions by forming cross-links between adjacent cellulose polymer chains. These give cotton some elastic and resiliency properties.

8.2.3 Fiber strength parameters

Although most fibers belong to the category of polymer material, their molecular structure and deformational behavior provide them with unique values for strength parameters. In this regard, it will be useful to divide fibers into two main categories: conventional fibers and speciality fibers. The first category represents fibers that are commonly used for traditional fibrous products. These include natural fibers and common synthetic fibers (e.g. nylon, polyester, polypropylene and acrylic fibers). The second category represents fibers that are commonly used for function-focus fibrous products (e.g. aramid fibers, carbon fibers and glass fibers). Values for strength-related parameters of conventional fibers are listed in Table 8.1 to Table 8.3, and those of speciality fibers are listed in Table 8.4. The units

Table 8.1 Values of some strength parameters of natural fibers[3,6,7]

Fiber type	Tenacity (g$_f$/denier)		Breaking elongation (%)	Flexural rigidity (g$_f$/denier)	Elastic recovery (%) at 1%	Elastic recovery (%) at 3%
	Dry	Wet				
Cotton	2.7–4.0	3.5–5.5	4–6	60–70	50	35
Wool	1.8–2.0	1.3–1.4	25–45	4–6	60	
Flax	6–7	7–9	1.5–3.5	160–180	80	65
Silk	4–5	3–4	20–25	50–120	45	40

Table 8.2 Values of tenacity and elongation of synthetic fibers in dry and wet conditions[3,6,7]

Fiber type	Tenacity (g$_f$/denier)				Breaking elongation (%)			
	Dry		Wet		Dry		Wet	
	RT	HT	RT	HT	RT	HT	RT	HT
Polyester (filament)	2.8–5.6	6.8–9.2	2.8–5.6	6.8–9.2	24–42	10–25	24–42	10–25
Polyester (staple)	2.4–7.0	5.7–6.9	2.4–7.0	5.7–6.9	12–55	20–34	12–55	20–34
Polyester (partially oriented)	2.0–3.0		2.0–3.0		120–170		120–170	
Nylon 6 (filament)	4.0–7.2	6.5–9.0	3.7–6.2	5.8–8.2	17–45	16–24	20–47	19–33
Nylon 6 (staple)	3.5–7.2				30–90		42–100	
Nylon 6,6 (filament)	2.3–6.0	5.9–9.8	2.0–5.5	5.1–8.0	25–65	15–28	30–70	18–32
Nylon 6,6 (staple)	2.9–7.2		2.5–6.1		16–75		18–78	
Lyocell (Tencel™)		4.8–5.0	3.8–4.2			14–16	16–18	
Rayon (Fibro™) (filament and staple)	2.3	3.0	1.1	1.5	18–22		18–22	
Acetate (filament and staple)	1.2–1.4		0.8–1.0		25–45		35–50	
Acrylic (Acrilan™) (staple)	2.2–2.3		1.8–2.4		40–55		40–60	
Olefin (Spectra™)		30–35		30–35		27–36		27–36
Olefin-polypropylene (filament)	2.5–5.5		2.5–5.5		30–100		30–100	
Olefin-polypropylene (staple)	2.5–5.5		2.5–5.5		30–150		30–150	
Spandex (Lycra®)	0.8–1.0				400–800			

Table 8.3 Values of stiffness, toughness and elastic recovery of synthetic fibers[3,6,7]

Fiber type	Stiffness (g_f/denier)		Toughness (g_f.cm)		Elastic recovery (%)		Specific gravity	
	RT	HT	RT	HT	RT	HT	RT	HT
Polyester (filament)	10–30	30	0.4–1.1	0.5–0.7	76/3	88/3	1.38	1.39
Polyester (staple)	12–17		0.2–1.1		81/3		1.38	
Polyester (partially oriented)			1.3–1.8				1.34	
Nylon 6 (filament)	18–23	29–48	0.67–0.90	0.75–0.84	98–100/1–10	99–100/1–8	1.14	
Nylon 6 (staple)	17–20		0.64–0.78		100/2		1.14	
Nylon 6,6 (filament)	5–24	21–58	0.8–1.25	0.8–1.28	88/3	89/3	1.14	1.14
Nylon 6,6 (staple)	10–45		0.58–1.37		82/3		1.14	
Lyocell (Tencel™)		30		0.34				1.56
Rayon (Fibro™) (filament and staple)							1.53	
Acetate (filament and staple)	3.5–5.5		0.17–0.30		48–65/4		1.32	
Acrylic (Acrilan™) (staple)	5–7		0.4–0.5		99/2 89/5		1.17	
Olefin-polypropylene (filament)	12–25		0.75–3.0		95/5 85/10			
Olefin-polypropylene (staple)					93/5 85/10		0.91	
Spandex (Lycra®)					97/50 to 99/200		1.2	

Table 8.4 Values of strength parameters of some specialty fibers[8,9]

Fiber type	Specific gravity (g/cc)	Tenacity (N/tex)	Initial modulus (N/tex)	Breaking elongation (%)
Kevlar®29	1.44	2.030	49	3.6
Kevlar®49	1.44	2.080	78	2.4
Kevlar®149	1.44	1.680	115	1.3
Twaron®	1.44	2.100	60	3.6
Nomex®	1.46	0.485	7.5	35
HPPE (Dyneema® SK60)-1 dpf	0.97	2.800	91	3.5
HPPE (Dyneema® SK65)-1 dpf	0.97	3.100	97	3.6
HPPE (Dyneema® SK71)-1 dpf	0.97	3.500	122	3.7
Spectra 900-10 dpf	0.97	2.600	75	3.6
Spectra 2000-3.5 dpf	0.97	3.400	120	2.9
PBO poly(*p*-phenylene benzobisoxazole	1.56	2.54	177	
Carbon fiber-PAN® (Toray)	1.80	2.0–6.0	180–450	0.7–2.0
E-glass	2.55	1.5–2.5	54	1.8–3.2
S-glass	2.50	2.0–3.0	62	4.0
Steel wire	7.85	0.18	26	1.5

used in these tables are those used commonly among fiber and polymer engineers.

Examination of the values of strength-related parameters reveals a number of important points:

- As shown in Fig. 8.3, tenacity values of conventional fibers range from about 140 to 850 MPa (0.1 N/tex to about 0.65 N/tex). Some high-tenacity fibers can reach up to about 1200 MPa (0.9 N/tex). Note that synthetic fibers such as nylon and polyester can be produced in a wide range of tenacity depending on the intended applications.
- As shown in Fig. 8.4, some specialty fibers exhibit very high tenacity values that easily fit within the range of metals and ceramic materials. The level of tenacity of this category of fiber will depend on the emphasis of the application. For example, Nomex fiber, which is essentially a flame retardant meta-aramid, has a tenacity of 0.7 GPa. Kevlar fiber, which is a strength-oriented product, can have a tenacity of about 3 GPa.
- Most conventional fibers, except cotton and flax, have high breaking elongation values (exceeding 15%). Speciality fibers, on the other hand, exhibit low breaking elongation as shown in Fig. 8.5.

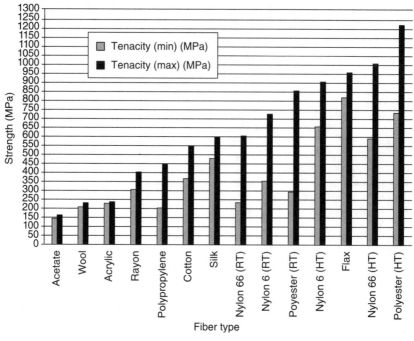

8.3 Typical values of strength for different conventional fiber types.[3,6,7]

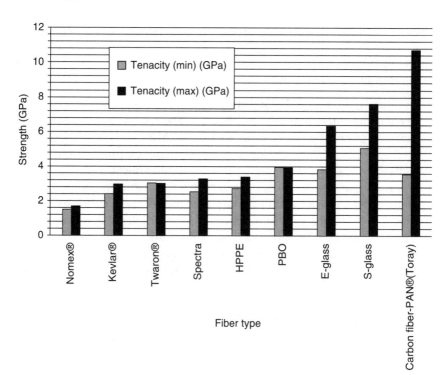

8.4 Typical values of strength for different speciality fibers.[5–9]

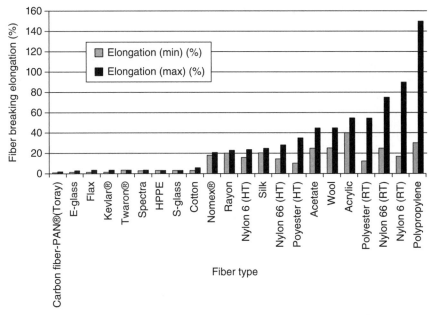

8.5 Typical values of elongation for different fiber types.[2,3,5-9]

• As indicated earlier, the most unique characteristic of fibers is flexibility. This is the characteristic that yields the comfort and fit of traditional fibrous products and the easy assembly and shape manipulation of function-focus fibrous products. The most common measure of flexibility is the initial modulus, known as the Young's modulus. Figure 8.6 shows another one of the Cambridge material charts[1] in which values of Young's modulus for polymers range from about 0.08 to 10 GPa. These values are clearly below those of metals, ceramics and composite materials. Most conventional fibers exhibit values within this range, as shown in Fig. 8.7. Only flax fibers can exceed this range.
• Specialty fibers exhibit Young's modulus values approaching those of metals and ceramic materials. This is clearly illustrated in Fig. 8.8.

8.2.4 Fiber thermal conductivity

Another parameter that distinguishes fibers from other materials is thermal conductivity. This is the ability of material to conduct heat. The general expression for thermal conductivity is as follows:

$$k = \frac{\Delta Q}{A \times \Delta t} \times \frac{x}{\Delta T} \qquad (8.1)$$

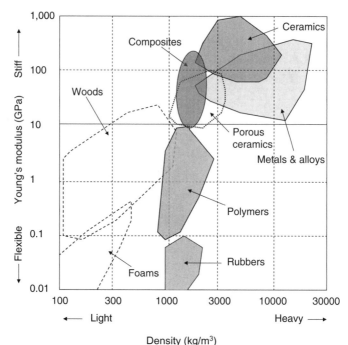

8.6 Young's modulus versus density for different materials.[1]

where $\dfrac{\Delta Q}{\Delta t}$ is the rate of heat flow, A is the total surface area of conducting surface, ΔT is temperature difference and x is the thickness of conducting surface separating two temperature levels.

The above expression indicates that thermal conductivity is a direct function of the quantity of heat, ΔQ, transmitted during time, Δt, through a thickness, x, in a direction normal to a surface of area A, owing to a temperature difference, ΔT, under steady state conditions and when the heat transfer is dependent only on the temperature gradient. In SI units, thermal conductivity is therefore expressed in $W\ m^{-1}\ K^{-1}$.

Figure 8.9 shows typical values of thermal conductivity for some conventional and specialty fibers. Figure 8.10 shows thermal conductivity values for other common materials. Values in both figures should be taken only for general comparative purposes. Figure 8.11 shows values of thermal conductivity for highly conductive materials. In general, most fibers exhibit low thermal conductivity, which makes them useful in many heat insulation applications. Specialty fibers such as carbon and glass fibers can have exceptionally high thermal conductivity.

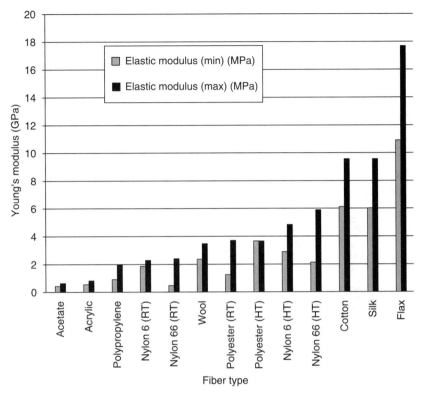

8.7 Typical values of Young's modulus for conventional fiber types.[2,3]

8.2.5 Fiber aspect ratio

A key characteristic that distinguishes fibers uniquely from other materials is the high aspect ratio (length/diameter ratio). Typical values of aspect ratio for fibers may range from 200 to several thousands. From a design viewpoint, fibers with a high aspect ratio should result in stronger yarns than those with a low aspect ratio by virtue of the long inter-fiber contact and the possibility of placing more fibers in the yarn cross-section.[10] More importantly, high aspect ratio can result in a significant improvement in fiber flexibility by virtue of the fact that the bending rigidity of fiber is proportional to d^4, where d is the fiber diameter. As discussed in Chapter 7, material options should be entertained in view of the design problem statement as well as both the performance and the manufacturing criteria associated with the material. When flexibility is the main performance criteria, a design engineer can have multiple options for producing high flexibility, as shown in Fig. 8.12. At the fiber level,

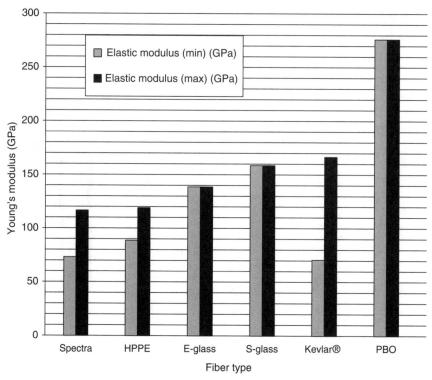

8.8 Typical values of Young's modulus for specialty fiber types.[5-9]

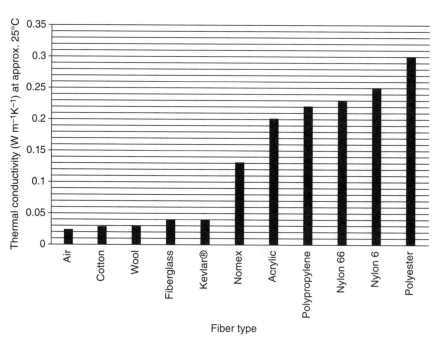

8.9 Typical values of thermal conductivity for some fibers.[2,4,5]
1 Btu ft/(h ft^2 °F) = 1.730 735 W m^{-1} K^{-1}.

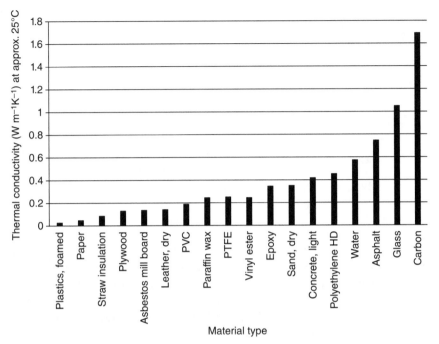

8.10 Typical values of thermal conductivity for some materials.[2,4,5]

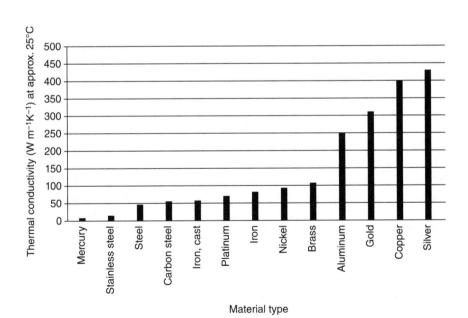

8.11 Typical values of thermal conductivity for highly conductive materials.[2,4,5]

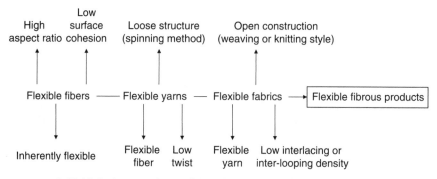

8.12 Main factors determining flexibility of fibrous products.

high flexibility in fibers should result in high flexibility in end products provided that the methods used to bind the fibers together into a yarn or fabric can preserve this flexibility or at least minimize its inevitable reduction. In this regard, the material choices will be an inherently flexible fiber, natural staple fibers with high aspect ratios or continuous filaments cut to various lengths, each associated with a different aspect ratio. At the yarn level, fiber flexibility can be transferred into yarn flexibility using an appropriate spinning method or low twist levels in the yarn. Similarly, fabrics with open constructions using flexible yarns can produce flexible end products.[10] When manufacturing criteria are considered, optimum values of fiber flexibility have to be used to avoid manufacturing problems as explained in Chapter 7 using the example of micodenier fibers.

8.2.6 Fiber surface characteristics

Different material categories are expected to have different surface structures by virtue of their atomic structure. For example, metals, being largely crystalline structures, are characterized by surfaces in which atoms are arranged in a largely regular manner.[11] Ceramics, being essentially a combination of one or more metals and a non-metallic substance chemically bonded together, are expected to exhibit different surface textures.[12] For composite products, surface characteristics of both the matrix and the reinforcing components represent a key structural aspect as they determines both the integrity and the protection of these products. In recent years, a great deal of development of surface physics has been witnessed as a result of the development of vacuum technology, new surface sensitive probes, atomic microscopes and powerful analytical methods.

The surface morphology of fibers represents a key design aspect in many product development applications. Different natural fibers exhibit different inherent surface morphologies. For example, cotton fiber has a twisted-ribbon shape along the length of the fiber and a kidney-shaped cross-section.[13] Wool fibers on the other hand exhibit a scaly surface, similar to that of human hair, leading to directional frictional effects.[14] Synthetic fibers stem their surface characteristics from various sources including[15] (a) the polymeric surface structure, (b) molecular orientation, (c) fiber cross-sectional shape, (d) fiber crimp and (e) surface finish treatments. This provides a wide range of key characteristics such as surface cohesion, surface morphology, roughness, moisture management and cleanability. In addition, the processing performance of most fibers is influenced by surface friction and fiber cohesion. During processing, fibers are subject to repeated rubbing between each other and against machine parts (wires, rollers, etc). As a result, an optimum cohesion level should be determined for each fiber type. In this regard, high enough cohesion may be required to maintain the bulk integrity of the fiber strand. Meanwhile, too low fiber cohesion may not be desirable as it can result in an uncontrollable flow of material during manufacturing. In the end product, fibers may be blended with other fibers, or they may be bonded with non-fibrous materials. This requires an optimum surface cohesion of fibers. In Chapter 12, more discussion of fiber surface characteristics will be presented in the context of the performance of sportswear.

In closing this section, it is important to point out that the fiber-related aspects discussed above represent key design factors that should be taken into consideration in any product development project involving fibrous products. Understanding these aspects can assist a great deal in exploring possible solutions to design problems. In addition to these aspects, design engineers should refer to the plausible ranges of various fiber characteristics available in the literature or in fiber producer documents. Figure 8.13 provides a summary of the key fiber attributes that can influence the performance characteristics of fibrous products. In recent years, nanotechnology has introduced a new category of fibers called 'nanofibers'. In comparison with the microdenier fiber, which has a diameter of few micrometers, a nanofiber will have a diameter of less than one micrometer. This three to four atoms thick fiber will add new dimensions to the uniqueness of fiber attributes in comparison to those of other categories of material. In addition, it will allow a dominant use of fibers in critical applications such as precision medical products, electronics, precision filters, smart garments, composite, insulation, aerospace, capacitors, transistors, drug delivery systems, battery separators, energy storage, fuel cells and information technology.[16,17] More discussion of nanotechnology will be presented in Chapter 11.

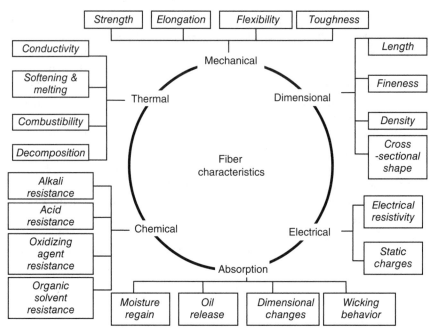

8.13 Common performance-related fiber attributes.

8.3   Review of different fiber types

Reviews of fiber types and their associated properties can be found in numerous literatures sources. For conventional fibers, the author recommends the classic book entitled *Physical Properties of Textile Fibers* by Morton and Hearle.[3] This book covers many of the fundamental aspects associated with both the internal structure and the properties of conventional fibers. For publications focusing on synthetic fibers, technology and properties, there are many sources of information including the ones listed in the references of this chapter.[6,7,18–20] For information regarding specialty fibers, producers' technical reports obtained from their websites are recommended. In addition, new books have been published that deal specifically with speciality fibers.[8,9] In this section, only a brief review of different fibers is presented as a quick reference for the reader.

8.4   Natural fibers

Natural fibers are those from natural sources, vegetable or animal. Examples of the former include cotton, flax and jute. Examples of the latter include wool and camel hair. Key points in dealing with these fibers in a product development project are as follows:

- Owing to the nature of this category of fibers, high inherent variability in their properties should be expected. As a result, they should be described by both average values and associated variability measures, such as range or variance.
- Inherent variability cannot be eliminated except by extracting fibers that have extreme values (e.g. carding or combing to remove short fibers).
- Inherent variability can be manipulated through many techniques of fiber selection and appropriate blending methods.[21] The purpose of this manipulation is to produce a consistent fiber mixture exhibiting a consistent pattern of inherent variability.

8.4.1   Cotton fibers

Cotton is a natural cellulosic fiber made from long chains of natural cellulose (carbon, hydrogen and oxygen). It exhibits a long linear molecular chain of over 10 000 cellulosic units held together by strong intermolecular forces.[3] As a result, it is reasonably strong within its use boundaries. The length to diameter ratio of the fiber is in the order of several thousand, which provides a great deal of flexibility, processing ease and good wearing performance.[10,21] A typical cotton fiber length may range from 0.90 to 1.5 inch (23–38 mm) and fiber fineness may range from 100 to 200 millitex (0.9 to 1.8 denier). Most cotton fibers exhibit a range of fiber bundle tenacity from 25 to 35 g_f/tex (or 2.7 to 3.9 g_f/denier). However, some cotton varieties exhibiting up to 40 g_f/tex (4.4 g_f/denier) tenacity can be found particularly in the extra long staple (ELS) category. Cotton fiber elongation typically ranges from 4 to 6%.

The majority of cotton fibers are used in traditional fibrous products, particularly apparel. This is due to their natural appeal and aesthetic properties.[21] Cotton fibers are also used widely for household and interior products such as towels, sheets, pillowcases, bedspreads, tablecloth and upholstery. In recent years, efforts have been made to use cotton in function-focus fibrous products such as fire-resistant clothing and commercial carpets. In these applications, cotton fibers are treated with chemicals to enhance their performance. For example, chemical treatments such as Proban and Pyrovatex are used to make cotton fire-retardant products.[22] In addition, cotton may be blended with wool and low-melt polyester fiber for utilization in carpets without the need for any topical finish.

8.4.2   Bast fibers

Another group of cellulosic fibers is bast fibers or long-vegetable fibers. These include[23] flax, hemp, jute and ramie. These fibers are relatively coarse and they exhibit high levels of durability. Flax (also called linen) is the most commonly used fiber of this group. It is used to make a variety of apparel products including dresses, skirts, blouses, suits, coats and hats. It is also used in interior and home furnishing products such as draperies and upholstery fabrics. Although flax exhibits approximately the same degree of polymerization as cotton, it is stronger than cotton as a result of its higher molecular orientation and higher crystallinity (see Table 8.1 and Fig. 8.3). It has a tenacity of about 6 g_f/denier and elongation at break from 1.5 to 3.5%. Like cotton, flax is stronger in wet conditions than in dry conditions (approximately 20% stronger).

Hemp is another long-vegetable fiber that was once the principal fiber used for marine cordage until replaced by abaca and sisal. It is still used extensively for twine and for many of the same products as linen. The fiber is usually about 4–7 ft (1–2 m) in length, and 22 micrometers diameter. It is typically spun into coarser and strong yarns. The ends of the fibers are blunt and very thick-walled, and show some branching. This branching distinguishes hemp from linen under the microscope.

Jute fiber is commonly used for making strong and bulky fabric or twine that is used in wrapping or bag materials (e.g. hessian, known as burlap in the USA and a heavier-weight fabric known as sacking). Jute fiber is usually 5–10 ft (1.5–3 m) or more in length and 20 micrometers in diameter. Synthetic fibers, especially polypropylene, have made substantial inroads into the markets for jute. Polypropylene bags and prime back for tufted carpeting have displaced large quantities of jute. Also, bulk handling has eliminated much of the former market for grain bags, especially in the USA. Another fiber competing with jute, particularly in products such as sacks, bags and paper, is kenaf. This is a long-vegetable fiber that has a similar appearance to jute, although the fiber is somewhat lighter in color. It also has similar or slightly lower strength than jute.

Ramie fiber is produced in length ranges from 28–60 inch (70–150 cm) and it can be up to 50 micrometers in diameter. Ramie fabrics are typically very strong and stiff. They also gain strength when wet and are highly resistant to mildew and rot.

8.4.3 Wool fibers

Wool fiber is another common type of fiber used in traditional fibrous products. It is typically more expensive than cotton and also in more limited supply as it primarily comes from the fleece of sheep. As an animal fiber, the main component of wool is protein (called keratin), which has a polypeptide chain with amino acid side chains. Keratin has a helical chain structure with strong hydrogen bonding.[3] The surface of the fiber has a very unique surface morphology characterized by overlapping scales extending lengthwise from the cuticle. These scales provide a 'differential frictional effect' with friction against the scales being significantly higher than that in the direction of the scales. Wool fibers may have fineness (diameter) ranging from 10–30 micrometers and fiber length ranging from 1–3 inch (2.5–7.6 cm). Coarse wool fibers can be very stiff, causing irritation and a prickling effect when projecting from the fabric against human skin. When wool fibers are used in carpets, this stiffness may provide a positive effect. The tenacity of wool fibers can vary significantly depending on the type of wool used; a typical range may be from 1.8–2.0 g_f/denier. Fiber elongation also varies from 25–45%. This high extensibility has a positive effect on its comfort characteristics. Like cotton, wool fibers are mostly used in apparel products such as men's suits, women's suits and dresses, coats, casual shirts, and scarves or hats. More than 10% of the wool fibers are used for carpets and rugs. It also holds a solid place in household products such as blankets and upholstery.

8.4.4 Silk fibers

Silk is another protein natural fiber used in making traditional fibrous products, particularly luxury apparel products. It is produced by the silkworm and is the only naturally produced continuous filament fiber but it can be used in staple form. In the context of durability, silk is stronger than wool but is also stiffer. This strength is a direct result of high degree of polymer orientation and high crystallinity enhanced by strong hydrogen bonding. Although silk has long-chain molecules as its backbone, there are various sorts of side chains attached to these. One of the major problems of silk, which directly influences the performance criteria of silk products, is its low crease resistance. This problem is typically handled by epoxide treatments.[24]

8.5   Regenerated synthetic fibers

Another series of fibers used in traditional fibrous products are regenerated fibers. These include rayon and acetate fibers. In Chapter 2, the story of rayon development was presented; it marked the beginning of unlimited development of synthetic fibers. In comparison with cotton, conventional viscose rayon exhibits inferior physical properties as a result of its lower degree of polymerization and lower crystallinity. Regular viscose rayon fibers have medium strength, low modulus and high elongation. New developments resulted in many derivatives of regenerated fibers with better physical properties (e.g. high tenacity and high wet modulus viscose). Acetate fibers are made from cellulose acetate polymer solution. This fiber is relatively less durable than other fibers as it exhibits poor strength and poor abrasion resistance. Regenerated fibers are mostly used in apparel products such as blouses, shirts, dresses, shirts and slacks, and lining fabrics for suits and coats.

8.6   Synthetic fibers

Finally, a wide range of synthetic fibers (oil and coal based) are used in making traditional fibrous products.[6,7] The most commonly used fibers are polyester, nylon, acrylic and polypropylene fibers. These fibers offer a wide range of values with different measures of durability giving unlimited opportunities for designers of traditional fibrous products to select desirable levels of mechanical properties and other important characteristics. For example, melt-extruded fibers such as nylon 6.6 and nylon 6 are made in a wide variety of fiber fineness and cross-sectional shapes that are suitable for different types of traditional fibrous products. These fibers can be drawn at different draw ratios leading to different levels of strength and elongation. They are superior in elastic recovery and dimensional stability to many of the natural or regenerated fibers. A significant amount of nylon fiber is used for carpet and rug products; a relatively small amount is used for apparel products such as hosiery and socks, underwear and nightwear. Nylon fibers can be made with a regular tenacity suitable for apparel products or a high tenacity suitable for the carpet market (see Tables 8.2 and 8.3).

Another popular synthetic fiber, also produced by melt spinning, is polyester, defined as any long-chain synthetic polymer composed of a least 85% by weight of an ester of a substituted aromatic carboxylic acid, including but not restricted to substituted terephthalate units and parasubstituted hydroxybenzoate units. Again, this fiber enjoys excellent strength and good elongation properties. The fact that this fiber can be effectively and efficiently blended with cotton provided a great market share in apparel and household products. Polyester has lower elastic recovery than nylon when elongated at 3%. As a result, nylon is more appealing for products such as women's sheer hosiery and polyester is more appealing in durable clothing.[23]

Acrylic fibers exhibit wool-like properties which make them attractive in the apparel market. They are known as manufactured fibers in which the basic substance is a long-chain synthetic polymer composed of at least 85% by weight of acrylonitrile units. These are spun into fibers by dry or wet spinning methods.[6,7] Acrylic filaments are commonly converted into staple fibers for apparel products. They exhibit slightly higher tenacity and higher toughness than wool fibers. They also exhibit higher elastic recovery, which adds to their appeal in the apparel market. Indeed, over 75% of acrylic products are in the apparel market (e.g. sweaters, high-pile fleece for coats and hosiery). Acrylic fibers also hold a solid place in the household and interior markets with products such as blankets, draperies and upholstery.

Polyolefin fibers are commonly known as manufactured fibers in which the basic component is any long-chain synthetic polymer composed of at least 85% by weight of ethylene, propylene or other olefin units.[7] They include polyethylene and polypropylene made by addition polymerization of ethylene and propylene and subsequent melt extrusion, respectively. Polyethylene did not perform well in the traditional market owing to its low tenacity and low melting point (110°C). They are available in coarse deniers for particular applications such as blinds, awnings, curtains and car interiors. Polypropylene fibers, on the other hand, are widely accepted in the traditional market. They have a high melting point of about 170°C, an average tenacity of 4 g_f/denier and high breaking elongation. These properties enhance the polyolefin position in the carpet and rug industry and in many industrial applications.[6,7,23]

Finally, a special category of fibers that has received increasing popularity in recent years is the elastomeric fiber. This is defined as a material which at room temperature can be stretched repeatedly to at least twice its original length, and upon immediate release of the stretch, will return to its approximate original length.[23] With at least 85% segmented polyurethane in their structure, elastomeric fibers are never used as the sole component of a product; instead they typically play a supporting role by adding elastic features to fancy yarns and apparel products. The most common type of elastomeric fiber is spandex (commercially known as Lycra®). This fiber is typically in a filament form with deniers ranging from 20 to 5400 denier. In the context of durability, spandex has a lower tenacity than most fibers but its high elongation and superior elastic recovery compensate for its low strength.

8.7   Fibers for high-performance applications

High-performance fibers represent a special category of fibers that is primarily made for function-focus products, but can be used for some traditional fibrous products (e.g. some apparel support, upholstery and floor coverings). From a design viewpoint, high-performance fibers provide unlimited design-oriented criteria that can promote fibers to virtually any field from ground transportation to aerospace and from biotechnology to computer and communication applications. Some of these applications are discussed later, in Chapter 13. The key criteria of high-performance fibers are[8,9,25–30] extremely high strength, exceptionally high temperature resistance and unique geometrical characteristics (surface morphology and cross-sectional shapes). These superior criteria have been a result of major advances in polymer and fiber technology. Examples of high-performance fibers are discussed below. The reader should refer to Table 8.4 and Figures 8.1, 8.4, 8.5, 8.8 and 8.10 to support this discussion.

8.7.1   Aramid fibers

Aramid fibers are polyamides, where each amide group is formed by the reaction of an amino group of one molecule with a carboxyl group of another. However, the presence of aromatic rings makes them more stable than with the linear arrangements of atoms. This provides great strength and heat stability. Two of the traditional aramid fibers are Kevlar®, a high-strength fiber, and Nomex®, a heat-resistant fiber.

Kevlar® was first produced by E.I. Du Pont de Nemours & Company, in the early 1970s as a candidate fiber for reinforcing tires and some plastics. Shortly afterwards, this light weight, strong and tough fiber was incorporated into many products including composites, ballistics, tires, ropes, cables, asbestos replacement and protective apparel.[25] Some radial automobile tires reinforced with Kevlar® cords are similar to those reinforced with steel. Over the years, other Kevlar-like fibers have been introduced, such as Twaron by Accordis BV and Technora by Teijin, Ltd.

Kevlar® is commonly called para-aramid or poly(p-phenylene terephthalamide). It belongs to a class of materials known as liquid crystalline polymers.[8,9,25,26] In contrast with conventional flexible polymers, which in solution can easily bend and entangle (forming random coils), Kevlar polymers are very rigid and rod-like. As a result, in solution they aggregate to form ordered domains in parallel arrays. Figure 8.14 shows a comparison of fibers from conventional flexible polymers and rigid polymers. When the polymer solution (in concentrated sulfuric acid solvent) is extruded through a spinneret and drawn through an air gap during fiber production, the liquid crystalline domains are oriented and aligned in the direction of flow, yielding an exceptional degree of alignment of long, straight polymer chains parallel to the fiber axis. The final structure is an anisotropic, high strength, high modulus along the fiber axis. It is also fibrillar, which has a profound effect on fiber properties and failure mechanisms. Subsequent high-temperature processing under tension can further increase the orientation of the crystalline structure and result in a higher fiber modulus. Kevlar® fiber can be in a continuous filament form or staple fiber form. This makes it formable into spun yarns, woven or knit fabric forms, textured yarn, needle-punched felts, spunlaced sheets and wet-laid papers.

Nomex® is the registered brand name of a flame retardant meta-aramid material. This m-aramid or poly(m-phenylene isophthalamide) was first discovered by DuPont in the 1970s. The DuPont scientist responsible for discoveries leading to the creation of Nomex® is Dr Wilfred Sweeny. This

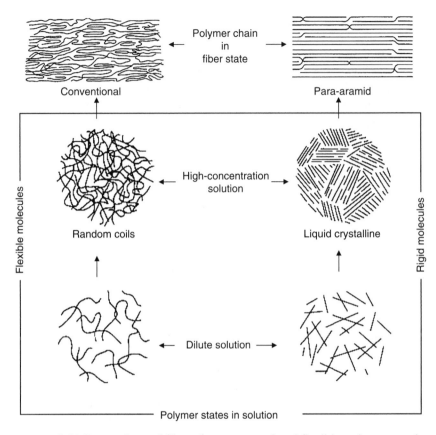

Polymer chain
in
fiber state

Conventional

Para-aramid

Flexible molecules

Rigid molecules

High-concentration
solution

Random coils

Liquid crystalline

Dilute solution

Polymer states in solution

8.14 Comparison of fibers from conventional flexible polymers and rigid polymers.[8,25]

unique material can be made in either sheet or fiber form. Nomex® sheet is actually a calendered paper commonly used for electrical insulation applications such as circuit boards and transformer cores as well as fire-proof honeycomb structures where it is saturated with a phenolic resin.[8] These honeycomb structures, as well as mylar–nomex laminates are used extensively in aircraft construction. In addition, both the firefighting and vehicle racing industries use Nomex® fibers to design clothing and equipment that can withstand intense heat.

Although Nomex® belongs to the class of aramid fibers like Kevlar® and both are largely heat and flame resistant, Kevlar® has a para-orientation that can be molecularly oriented and aligned to provide exceptionally high strength. Meta-aramid, or Nomex®, on the other hand, cannot be aligned during filament formation. As a result, it has a lower strength and modulus than Kevlar® (see Table 8.4). Under exposure to extreme heat, Nomex® structures consolidate and thicken. When blended with a small percentage of Kevlar®, a combination of high temperature resistance and strength enables these swollen fabrics to remain intact. This is referred to as non-break-open protection.[8,9] Another key attribute of Nomex® is its light sheet weight (as low as 1.5 oz per square yard or 48 g m^{-2}). This makes it possible to design fabrics and protective garments and gloves for various applications. It also extends its applications to areas such as reinforcement fabric in diaphragms and hose constructions. The fiber is also highly resistant to hydrolysis, alkali and oxidative substances. This makes it an excellent candidate for a premium class of fabrics used for rubber reinforcement. Additional good attributes of Nomex® are low shrinkage and good dye affinity.

Examples of Nomex® products include Nomex® hoods used as a common piece of firefighting equipment. This piece is typically placed on the head on top of a firefighter's face mask to protect the portions of the head not covered by the helmet and face mask from intense heat and fire. Racing car drivers use a similar hood to protect them in the event of fire engulfing their cars. Military pilots and aircrew wear one-piece coveralls (or flight suits) made of about 90% Nomex® to protect them from the possibility of cockpit fires. The remaining 10% is usually Kevlar thread used to hold the fabric together at the seams. The US space program also used its share of Nomex® and Kevlar®, particularly for the extravehicular mobility unit and ACES (advanced crew escape suit) pressure suit. These items require high fire resistance and high protection against extreme environmental conditions such as water immersion to near vacuum. Other Nomex® products include thermal blankets on the payload bay doors, fuselage and upper wing surfaces of the space shuttle orbiter.

8.7.2   Gel-spun polyethylene fibers

This category of fibers represents ultra-strong and high-modulus material derived from the simple and flexible polyethylene molecule. They are commonly called high-performance polyethylene (HPPE) fibers, high-modulus polyethylene (HMPE) fibers, or sometimes extended chain polyethylene (ECPE) fibers.[27] The molecular structure of this category of fibers is different from that of para-aramid fibers in that it is not a rod-like structure which needs to be oriented in one direction to form a strong fiber. Instead, polyethylene has much longer and flexible molecules which, by physical treatments, can be forced to assume straight (extended) conformation and orientation along the fiber axis.[28]

In the real world, polyethylene is probably the most commonly used polymer (or plastic) that humans use in daily life. This is the polymer that is used to make grocery bags, shampoo bottles and children's toys. For such diverse applications, the normal polyethylene polymer has a very simple structure, the simplest of all commercial polymers. It is nothing more than a long chain of carbon atoms, with two hydrogen atoms attached to each carbon atom. In this simple form, the molecules are not oriented and can easily be torn apart. As a result, the formation of a high-strength polymer requires stretching, orienting and forming a highly ordered (crystalline) structure from a long molecular chain. Accordingly, the starting substance should be polyethylene with an ultra-high molecular weight (UHMW-PE). This, in turn, creates a problem in spinning since extrusion of a very high-viscosity melt is extremely difficult. Drawing the filaments for high orientation is also difficult as a great deal of entanglement of the molecular chains is to be expected with this type of polymer. These two problems were solved using so-called 'gel spinning'.[27]

In principle, gel spinning operates on the basis of dissolving the molecules in a solvent prior to extrusion through a spinneret.[27,28] This only allows smooth extrusion, but molecular entanglement has to be resolved through super-drawing of the gel material after spinning. Figure 8.15 shows the basic difference between normal PE and HPPE.

The most common types of high-performance polyethylene fibers are Dyneema® and Spectra®. These are produced as multifilament yarns with a wide range of denier per filament (dpf) from 0.3 to 10 dpf. The tenacity of a single filament can be greater than 3 N/tex and the modulus can be greater than 120 N/tex (see Table 8.4). Since fiber density is less than one, the fiber can float on water. The combination of high tenacity and low density makes the specific strength of this fiber higher than steel (10 to 15 times that of good quality steel). The modulus of high-performance polyethylene fiber is second only to that of special carbon grades. Elongation at break is relatively small, which is common in most high-performance fibers. However, the high tenacity makes these fibers extremely tough, as indicated by their high values of energy to break.[28] The fibers are also highly resistant to abrasion, moisture, UV rays and chemicals.

The Dyneema® fiber, developed by the Dutch company DSM, has been used in many high-strength, low-weight applications. These include bulletproof armor and protective clothing for law enforcement and military

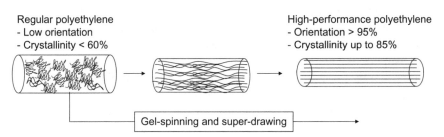

Regular polyethylene
- Low orientation
- Crystallinity < 60%

High-performance polyethylene
- Orientation > 95%
- Crystallinity up to 85%

Gel-spinning and super-drawing

8.15 Comparison of molecular structures of normal PE and HPPE.[26,27]

personnel. Some body armors have been designed from a combination of Dyneema high-performance polyethylene fiber and Steelskin steel cord material to provide stab protection against edged weapons, an application that traditionally required heavier and stiffer outfits which were uncomfortable to use. The underlying design principle of this product is based on blunting and damaging a blade with each thrust, using the steel wire, while further absorbing the impact energy by the super-strong fiber to stop the damaged knife and minimize trauma.

The Spectra® fiber produced by Honeywell has also been used in protective clothing (or shield technology). Shield technology lays parallel strands of synthetic fiber side by side and holds them in place with a resin system, creating a unidirectional tape. Two layers are then cross-plied at right angles (0°/90°) and fused into a composite structure under heat and pressure. The pre-consolidated cross-plied material is then packaged as rolls to make a ready-for-use product. This technology can use Spectra® fiber or other fiber types such as aramid fiber. The distinct advantage of shield technology, especially in armor applications, is that it preserves the strength of the Spectra® fiber or any other fiber encased within the resin matrix. Because the fibers are not crimped as they are in a traditional weaving process, the energy of the projectile is allowed to dissipate rapidly along the length of the fiber. Normally, the weaving process induces a great deal of bending and flexing, which reduces the characteristic molecular alignment of the high-performance polyethylene fiber.

8.7.3 Carbon fibers

Carbon fibers contain at least 90% carbon by weight. They are commonly derived from several organic polymers, such as rayon and PAN (polyacrylonitrile). The first commercial carbon fiber was rayon-based and was introduced in 1959. This fiber found its applications primarily in military products. Since 1970, PAN-based fibers have largely replaced rayon-based fibers in most applications. This was attributed to their superior tensile strength. PAN-based carbon fibers are now used in a wide array of applications such as aircraft brakes, space structures, military and commercial planes, lithium batteries, sporting goods and structural reinforcement in construction materials.[8,9,29] They can be woven into sheets, tubes, or other desired structures and they are often used in making carbon-fiber composites using epoxy resins or other binders.

Another more advanced carbon fiber is the so-called pitch-based carbon fiber. This is relatively very expensive, but it is unique in its ability to achieve ultra high Young's modulus and thermal conductivity. The high cost of this type of fiber has restricted their use to some specific military and space applications. A lower modulus, non-graphitized mesophase-pitch-based fiber, at lower cost, was also introduced for extensive use in products such as aircraft brakes. In addition to strength and modulus characteristics, it also exhibits good thermal and electrical conductivity.

PAN-based carbon fibers are made using many different methods.[29] The polymer is made by free radical polymerization, either in solution or in a solvent–water suspension. The polymer is then dried and redissolved in another solvent for spinning, either by wet spinning or dry spinning. Melt-spinning PAN plasticized with water or polyethylene glycol is another spinning possibility that is under development. The preferred method for high-strength carbon fiber (i.e. 80% strength improvement over conventional carbon fibers) is wet spinning using clean room conditions followed by heat treatment. Examples of carbon fibers produced by these methods include Toray® T800 and T1000. To make carbon fibers, the polymer is stretched into alignment parallel to what will eventually be the axis of the fiber. Then, an oxidation treatment in air between 200 and 300°C transforms the polymer into a non-meltable precursor fiber. This precursor fiber is then heated in a nitrogen environment. As the temperature is raised, volatile products are given off until the carbon fiber is composed of at least 90% carbon. The temperature used to treat the fibers varies between 1000°C and 2500°C depending on the desired properties of the carbon fiber. Under these high temperatures, carbon fibers with diameters ranging from 6 to 10 μm can be produced. With a fiber density ranging from 1.75 to 2.0 g cm^{-3}, fiber strength can range from 3–7 GPa, modulus from 200–500 GPa, compressive strength from 1–3 GPa and shear modulus from 10–15 GPa. Carbon fibers made from pitch can have modulus, thermal and electrical conductivities as high as 900 GPa, 1000 W m^{-1} K^{-1} and 106 S m^{-1}, respectively.

Graphite fibers can be considered as derivatives for carbon fibers. If during the treatment process for carbon fibers, the temperature is raised above 2500°C, graphite will be formed instead of carbon fibers. However, most of the graphite used in industry is manufactured by heating petroleum by-products to about 2800°C. The petroleum by-products are similar to the polymers used in the carbon fiber process in that both contain chains of carbon atoms. One may consider graphite as a soft form of carbon, which exhibits a unique combination of very low density and high elastic modulus with mechanical strength increasing with increasing temperature. Indeed, this unique material is capable of mechanical service at temperatures of 2200°C (4000°F) or higher. The main problem with graphite is its vulnerability to oxidation since it is essentially a form of carbon.

8.7.4   Glass fibers

The art of heating sand and limestone to form a molten liquid has been known from the time of the ancient Egyptians. Since the 1930s, glass has been spun into a fiber that is sufficiently pliable to be woven into fabrics. Some believe that Napoleon's funeral coffin was decorated with glass fiber textiles.[2] Glass fibers are prepared by melt spinning previously formed glass marbles, and the molten filaments are drawn down to very fine dimensions. Glass fibers are inherently stiff, but their flexibility stems from fiber fineness. Indeed, the fibers are so stiff that when broken they can penetrate human skin, making them unsuitable for use in apparel or upholstery. Because of their good resistance to the degrading effects of sunlight and their flame resistance, they can be used for curtains and drapery. They also provide a non-rotting, non-settling insulating material for homes and industrial uses.

In Chapter 7, it was indicated that glass fibers have been used as a substitute for asbestos fibers in high-temperature applications because of their high heat resistance (up to 450°C). From a design viewpoint, the thermal performance of glass fibers is attributed to two key factors: the thermal conductivity of the glass itself (see Fig. 8.10) and the structural features of a fiberglass product (structure density, fiber dimensions and air/fiber volume ratio). Fiberglass is also commonly used for filtration purposes. For these applications, the surface area of fibers and the structural pore size represent the key design parameters. In this regard, fiberglass can be made of super-fine to medium or coarse diameters (0.05–25 µm). For apparel applications, fiberglass has poor esthetic characteristics, high densities and is difficult to process. Another critical field of fiberglass applications is electronics and communication. For these applications, optical fibers made from fiberglass are extremely effective.

Since silica is an excellent glass base, inorganic glasses are all made from silica. The problem is that silica $(SiO_{4/2})_n$, being a three-dimensional network, cannot be converted into a liquid form easily. It fails to exhibit a sharp melting point and starts to soften at 1200°C but even at this point is not fluid enough for extrusion into filaments; this requires a higher temperature of up to 2000°C. Some additives may be used to reduce the melting point of silica.[30] The majority of continuous glass fibers are spun from the so-called E-glass formulation. E-glass fibers can be used as reinforcing materials for resins, rubber or polymer composites. Some E-glass fibers are used for fire-resistant applications. Another type of glass fiber is the so-called A-glass. This is more economically attractive as it utilizes plate glass scrap, made in a re-melt process rather than difficult direct melting. A-glass normally exhibits half the strength of E-glass and it is commonly used as an insulation material in many thermal or acoustic applications.[30] Another glass formulation is the so-called C-glass, which commonly used to be substituted for the deficiency of E-glass in applications requiring better acid and alkalis resistance. C-glass may also be used in place of E-glass for the reinforcement of bitumen for roofing mats. For high strength composite structures, S-glass, developed by Owens-Corning Fiberglass, is commonly used. Other common applications for glass fibers include electrical circuit boards, automobile water-pump housings and durable pipes.

8.7.5 Metallic and ceramic fibers

Metallic and ceramic fibers represent a special category of fibers that aims to use the outstanding features of metal alloys and ceramic material in a semi-flexible form to allow easier manipulation of these materials in applications requiring their precise and directed incorporation in various structures. Metallic fibers of silver and gold have been used for millennia to decorate fabrics. Today metallic fibers serve functional as well as decorative purposes. These fibers are formed by drawing metal wires through successively finer dies to achieve the desired diameter. Although gold and silver are the easiest to draw, modern methods have allowed the manufacture of steel, tantalum and zirconium fibers.[6] Because they are electrical conductors, metal fibers have been blended into fabrics to reduce the tendency to develop static electrical charges.

The idea of developing ceramic fibers was largely driven by the need to reinforce ceramic matrix composites in applications where temperatures can be elevated to above 1000°C. This type of application required ceramic structures that are fine and flexible for easy manipulation and directed functions. Now, many ceramic fibers (oxide and non-oxide) are available in the market with diameters ranging from 10–20 µm. In addition, larger diameters ceramic fibers (over 100 µm) are also available for applications such as gas turbines, heat exchangers and contaminant walls for fusion reactors.[5]

In comparison with organic fibers or even glass fibers, ceramic fibers are uncontested in withstanding high temperature applications. In comparison with carbon fibers, ceramic fibers will prevail on the basis of the oxidizing and corrosive issues associated with carbon fibers. Carbon fibers will also degrade at temperatures above 300°C. The inherent stiffness of ceramic fibers can limit their formation into fabrics as they must be woven to meet this requirement. However, this manufacturing limitation can be largely overcome by using finer ceramic fibers (about 10 µm) to allow enough flexibility for fabric formation.

8.8   References

1. CAMBRIDGE MATERIAL CHARTS, http://www-materials.eng.cam.ac.uk/mpsite/interactive_charts, Department of Mechanics, Materials, Design, United Kingdom, 2002.

2. CARRAHER C E JR, *Polymer,* in AccessScience@McGraw-Hill, http://www. accessscience.com, DOI 10.1036/1097-8542.535100, last modified: March 12, 2004.

3. MORTON W S and HEARLE J W S, *Physical Properties of Textile Fibers,* The Textile Institute–Butterworths, Manchester and London, 1962.

4. SEYMOUR R B, *Engineering Polymer Sourcebook,* McGraw-Hill, NY, 1990.

5. SMITH W F, *Materials Science and Engineering,* in AccessScience@McGraw-Hill, http://www.accessscience.com, DOI 10.1036/1097-8542.409550, last modified: May 4, 2001.

6. BLOCK I, *Manufactured Fiber,* in AccessScience@McGraw-Hill, http://www. accessscience.com, DOI 10.1036/1097-8542.404050, last modified: May 6, 2002.

7. KLEIN W, *Man-made Fibers and Their Processing.* The Textile Institute, Manchester, 1994.

8. HEARLE J W S (ed), *High-Performance Fibers,* Woodhead Publishing Limited, Cambridge, UK, 2001.

9. HONGU T and PHILLIPS G O, *New Fibers,* 2nd edition, Woodhead Publishing Limited, Cambridge, UK, 1997.

10. EL MOGAHZY Y and CHEWNING C, *Fiber To Yarn Manufacturing Technology,* Cotton Incorporated, Cary, NC, USA, 2001.

11. KIEJNA A and WOJCIECHOWSKI K, *Metal Surface Electron Physics,* Elsevier Science, Oxford, UK, 1996.

12. OSTERMANN M, *The Ceramic Surface,* A & C Black Publishers, London, UK, 2002.

13. EL MOGAHZY Y, 'Friction and surface characteristics of cotton fibers', *Friction in Textile Materials,* Gupta B.S. (ed), Chapter 6, Woodhead Publishing Limited, Cambridge, UK, 2008.

14. RIPPON J A, 'Friction, felting and shrink-proofing of wool', *Friction in Textile Materials,* Gupta B.S. (ed), Chapter 7, Woodhead Publishing Limited, Cambridge, UK, 2008.

15. EL MOGAHZY Y, 'Friction and surface characteristics of synthetic fibers', *Friction in Textile Materials,* Gupta B.S. (ed), Chapter 8, Woodhead Publishing Limited, Cambridge, UK, 2008.

16. NALWA H S, *Handbook of Nanostructured Materials and Nanotechnology,* Academic Press, San Diego, 2000.

17. HUANG Z M, ZHANG Y Z, KOTAKI M and RAMAKRISHNA S, 'A review on polymer nanofibers by electrospinning and their applications in nanocomposites', *Composites Science and Technology,* 2003, **63**, 2223–53.

18. ROBINSON J S, (ed), 'Fiber-forming polymers–recent advances', *Chemical Technology Review No. 150,* Noyes Data Corporation (ndc), Park Ridge, NJ, 1980.

19. ROBINSON J S, (ed), 'Spinning, extruding and processing of fibers–recent advances', *Chemical Technology Review No. 159,* Noyes Data Corporation (ndc), Park Ridge, NJ, 1980.

20. ROBINSON J S, (ed), 'Manufacture of yarns and fabrics from synthetic fibers', *Chemical Technology Review No. 163,* Noyes Data Corporation (ndc), Park Ridge, NJ, 1980.

21. EL MOGAHZY Y, *Fiber Selection and Blending: structural, interactive, and appearance modes.* Video-lecture produced by Quality-Business Consulting, Auburn, AL, 2002, http://www.qualitybc.com/.

22. KANDOLA B K and HORROCKS A R, 'Complex char formation in flame-retardant fiber-intumescent combinations-II. Thermal analytical studies', *Polymer Degradation and Stability*, 1996, **54**, 289–303.
23. HATCH K L, *Textile Science*, West Publishing Company, Minneapolis, NY, 1999.
24. TSUKADA M, SHIOZAKI H and URASHIMA H, 'Changes in the mechanical properties of silk fabrics modified with epoxide and their relation to the fabric construction', *Journal Textile Institute*, 1989, **80**, 4.
25. YANG H H, *Kevlar Aramid Fiber*, John Wiley & Sons, New York, 1993.
26. FLORY P J, *Molecular Theory of Liquid Crystals*, in Advances in Polymer Science series, Vol 59, Springer-Verlag Berlin, 1984, 1–36.
27. JACOBS M J M and MENKE J J, 'New technologies in gel-spinning the world's strongest fiber', *Techtextil-Symposium*, Lecture 213, Frankfurt, Techtextile, 1995.
28. YASUDA H, BAN K and OHTA Y, *Advance Fiber Spinning Technology*, Nakajima T. (ed), Woodhead Publishing Limited, Cambridge, UK, 1994.
29. GUIGON M, OBERLIN A and DESARMOT G, 'Microtexture and structure of some high tensile strength, PAN-base carbon fiber', *Fiber Science and Technology*, 1984, **20**, 55–72.
30. LOEWENSTEIN K, *The Manufacturing Technology of Continuous Glass Fibers*, 3rd edtion, Elsevier, Amsterdam, 1993.

9

Structure and types of yarn for textile product design

Abstract: In the design of fibrous products in which fabrics are made from yarns, the choice of appropriate yarn type and yarn characteristics is critical. Commonly, design engineers focus more on fabric construction and finishing techniques to meet the performance requirements of fibrous products. Despite the benefits of such approaches, they can be very costly if not coordinated with the choice of appropriate yarn. This chapter focuses on yarn and yarn structure from a product-design viewpoint. It begins with the basic structural features of yarns and moves into detailed discussions on the different yarn types. The discussion of yarn structure in this chapter is intended to direct the reader's attention to yarn structural features which can result in desired levels of product performance. Accordingly, key yarn design-related aspects are discussed, including fiber compactness in the yarn, fiber arrangement and fiber mobility.

Key words: yarn classification; yarn specification; specialty yarns; compound yarns; ropes; yarn count; yarn twist; wrap spinning; hollow spindle; idealized yarn structure; yarn hairiness.

9.1 Yarn classification

In classic terms, a yarn may be defined as a long fine fiber strand, consisting of either twisted staple fibers or parallel continuous filaments, which is capable of being interlaced into a woven structure, intermeshed into a knit structure, or inter-twisted into braids, ropes or cords. This definition implies that there are two main types of yarn: continuous filament yarns and spun yarns. A continuous filament yarn represents a simple structure in which multiple filaments are laid side by side in a parallel arrangement. This type of yarn is typically made by extruding polymer liquid through a spinneret to form liquid filaments that are solidified into a continuous fiber strand. As shown in Fig. 9.1, a continuous filament yarn is commonly called a monofilament yarn when it consists of a single filament or a multifilament yarn when it consists of many filaments. Continuous filaments can also be converted into other structural derivatives by deliberate entanglement or geometrical reconfiguration, using a process called texturizing for the purpose of producing stretchy or bulky yarns.[1,2] Spun yarns, on the other hand are produced from staple fibers of natural or synthetic sources using a number of consecutive processes such as blending, cleaning, opening,

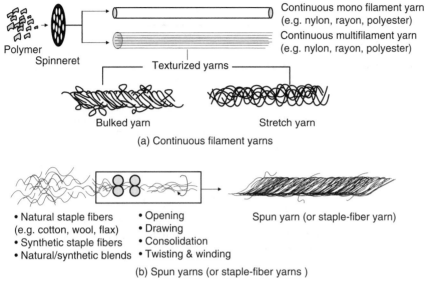

Polymer   Spinneret

Continuous mono filament yarn
(e.g. nylon, rayon, polyester)

Continuous multifilament yarn
(e.g. nylon, rayon, polyester)

Texturized yarns

Bulked yarn   Stretch yarn

(a) Continuous filament yarns

• Natural staple fibers   • Opening   Spun yarn (or staple-fiber yarn)
  (e.g. cotton, wool, flax)   • Drawing
• Synthetic staple fibers   • Consolidation
• Natural/synthetic blends   • Twisting & winding

(b) Spun yarns (or staple-fiber yarns)

9.1 Continuous filament yarns and spun (staple-fiber) yarns.

drawing and spinning to align the fibers and consolidate them into a yarn via twisting or other means.[3]

In addition to the above main classification, yarns can be classified in many different ways. They can be classified by structural complexity into single yarns, plied yarns and cabled yarns as shown in Table 9.1, by the method of fiber preparation into carded, combed, worsted and woolen yarns as shown in Table 9.2 and by the method of spinning into ring-spun, rotor-spun, air-jet and friction-spun yarn as shown in Table 9.3. Details of the spinning methods used to produce these yarns are outside the scope of this book and the reader may refer to previous literature dealing with this subject.[1-6]

The point of classifying yarn into different categories is to emphasize the fact that there is an infinite variety of yarn structures leading to numerous design options depending on the desired physical properties and performance characteristics required in fibrous products. These structures can be produced using innovative polymer technologies, special fiber blends, well-established technologies of spinning preparation and innovative fiber consolidation techniques. In addition, special yarn types can be produced to meet highly specific functions as will be discussed later in this chapter.

9.2   Yarn specifications

In practice, yarns can be described by many specifications depending on the desired yarn performance, fabric type and end product performance.

Table 9.1 Yarn types: by structural complexity[1-4]

Yarn type	General features
Single yarn	• A single yarn can be made from continuous filaments, monofilaments or staple fibers. • Continuous filament yarn basically consists of a monofilament or multifilament arrangement. • Staple-fiber (or spun) yarns are made by twisting fibers or by a false-twist technique in which other forms of binding fibers such as wrapping can be used.
Plied yarn	• Plied yarn commonly consists of two single yarns twisted together to form a thicker yarn. • A ply yarn may be twisted in the same direction as the single yarn or in opposite direction to add visual and appearance effects.
Cabled (multi-folded) yarn	• Several plied yarns can be twisted together to form cabled yarns. • Cabled yarns are typically used for heavy-duty industrial applications such as mooring and heavy-weight lifting.

Figure 9.2 illustrates different types of yarn specifications. Normally, the first specification is yarn type. As indicated above, yarns may be classified as spun yarns or continuous-filament yarns. Within each category, there are many types that can also be specified, each of which exhibits unique characteristics in relation to the end product as demonstrated in Tables 9.1–9.3. Any yarn type must be associated with the fiber type used or fiber content. This may be 100% of a single fiber type (e.g. 100% cotton or 100% wool), or a blend of two or more fiber types (e.g. 50% cotton/50% polyester yarn). Yarns may also be made of multiple structures such as ply, cords, ropes, compound and fancy yarns. Yarn type may also be specified by some desired chemical treatment such as mercerization and slack-mercerization in the case of cotton yarns, shrink resistance for wool yarns, degumming

Table 9.2 Yarn types: by preparation[1-4]

Yarn type	General features
Carded yarn 	• A carded yarn is a single yarn made using cotton processing equipment in which fibers are opened, cleaned, carded and drawn prior to spinning. • It may be considered as an economy yarn as a result of its lower manufacturing cost than a combed yarn (presented below). • The yarn lacks good fiber orientation, has possible moderate to high trash content (when made from cotton), and relatively high neps. • It is typically used for coarse to medium yarn counts that can be woven or knitted into a wide range of apparel from heavy denim to shirts and blouses.
Combed yarn 	• A combed yarn is a single yarn made using the same cotton processing equipment for carded yarn, but with the addition of the so-called combing process. • The combing process substantially upgrades the fiber quality and, consequently, the quality of the yarn produced. • Combing removes short fibers (≤0.5 inch; 1.3 cm), removes neps and reduces trash content to nearly zero. It also results in a superior fiber orientation in the yarn, leading to smoother and softer yarns. • Combed yarns are typically upper medium to fine count yarns. They are also more expensive than carded yarns. • Combed yarns are typically used for high-fashion apparel with comfort and durability representing a unique performance combination.
Woolen yarn 	• Woolen yarns are made on the so-called woolen system using basic steps such as fiber selection, dusting, scouring, drying, carding and spinning • Products produced from woolen yarns include blankets, carpets, woven rags, knitted rags, hand-knitting yarn and tweed cloth.
Worsted yarn 	• This is a high quality yarn made by a series of operations to yield stronger and finer wool yarn quality than that of woolen yarns. • Operations involved in worsted yarn include sorting, blending, dusting, scouring, drying-oiling, carding, combing, gilling operations and drawing. • Products produced from worsted yarns include high quality fashionable wool apparel.

Table 9.3 Yarn types: by spinning system[1-4]

Yarn type	General features
Ring-spun yarns  Carded Combed	• Yarns are made by the ring-spinning system. • Can be carded or combed • Fibers in the yarn exhibit largely true twist and take a helical path crossing the yarn layers. • Some fiber points can be in the core of the yarn and others can be in intermediate or outer layers owing to the phenomenon of fiber migration. • They can be made in a wide range of yarn count and twist. • The strongest yarn of all spun yarns • It can exhibit high hairiness and high mass variation. • The most diverse yarn type as it can be used in all types of fabric from knit to woven.
Compact-spun yarn	• Yarns are made by compact spinning which is a modified ring-spinning system in which fibers are aerodynamically condensed to reduce hairiness and improve yarn strength. • They is commonly used for fine yarns and high-quality apparel.
Rotor (open-end) spun yarns 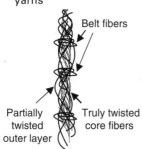 Belt fibers Partially twisted outer layer Truly twisted core fibers	• Yarns are made by rotor spinning. • They can be carded or combed. • Three-layer structure: truly twisted core fibers, partially twisted outer layer and belt fibers • Limited to coarse-to-medium yarn count and requires higher twist than ring-spun • Relatively weaker than ring-spun yarns but has lower mass variation • They can be used for many knit or woven apparels, but greatest market niche is denim fabrics.
Air-jet spun yarn 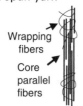 Wrapping fibers Core parallel fibers	• Yarns are made by air-jet spinning, the fastest spinning system. • They consist of two layers, core-parallel fibers and wrapping fibers • Yarns exhibit no twist and the source of strength is the wrapping fibers. • Used for many apparel and home products particularly sheets and bed products
Friction-spun yarn 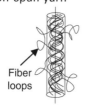 Fiber loops	• Yarns are made by friction spinning. • Yarn exhibits true twisted fibers but a great deal of fiber loops are presented. • They can only be made in very coarse yarn counts used for industrial applications. • Largely used for making industrial yarns of different structures including core/sheath yarns • They can be made from a blend of raw fibers and waste fibers.

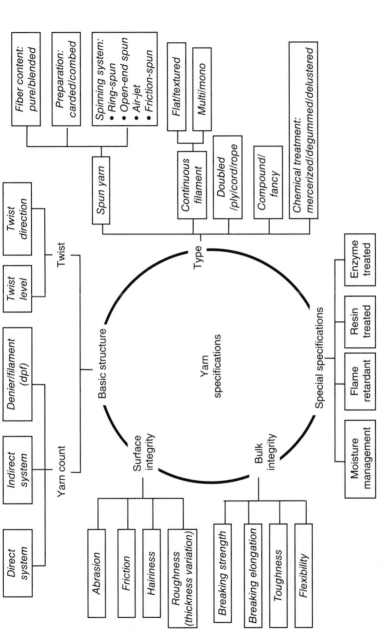

9.2 Common yarn specifications.

Table 9.4 Yarn count systems and units[2,3,6]

System	Term used	Common use	Symbol	Unit of count	Equivalent tex
Direct	Tex	Yarns	tex	g/km	1
	Decitex	Yarns	dtex	g/10 km	0.1
	Millitex	Fibers	mtex	g/1000 km	0.001
	Kilotex	Slivers	ktex	g/m	1000
	Denier	Yarns & fibers	den	g/9 km	0.1111
	Grains	Slivers	gr	gr/yd	70.8
Indirect	English (or cotton count)	Yarns & slivers	N_e	840 yd hanks/lb	590.5
	Worsted count	Yarns	Ne_w	560 yd hanks/lb	885.5
	Woolen	Yarns	N_y	256 yd hanks/lb	1938
	Metric	Yarns	Nm	Km/kg	1000

for silk yarns, delustering for synthetic yarns and texturing for continuous-filament yarns.

Basic structural parameters of yarns must also be specified. The two structural parameters that are commonly specified for any yarn are count and twist. Ideally, yarn count, also known as yarn size or yarn fineness, should be described by yarn diameter or thickness. However, the difficulty associated with specifying a yarn diameter, owing to the high variability in yarn thickness and the deviation from complete roundness in yarn cross-section, resulted in using the yarn linear density as an indirect measure of fineness. In general, linear density is defined as the mass per unit length of a fiber, sliver or yarn. In the case of spun yarns, two systems of linear density (or yarn count) are commonly used (see Table 9.4): the direct system, which is based on weight per unit length, and the indirect system, which is based on length per unit weight. Examples of the direct system are denier and tex. As discussed earlier, the tex system expresses linear density using the weight in grams of 1000 meters of yarn, and the denier expresses it using the weight in grams of 9000 meters of yarn. Examples of the indirect system are the cotton and worsted yarn systems. The cotton system is based on the number of 840-yard lengths per pound of yarn and the worsted system uses 560-yard lengths per pound.

The number of filaments in a continuous-filament yarn is typically given in conjunction with the yarn count. For example, the designation '80/32' can be used to imply a yarn count of 80 dtex consisting of 32 filaments.

This designation indicates that the dtex per filament is 80/32 or 2.5. The tex system is considered by the ASTM as the universal yarn count system. The millitex and the decitex units are extensions of the tex system which are usually used to describe fiber or filament fineness. For intermediate heavy products such as slivers, the 'kilotex' is commonly used.

Yarn twist is typically described using two parameters: twist level and twist direction. Twist level is commonly expressed by the number of turns of twist per unit length. Another way to express twist level is by using the twist multiplier. These two parameters will be discussed in detail later in the context of yarn structure. In general, fibers are twisted to produce some integrity and strength for the yarn. Different yarn types will require different levels of twist by virtue of their desired performance and end use applications. Continuous filament yarns typically require no twist to impart strength. Nevertheless, a small amount of twist (one or two turns per inch 2.5 cm) may occasionally be inserted in this type of yarn merely to control the fibers and prevent them from splitting apart. Twist may also be inserted in continuous filament yarns to avoid ballooning out as a result of accumulating electrical charges.[3,6] In some situations, a fairly high amount of twist may be inserted in continuous filament yarns to break up luster of the yarn or to impart some other effect or fancy attribute.[6] However, high twist levels in these yarns can result in deterioration of their strength.

In spun yarns, twist is necessary to maintain yarn integrity and provide strength. Different spun yarns may require different levels of twist. For example, warp yarns used for weaving normally have higher twist levels than weft yarns because of the higher strength required in these yarns. Knit yarns typically have lower twist than woven yarns to provide better softness.[3] Some spun yarns may exhibit very low twist to produce lofty structures. This is typically the case for weft yarns that are to be napped by teasing out the ends of the staple fibers to create soft, fuzzy surfaces. Other spun yarns may exhibit very high twist levels to meet their application needs. These include voile and crepe yarns. Voile yarns are typically made with high twist to create an intentionally stiff feel in the fabric by plying yarns in the same twist direction as the single yarns to increase the total twist. This provides a light weight stiff furnishing or curtain fabrics that can be made from 100% cotton or cotton blends with linen or polyester. Crepe yarns, also known as unbalanced yarns, have the highest levels of twist. The idea is to create a yarn with the high liveliness desired in some apparel products.

In a spun yarn, fibers may be twisted in clockwise or counter clockwise directions. Commonly, these are described as Z-twist or S-twist. A single yarn has 'Z' twist if, when it is held in the vertical position, the fibers inclined to the axis of the yarn conform in direction of slope to the central portion of the letter Z. On the other hand, a single yarn has 'S' twist if, when it is held in the vertical position, the fibers inclined to the axis of the yarn conform in direction of slope to the central portion of the letter S.

In practice, the importance of twist direction is realized when two single yarns are twisted to form a ply yarn. Ply twist may be Z on Z, or S on Z depending on appearance and strength requirements of the ply yarn. When the yarn is woven or knitted into a fabric, the direction of twist influences the appearance of the fabric. When a cloth is woven with the warp threads in alternate bands of S and Z twist, a subdued stripe effect is observed in the finished cloth owing to the difference in the way the incident light is reflected by the two sets of yarns. In twill fabric, the direction of twist in the yarn largely determines the predominance of the twill effect. For right-handed twill, the best contrasting effect will be obtained when a yarn with a Z twist is used; on the other hand, a left-handed twist will produce a fabric having a flat appearance. In some cases, yarns with opposite twist directions are used to produce special surface texture effects in crepe fabrics. Twist direction will also have a great influence on fabric stability, which may be described by the amount of skew or 'torque' in the fabric. This problem often exists in cotton single jersey knit where knitted wales and courses are angularly displaced from the ideal perpendicular angle. One of the solutions to this problem is to coordinate the direction of twist with the direction of machine rotation. With other factors being similar, yarn of Z twist is found to give less skew with machines rotating counter-clockwise. Fabrics coming off the needles of a counterclockwise rotating machine have courses with left-hand skew and yarns with a Z twist yield right-hand wale skew. Thus, the two effects offset each other to yield less net skew. Clockwise rotating machines yield less skew with S twist.

Almost all yarn specifications will involve yarn type, count and twist. In addition, other specifications may be stated depending on the fabric-forming system used and the intended application or end product. These include yarn bulk integrity, described by physical or mechanical parameters such as strength, elongation, toughness and flexibility, and yarn surface integrity described by yarn hairiness, abrasion resistance, friction and roughness. For function-focus fibrous products, yarn types may also be specified in terms of special treatments such as hydrophilic finish, hydrophobic finish, flame-retardant finish, resin treatment and enzyme treatment. These types of finish will be discussed in Chapter 11.

9.3   Specialty yarns

In addition to the common yarn types listed in Tables 9.1 to Table 9.3, many specialty yarns can be made to serve particular purposes in both traditional and function-focus products. These yarns are made through manipulation of the basic yarn structures to produce unique structural features.

In the following sections, three types of specialty yarns are discussed: compound yarns, ropes and fancy yarns.

9.3.1 Compound yarns (function-focus yarns)

A compound yarn typically consists of two or more different fiber strands in a core–wrap structure with the idea being to get a net effect combining the characteristics of the core and the cover components.[4,7,8] Many core–wrap yarns consist of core filament(s) wrapped or covered by a yarn. These may be categorized as covered-spun compound yarns (see Fig. 9.3a to 9.3c). This type of yarn can be produced using so-called wrap spinning with a hollow spindle.[4,7] The core being in the inner layer of the yarn typically provides specific bulk characteristics such as exceptional strength (e.g. polyester or nylon filament) or high elasticity (e.g. rubber or spandex). The sheath or the yarn wrap should have a good grip on the core to avoid fiber separation or core slippage. It should also protect the core filament from adverse external effects. This is particularly important when the core filament is elastomeric (e.g. rubber or spandex) as this type of filament can degrade by exposure to many chemicals or ultraviolet light. In addition, the wrap, being in contact with the wearer, should exhibit comfort characteristics such as softness, flexibility and porosity. The wrap yarn can be a spun or filament yarn. In some cases, two wrap yarns may be used with one twisted in the opposite direction to the other (Fig. 9.3c). This results in special appearance effects as well as a good balance against potential yarn snarling.

Another type of compound yarn can be made from a readily separable core surrounded by fibers or fiber strands. These are generally known as core–spun yarns. Different combinations of this type include[7] filament core/staple-fiber wrap (Fig. 9.3d), staple-fiber core/filament wrap (Fig. 9.3e) and staple-fiber core/staple-fiber wrap (Fig. 9.3f). Most common spinning systems such as ring spinning, air-jet spinning and friction spinning can be equipped with accessories to produce these types of yarn.

Among the types of core–spun yarns mentioned above, the most widely used are the filament core/staple-fiber wrap yarn. In these yarns, the core consists of continuous filaments surrounded completely by a wrap of staple fibers. It is often referred to as core/sheath yarn since the untwisted staple fibers surrounding the filament core act as a sheath or a casing holding the core. Potential products made from this yarn can range from traditional apparel and household fabrics to function-focus fibrous products. Examples of these products are as follows:[4,7–11]

- Some bed sheets and knit fabrics made from core–spun yarns with polyester-filament core and a sheath of combed cotton fibers; the

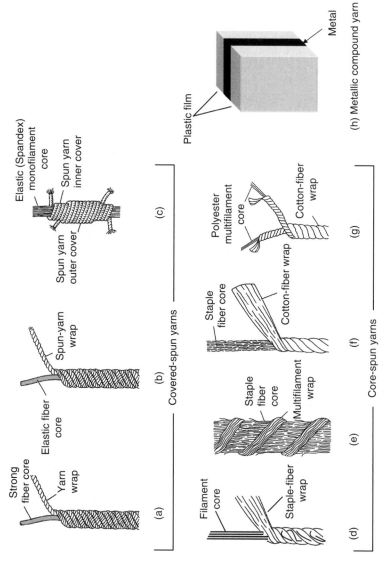

9.3 Examples of compound (or composite) yarns.[4,7]

obvious advantages of these products are high strength and durability provided by the polyester filament and comfort characteristics provided by the cotton sheath;

- Some sewing threads made from polyester core/cotton wrap fibers by plying and twisting together two core/sheath yarns (see Fig. 9.3g);
- Woven corduroy fabric and stretchable denim made from elastomeric core/staple-fiber wrap yarns made from spandex and cotton;
- Some cotton-knit swimsuits made from spandex core and cotton fiber wrap to provide better fit using a very small amount of spandex (as little as 1%); this yields the traditional performance of fabrics made from spun yarns with the additional advantages of good elongation and elastic recovery provided by the spandex component;
- Some secondary carpet backing made from polypropylene split or tape core/polypropylene fibers wrap;
- Flame-retardant and protective clothing made from fiberglass–filament core/aramid fibers wrap and carbon core/aramid fibers wrap;
- Protective gloves made with steel wire surrounded by aramid-fiber blends.

In light of the above examples, it is obvious that the idea that a yarn can be made from two or more different components has opened up a wide range of opportunities for creative products in many different fields including protective clothing, e-textiles and optical applications.[4,7–11] Now, compound yarns can be made in many different combinations such as high-performance fibers core/conventional fibers wrap, metallic wire core/fiber wrap and optical fiber core/conventional fiber wrap.

The design of compound yarns is likely to face many issues and design problems that should be clearly stated and identified. For example, in the design of staple-fiber core/filament wrap yarns (Fig. 9.3e), untwisted parallel staple fibers (typically 80–95% by weight) are wrapped by a multifilament strand. Obstacles facing this design typically result from the difficulty of maintaining continuous and consistent streams of staple fibers and the need for precise wrapping.[7] As a result, it is not as widely used as the filament core/staple-fiber wrap yarns discussed above. From a functional performance viewpoint, it is yet to be seen whether this type of yarn can provide better compound criteria than the filament core/staple-fiber wrap yarns. However, the design concept of untwisted parallel staple fibers wrapped with multifilament yarns reveals many advantages over the conventional blended yarns. For instance, the absence of twist in the staple fiber strand should automatically result in a more even yarn diameter than the conventional yarn. The strength and flexibility of the yarn will primarily depend on the wrapping effect. Other merits such as improved abrasion resistance, reduced hairiness and better pilling resistance can also be realized by virtue of the structure of the yarn.

Core–spun yarns made from staple-fiber core/staple-fiber wrap typically have a strand of parallel, untwisted staple fibers completely wrapped with another staple-fiber strand. The need for this type of yarn was a direct result of some of the performance problems associated with filament core/staple-fiber wrap yarns, particularly when normal filament (not elastomeric filament) is used. This structure can easily be produced using conventional spinning systems such as air-jet spinning and friction spinning.

The concept of compound yarn is not restricted to conventional spinning as many innovative techniques can be used to produce other types of compound structures that may be loosely called compound yarns. For example, compound structures can be made from laminated fabrics in which a central core of metal is sandwiched by two layers of plastic films such as nylon (see Fig. 9.3h). The three layers are then passed through squeeze rollers that bind the layers together. Strips of thin films or yarn are then cut and wound onto bobbins.[4,7] This metallic core/plastic cover yarn can be used in many applications, particularly when static charge buildup is to be avoided as the core component will easily conduct electricity. In addition, metallic core yarns with nylon coating have been used in carpeting.

Compound yarns can also be made directly from continuous filament yarns using multiple polymeric materials that are extruded using different arrangements. Figure 9.4 shows three examples of these arrangements, namely side-by-side, sheath-core and matrix arrangement.[4,5,7] Each one of these arrangements can provide pre-specified functions depending on the

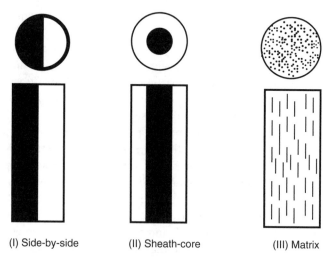

(I) Side-by-side   (II) Sheath-core   (III) Matrix

9.4 Heterogeneous compound filament yarns.[4,5,7]

polymeric material used. This design concept can obviously provide unlimited options using different types of polymeric materials, different percentages of each type and innovative polymer arrangements.

In light of the above review of compound yarns, it is the author's opinion that this category of yarn should be regarded as 'function-focus yarn'. This term is more precise, particularly in view of the fact that another category of yarn called 'fancy yarns', which will be discussed shortly, shares a great deal of the structural features of compound yarns, except that it is primarily intended for visual esthetic applications and not necessarily for specific or technical functions. A function-focus yarn is a yarn primarily intended for specific technical applications via the choice of appropriate fibers or filament and the formation of structural arrangements that directly serve the purpose of the technical application in question. This category of yarn may primarily be driven by the unique functional features of the fibers or filaments used, or it may represent a combination of fiber type and yarn structural arrangement targeted toward specific functional performances such as protective clothing[10] or thermoplastic composites.[11]

9.3.2 Ropes

The general structure of a rope is shown in Fig. 9.5a. The basic component is a single yarn made from staple fibers or continuous filaments; single yarns are plied and twisted together to produce rope yarns; a number of rope yarns are arranged in helical layers to produce a rope strand; finally, groups of strands are laid together to form the finished rope. The term 'laid' indicates that the individual strands are not simply twisted together but rather are revolving in a planetary pattern to form the final rope. Ropes and cords have gained significant interest in recent years as a result of the need for lighter, stronger and easier-to-handle rope structures than those made from metallic wires for heavy-duty applications such as civil and ocean engineering. In the context of product development, building a prototype rope and testing it to determine its performance, and then building another rope in an attempt to improve that performance can be very expensive. In the past, this method was affordable with small conventional ropes. When very large ropes made of polyester or high-modulus aramids are to be developed, conventional methods will be too expensive and too time consuming. Alternatively, computer modeling of the rope mechanical behavior as a function of construction, fiber type and surrounding conditions can provide great assistance to design engineers.[12–14]

One of the key design parameters of ropes is the rope construction. Ropes can take many different constructions that serve different applications and loading demands.[13,14] Examples of these constructions are shown in Fig. 9.5b and 9.5c. The rope construction is commonly defined by the

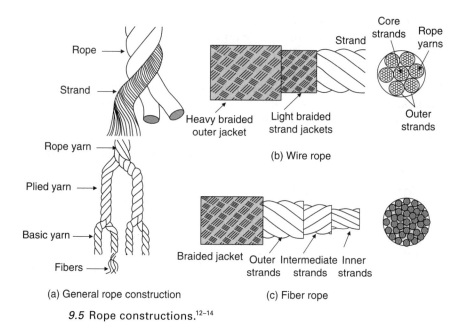

9.5 Rope constructions.[12-14]

number of strands involved and the structures of different rope layers. In case of wire ropes and some aramid ropes, the 7-strand form shown in Fig. 9.5b is commonly used. The rope typically consists of a heavy-braided outer jacket, a light-braided strand jacket and the rope strands. More strand layers of up to 18 strands can also be used for this type of rope. When larger ropes are required, up to 36-strand construction can be formed, as shown in Fig. 9.5c. In this case the rope strands are typically smaller in size. A compact construction of this type of rope can be achieved by using the same lay length and direction for all strands. A torque-free rope of this type can be achieved when the direction of lay of the outer strand layer opposes that of the inner two layers (see Fig. 9.5c). A torque-free rope means that the rope does not tend to rotate when tension is applied. However, this form of rope is not tolerant to torque and a small amount of rotation can upset the balance of loads between the inner and outer strand layers. Braiding is a common approach in constructing fiber ropes as it typically results in torque-free ropes. But again, they may not be torque tolerant, because applied rotation will transfer tension to only half of the strands.[14] However, they are not as easily affected by torque as the above mentioned 36 strand rope. Other construction forms include the 12-strand braided rope, which is widely accepted in conventional marine applications and the 8-strand rope (sometimes called plaited rope), which is less compact.

9.3.3 Fancy yarns

Fancy yarns, also called novelty or effect yarns, represent another category of yarns in which deliberate irregularities, discontinuities and color variations are introduced into the yarn structure with the primary intention being to produce enhanced esthetic and fashionable impressions.[4,7,15] Accordingly, fancy yarns are not necessarily intended for function-focus fibrous products.

Many fancy yarns are produced using a conventional staple-fiber spinning systems equipped with special accessories. Others are produced using a special spinning preparation in which blends of various fiber colors or fiber types that form fancy slivers are used. Texturized yarns made exclusively from continuous filaments can also be considered as a form of fancy yarn.

In Chapter 3, a number of important business and marketing issues related to product development were discussed. In addition, the importance of stimulating and attracting consumers was addressed. The development of fancy yarns represents a model example of how a market can be dynamically stimulated using creative ideas that can attract consumers and draw attention to the products with minimum added cost. Indeed, most fancy yarns merely represent random effects introduced to the traditional yarns using simple accessories mounted on conventional spinning machinery, such as ring spinning or rotor spinning.[7] The manufacturing cost that may be added as a result of using these accessories is largely counterbalanced by many other factors that can indeed result in a net cost reduction. For example, fancy yarns that are a result of induced irregularities and discontinuities require quality constraints and testing procedures that are much less than those implemented for traditional yarns. In addition, the possibility of using a wide range of fibers including low quality fibers and a variety of fiber wastes also adds to the overall reduction in manufacturing cost.

In general, fancy yarns may be divided into two major categories:[4,7,15] mono-fancy yarns and compound fancy yarns. The former are made from a single staple-fiber strand or multifilament yarn in which irregularities are introduced during the conversion from fibers to yarns; the latter consist of two or more fiber strands used to provide the desired combined effects. Figure 9.6a and 9.6b represents two examples of the first type of fancy yarn, namely slub yarns and thick and thin yarns. A slub yarn is basically a spun yarn (ring or rotor-spun yarn) in which thickness variations are deliberately induced by variation in the yarn twist combined with control of the fiber feed so that thick areas will exhibit lower twist than thin areas. Key structural parameters associated with slub yarns include slub size with respect to yarn diameter, slub length and slub spacing or frequency.

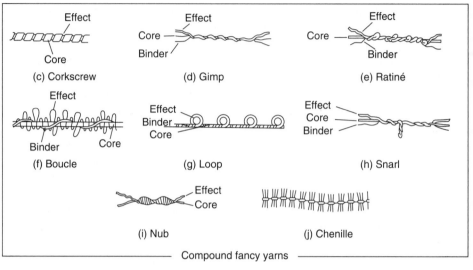

9.6 Examples of fancy yarns.[4,7,15]

Typically, slubs are randomly and irregularly spaced along the yarn axis. A thick-and-thin yarn is typically a multifilament yarn in which diameter variations are deliberately induced by variation in the filament flow rate through the spinneret during extrusion. This type of yarn provides a unique surface texture to the fabric made from the yarns similar to that of linen fabrics.

Compound fancy yarns can be found in an unlimited number of varieties and textures. The idea of a compound fancy yarn is to combine two or three strands together so that one of them acts as the fancy or effect strand, another acts as the yarn core and perhaps a third one acts as a binder to secure the effect strand to the core strand. The binder strand is usually small and is often not readily visible. The differences in the appearance of the various yarns are largely dependent on the extent of prominence of the effect strand. This is typically controlled by a number of structural factors including the size of the effect strand, the configuration assumed by the effect strand (wavy, looped, snarled, etc.) and the core-effect arrangement used.[7,15] The compound fancy yarns shown in Fig. 9.6c to 9.6j are listed in the order of their prominence if they are visualized side-by-side in a piece of fabric. Devices used to create these types of yarn have been available since the 19th century. These devices are now available as optional accessories with most commercial spinning machines.

In the context of design, it is important to point out that in addition to the deliberate irregularities induced in fancy yarns, fiber strands with different physical characteristics can also be used to make fancy yarns. For example, the corkscrew yarn shown in Fig. 9.6c typically has a fine strand spiraling around a soft and bulky strand. When this situation is reversed (i.e. a soft and bulky strand spiraling around a fine strand), the yarn is commonly called a spiral fancy yarn. Both gimp and ratiné yarns (Fig. 9.6d and 9.6e) exhibit a slightly wavy appearance. In both yarns, the effect strand is twisted around the core strand, but ratiné yarns are usually larger and wavier than gimp yarns owing to a larger effect strand. They both have an effect strand that lies closer to the core strand than is typical in boucle and loop.[4]

Boucle, loop and snarl yarns, shown in Fig. 9.6f, 9.6g, and 9.6h, respectively, have dramatic effect strands.[4,15] The effect strand in boucle yarns is usually the softest and bulkiest of this group and, typically, does not lie near the core strand. In loop yarns, the effect strand is usually made of long, rigid fibers that may also be lustrous, such as mohair or wool, or untwisted thick filament strand. In both boucle and loop yarns, the binder strand is necessary to hold the loops in place. The core strand in loop yarns is usually coarser and heavier than in boucle yarns. Boucle yarns are found in both woven and knit fabrics that are often called boucle fabrics. Loop yarns are frequently used in coating and suiting fabrics, for interior furnishings and for speciality fabrics, either knit or woven. A snarl yarn uses a twist-lively strand to form the projecting snarls. The twist of the effect strand is usually in the same direction as the twist that holds the effect and core strands; the binder strand is usually twisted in the opposite direction. Nub and slash yarns have the effect strand twisted around the core strand a number of times in a small area to form an enlarged bump or 'nub'. A binder strand may or may not be used. In a slash yarn, the enlarged area is longer and thinner than in a nub yarn.

Chenille yarns (Fig. 9.6j) resemble a caterpillar, hence the derivation of their name from the French word chenille, which means caterpillar. The effect is composed of tufts of yarn held between plied yarns that form the core of the chenille yarn. Chenille yarns are usually made from cotton, wool, rayon or polyester and are used in fabrics that are soft and velvety. The softness of chenille yarns makes them vulnerable to rubbing; consequently, fabrics containing them do not wear well because of a low abrasion resistance.[4,15] Chenille yarns are used as crosswise yarns in a woven fabric to create a velvet-like surface or occasionally placed in the filling to add surface effects. In woven fabrics, the yarns may be laid so that all the tufts lie on one side of the fabric or so that some tufts appear on both the face and the back. These fabrics are used for many products such as dresses, draperies and bedspreads. Chenille yarns are also used in knit fabrics to make bulky sweater knits. It should be pointed out that some fabric is called 'chenille' even though it does not contain chenille yarns. These fabrics are essentially tufted and they are often used for bathrobes and bedspreads.

9.4   Idealized yarn structure

In order to understand the role of yarn in design applications of fibrous products, it will be important first to understand yarn structure. Research on yarn structure can be traced back to the early work by Gégauff in 1907 followed by Platt's work in the late 1940s, then the work by Hearle and co-workers in the 1950s and 1960s. These contributions focused primarily on the mechanical behavior of yarn and the structural features that influence this behavior. Engineers involved in the design of yarn should read the book entitled *Structural Mechanics of Fibers, Yarns, and Fabrics*, published in 1969 and co-authored by Hearle, Grosberg and Backer.[16] This book provides an excellent review of these major contributions. Another excellent book by Zurek, entitled *The Structure of Yarn*, translated from Polish to English in 1975, provided a critical dimension of yarn structure, which is the nature of variability in fiber arrangement in different fiber strands.[17]

In the discussion below, an idealized twisted yarn model is described based on the work by Hearle *et al.*[16] This model is shown in Fig. 9.7. Understanding this basic model can assist a great deal in understanding the real yarn structure and its effects on yarn and fabric performance characteristics, as will be discussed later.

An idealized yarn structure is assumed to be in the form of a cylindrical body with well-know radius, composed of a series of concentric cylinders of differing radii with each fiber following a uniform helical path around one of the concentric cylinders, so that its distance from the yarn axis remains constant.[16] According to this structure, a fiber at the center will follow the straight line of the yarn axis; but, going out from the center, the helix angle gradually increases, since the number of turns of twist per unit length remains constant in all the layers. Under these ideal conditions, the length of one turn of twist, h, is given by $h = 1/T$, where T is the amount of twist in the yarn commonly expressed by the number of turns per unit length.

Using the opened-out surface of the ideal yarn, shown in Fig. 9.7, the following basic relationships can be derived:[16]

$$L^2 = h^2 + 4\pi^2 R^2$$
$$\tan \alpha = \frac{2\pi R}{h} = 2\pi R T$$

(9.1)

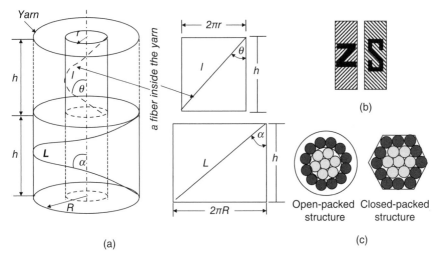

9.7 Twist in idealized yarn structure.[16] (a) Idealized twisted yarn structure, (b) twist direction, (c) idealized packing.

where R is the yarn radius, L is the length of the fiber in the yarn outer layer and α is the surface twist angle.

Using a general concentric layer of the yarn at a radius r, similar relationships for the fiber can be obtained (see Fig. 9.7):

$$l^2 = h^2 + 4\pi^2 r^2$$
$$\tan\theta = \frac{2\pi r}{h} = 2\pi r T \tag{9.2}$$

where θ is the helical angle of the fiber layer and l is the fiber length. The range of the angle θ is from zero at the center of the yarn to α at the yarn outer layer. This means that fibers at the center are straight and the helix angle reaches a maximum at the yarn outer layer. Thus, the yarn radius is $R = (2n - 1)r_f$, where r_f is the radius of the fiber and n is the nth fiber layer. In addition, all fibers in a given yarn layer have the same helix angle of twist.

One of the benefits of the idealized yarn structure stems from the possibility of defining twist in terms of key yarn structural parameters such as yarn diameter and yarn bulk density. From Equation (9.1), the twist angle is related to yarn diameter and the twist level by the following equation:

$$\tan\alpha = \frac{2\pi R}{h} = 2\pi R T = \pi D T \tag{9.3}$$

In a cylindrical yarn, the linear density of the yarn (mass/unit length) is given by:

$$\text{mass/unit length} = \pi R^2 \rho_y = \pi \frac{D^2}{4}\rho_y \tag{9.4}$$

where ρ_y is the yarn bulk density (g cm^{-3}) and R is the yarn radius (cm).

Using the tex unit of linear density, T_t, yarn tex can be expressed by the following equation:

$$\text{mass/unit length (tex)} = T_t = \pi \frac{D^2}{4}\rho_y \times 10^5 \tag{9.5}$$

This equation gives the following expression of yarn diameter:

$$D(\text{cm}) = \left(\frac{4T_t}{10^5 \pi \rho_y}\right)^{1/2} \tag{9.6}$$

From Equations 9.3 and 9.6:

$$\tan\alpha = \pi DT = \pi T\left(\frac{4T_t}{10^5 \pi \rho_y}\right)^{1/2} = 0.10112 \times T\left(\frac{T_t}{\rho_y}\right)^{1/2}$$

or

$$\tan\alpha = 0.0112(TM).(1/\rho^{1/2}) \tag{9.7}$$

The twist multiplier, TM, is a practical term used among spinners to express the level of twist in a spun yarn in terms of turns per inch and yarn count (tex), $TM = T.T_t^{1/2}$. Following the two systems of yarn count discussed earlier, there are two systems of twist multiplier: (a) the English system, $TM = T/\sqrt{N_e}$, where T is the twist in turns per inch, and (b) the tex system, $TM = T/\sqrt{tex}$, where T is expressed in turns per cm.

The idealized yarn model described above represents an important start in exploring key structural features of yarns. However, the limitations of this model should be considered in view of understanding the nature of a real yarn structure. In this regard, many points can be made:[2,3,6,16–18]

- A real yarn, spun or continuous filament, is not perfectly circular and most fibers do not have circular cross-sections.
- The twist level is not uniform along the yarn axis owing to variation in fiber properties and technological limitations that prevent a perfectly uniform twist.
- Upon twisting and owing to the lateral forces imposed, circular fibers tend to exhibit elliptical cross-sections.
- Fibers may take helical path in the yarn. However, this path is often distorted by the lack of full control on all fibers during spinning and the high variability in fiber length and fiber fineness.
- Fibers in the center of the yarns can be kinked or buckled.

- In a real spun yarn, many fibers will have ends projected outside the yarn body; this is known as yarn hairiness.

More limitations of the idealized yarn structure will be discussed later. However, and despite these limitations, the idealized yarn model still represents a key guideline in understanding the basic features of yarn structure. In the following section, key design-related factors of yarn are discussed in the context of yarn structure.

9.5   Design-related aspects of yarn structure

The yarn, being essentially an assembly of fibers, is largely governed by three key factors that not only influence its performance during weaving or knitting but also affect the functional performance of the fabric made from it. These factors are[2,3] (a) fiber compactness, (b) fiber arrangement and (c) fiber mobility.

9.5.1   Fiber compactness

The integrity of a yarn structure is maintained partially by some form of compactness of fibers or filaments in the yarn. In the case of a continuous filament yarn, filaments adhere together, by virtue of their large number and the long inter-filament contact under lateral forces imposed by the outer filaments pressing against the inner ones. This produces a highest compactness level as no air is allowed to penetrate between the filaments. When filament yarns are converted into textured yarn, a great deal of inter-filament space is created leading to lower compactness. In the case of spun yarns, fibers are compacted together by lateral forces imposed by the twist inserted to form the yarn. The discrete nature of the fibers provides for variable compactness with plenty of air pockets inside the yarn structure filling in the inter-fiber spaces.

The importance of fiber compactness stems from two important points that should be realized in the design of fibrous products. For a given fiber type, high fiber compactness is likely to result in low yarn compressibility (or low softness), high strength, low yarn flexibility, low yarn porosity and more moisture wicking along the yarn surface than along the fibers in the yarn. On the other hand, low fiber compactness is likely to result in high yarn compressibility (or high softness), high flexibility, high porosity and more moisture wicking along the fibers in the yarn. Between these two situations, yarns of different levels of fiber compactness can be produced to meet key fabric performance characteristics such as strength, hand, drape, moisture management and comfort. Furthermore, fiber compactness is a key factor in determining the yarn dimensional stability; a yarn can be considered dimensionally stable if the fiber compactness is the same in the relaxed state and under low levels of stress.[2]

The extent of fiber or filament compactness in a yarn structure can be expressed by two key factors: (a) the yarn bulk density (g cm^{-3}) or more precisely the specific volume, v_y, of yarn (cm^3 g^{-1}) and (b) the volume of still air inside the yarn. One can imagine a three-dimensional yarn structure consisting of fiber material and air pockets that are created by virtue of the discontinuities of fiber flow along the yarn axis in the case of spun yarns or separations between filaments in the case of continuous-filament yarn or texturized yarn. Accordingly, a yarn is essentially a porous structure. Based on the analysis of the ideal yarn model discussed above, the specific volume (cm^3 g^{-1}) of yarn can be derived from Equation (9.6) to yield the following expression:

$$v_y = \left(\frac{\pi R^2}{T_t}\right) \times 10^5 \qquad (9.8)$$

This equation suggests that the density of packing of fibers in the yarn remains constant throughout the model. This idealization resulted in further theoretical analysis in which idealized fiber packing forms in the yarn were suggested.[16] These include the open-packed yarn and the hexagonally closed-packed yarn shown in Fig. 9.7c. The effect of twist on the specific volume of yarn can be realized from the following equation derived from Equation (9.7):

$$v_y = \left(\frac{\tan^2 \alpha}{4\pi T_t T^2}\right) \times 10^5 \qquad (9.9)$$

The above relationship indicates that for a given yarn count, T_t, the effect of twist on yarn specific volume is primarily a function of the variation in yarn circumference caused by twisting. Note that, from Equation (9.3), that yarn circumference $\pi D = \tan\alpha/T$. Thus, according to the above equation if the yarn has a constant diameter, as suggested by the idealized model, the specific volume will also be constant. In a real yarn, variation in yarn diameter is greatly to be expected. An increase in yarn circumference, or yarn diameter, caused by twisting will result in a quadratic increase in its specific volume, or a quadratic reduction in bulk density. This situation can only hold if the increase in yarn diameter is achieved at a constant mass per unit length, T_t, or a constant yarn count. Another relationship developed by Neckar[19] indicates that yarn volumetric density has an approximately linear relationship with the product (twist.tex^{¼}) of spun yarn.

The true specific volume of yarn will depend on the volume occupied by the fibers and by the amount of inter-fiber space which is filled with air. This leads to a useful term called packing fraction, ϕ:[16]

$$\phi = \frac{v_{fiber}}{v_{yarn}} = \frac{v_{fiber}}{v_{fiber} + v_{air}} \qquad (9.9)$$

At the time the above equations were developed, ways to measure yarn density did not exist. In recent years, measures of yarn density were made possible by Uster® Technologies.[20] These measures make use of a combination of capacitive techniques that measure the yarn mass and optical techniques to measure the yarn thickness.[3] Figure 9.8 shows values of density for some different spun yarns. Some of the interesting points that can be revealed from this figure are as follows:

- Most traditional spun yarns exhibit volumetric density ranging from approximately $0.30–0.6\,g\,cm^{-3}$. This is a wide range from a yarn-performance viewpoint as it implies a wide range of yarn bulkiness, which could have a wide impact on fabric performance characteristics such as compressibility, softness, porosity and stiffness. Corresponding data for continuous-filament yarns are not available, but one can expect them to be significantly higher than those for spun yarns. Textured yarns can have a wide range of bulkiness depending on the method of texturizing used.
- Combed yarns are denser than carded yarns (see Fig. 9.8a). This is expected on the ground that the combing process results in better fiber alignment and better evenness in fiber length via the removal of short fibers and fiber hooks.[2,3,6]
- Ring-spun yarns have substantially higher density than rotor-spun yarns (Fig. 9.8b). This difference is directly attributed to the structural difference between the two yarn types summarized in Table 9.3.
- Compact yarn has the highest density of all traditional spun yarns (see Fig. 9.8c). This is a direct result of the deliberate compactness of fibers in the yarn,[3] making it a denser yarn (see Table 9.3).
- Yarns made for woven fabrics are denser than those made for knit fabrics (see Figures 9.8c and 9.8d). This is a direct result of the higher twist used for the former than the latter. Yarns made for woven fabrics are typically stronger than those made for knit fabrics and this requires higher twist for a given fiber type and a given set of fiber characteristics. On the other hand, yarns made for knit products are typically softer and more flexible; this requires low levels of twist.

In light of the above discussion, the key factors influencing fiber compactness in the yarn are spinning preparation (e.g. carded versus combed), spinning method (or yarn type) and yarn twist. Although the results in Fig. 9.8 show values of yarn density at different yarn counts, they should not be used as indications of the effects of yarn count on density. This is

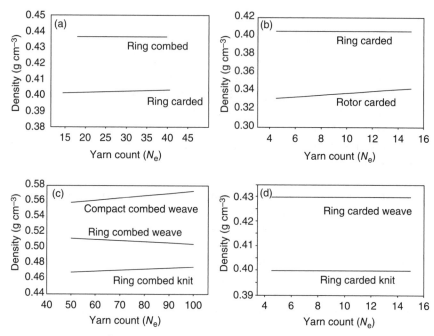

9.8 Uster® yarn density for different preparations and different spinning methods.[20]

due to the fact that these results are obtained from the Uster® Statistics data which are typically taken from various mills using various spinning equipment.

In addition to the above factors, fiber compactness in the yarn can be altered or controlled using many fiber-related parameters for a given yarn type and yarn count. These include fiber density, fiber length, fiber fineness, fiber crimp, fiber cross-sectional shape and the number of fibers that can be accommodated in the yarn cross-section. For example, fine and long fibers will normally result in higher yarn density than coarse and short fibers. In addition, fibers with circular cross-sectional shape are likely to form yarns with higher bulk density than those of triangular or trilobal shapes.

9.5.2 Fiber arrangement

The way that fibers are arranged in the yarn structure can have a great impact on a number of yarn and fabric performance characteristics such as yarn liveliness, yarn appearance, yarn strength, fabric dimensional stability and fabric cover. As illustrated in Fig. 9.1 and described in Tables 9.2 and 9.3, different yarn types exhibit different forms of fiber arrangement.

Obviously, the simplest fiber arrangement is that of a continuous-filament yarn where fibers (or continuous filaments) are typically arranged in a parallel and straight form. Deviations from this arrangement can be caused by slightly twisting the filaments or through a deliberate distortion in filament orientation as occurs in the texturizing process. In spun yarns, the fiber arrangement is quite different from the simple arrangement of continuous-filament yarns. The discrete nature of staple fibers makes it difficult fully to control the fiber flow in such a way as to produce a well-defined fiber arrangement. As a result, the analysis of the fiber arrangement in spun yarns typically begins with an idealized twisted yarn structure and then seeks ways to interpret or evaluate the deviation from this structure, as discussed earlier.

The idealized twisted yarn structure can be useful in exploring some of the important relationships between different yarn structural parameters. However, it does not fully describe the fiber arrangement in actual spun yarn. Indeed, a yarn following this idealized arrangement will neither be practically feasible nor useful. For example, the assumption that a yarn is composed of a series of concentric cylinders of differing radii and that each fiber follows a uniform helical path around one of the concentric cylinders, implies that the outer cylinders are held firmly by some cohesive forces without fiber interference. Obviously, this is not true as the fibers on the surface would be merely wrapped around and not gripped at all. In a real spun yarn, twisted fibers will still take some form of helical shape but they can migrate from one layer of the yarn to another with some points on one fiber being on the outer layer and others being in the intermediate layers or at the center of the yarn. This creates a self-locking structure that provides the necessary integrity to the yarn. This phenomenon, called 'fiber migration', was the subject of extensive research for many years.[16,17] Some yarn types may even have a greater deviation from the idealized twisted structure. For example, air-jet spun yarn exhibits no true twist with fibers in the core being largely parallel and outer fibers randomly wrapping around the core fibers. Thus, yarn structural integrity and fiber packing is largely a result of the continuity in the core fibers and the lateral pressure imposed by the wrap fibers.

9.5.3 Fiber mobility

Fiber mobility is a key design aspect of yarn as it determines the dimensional stability of both the yarn and the fabric woven or knitted from it. In case of continuous filament yarn, there is a small tendency for fiber movement in the yarn apart for the tendency of filaments to split apart under buckling effects imposed by processing factors or end use factors.[5] In the case of textured yarns, bulkiness or stretchability are both induced by some distortion in the filaments after which heat setting or self-locking of filament loops (as in air texturing) assist in providing dimensional stability to the yarn. Obviously, when such conditions are partially removed (e.g. by wearing, washing and drying), the filaments will have a greater tendency to mobilize.

In the case of spun yarns, the discrete nature of staple fibers can result in many modes of fiber mobility, depending on the external effects encountered. Under minimum yarn strain, fibers will tend to settle down and cohere. If the yarn is relaxed or slack, the fibers will be likely to be in a bending mode; if the yarn is tensioned, the fibers will follow the path of minimum length or a straight line between fiber ends; if the yarn is twisted, the fibers may be buckled or tend to snarl; and when a yarn end is set free, the fibers will tend to untwist. The point is that most spun yarns exhibit a great deal of fiber mobility which in one way or another can influence yarn performance during the conversion of yarn into fabrics and during the use of fabric in various applications.

Perhaps no greater fiber mobility is observed in a yarn than that of the fiber segments or filaments near the yarn surface. In untwisted continuous filament yarns, filaments in the outer layer will tend to snag or come loose, particularly under external rubbing conditions. For this reason, it is sometimes recommended that a slight twist be inserted in this type of yarn to avoid snagging as mentioned earlier. In spun yarns, an inevitable phenomenon is the existence of fiber segments protruding from the yarn body. This is called 'yarn hairiness' and it is measured by the number of hairs projecting from the yarn and their lengths. Typically, short hairs (less than 2 mm) adhered to the yarn body are acceptable as they provide a nice fuzzy feeling in apparel fabrics against the skin. This is particularly true when the protruding hairs are flexible so that they can bend easily against the skin. They may also create an opportunity to create tiny fiber pockets that can entrap air for good thermal insulation.[21] On the other hand, long hairs protruding from the yarn body can be of major adverse effect on yarn performance. During the conversion of yarns to fabrics, long hairs of different yarns will tend to entangle together creating fiber bridges that hinder a smooth flow of the yarn during processing. This phenomenon is frequently observed in the denim manufacturing process particularly in the stage of rope beaming.[22] One way to minimize or eliminate long hair is by using the singing process described in Chapter 11.

The level of hairiness in spun yarns will largely depend on a number of factors including yarn size, yarn twist, spinning preparation and spinning method. Figure 9.9 shows values of hairiness for different spun yarns.[20] In this case, yarn hairiness is measured by the so-called hairiness index, H, which is the length of hair per unit length of 1 cm. Thus, a hairiness index

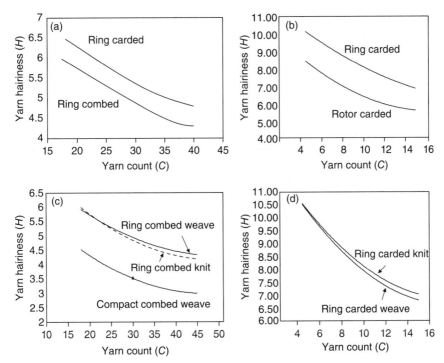

9.9 Uster® yarn hairiness for different preparations and different spinning methods.[20]

of 4 will simply mean that the accumulated length of hair of a certain yarn is 4 cm per unit length of 1 cm. Some of the interesting points that can be revealed from this figure are as follows:

- Most traditional spun yarns exhibit hairiness ranging approximately from 2.0 to 12. This is a wide range from a yarn-performance viewpoint as it can have a wide impact on fabric performance characteristics such as fabric texture, fabric hand or feel and fabric dimensional stability.
- In general, the finer the yarn, the lower the degree of hairiness. This is generally attributed to two reasons. The first reason is a statistical one, as finer fibers normally have a lower number of fibers per yarn cross-section leading to a lower chance of creating hairiness. The second reason is that fine yarns are typically made from long fibers or combed slivers with lower percentage of short fibers. Most hairiness is created by short fibers.[3,22]
- In line with the above point, combed yarns exhibit less hairiness than carded yarns (see Fig. 9.9a).

- Ring-spun yarns have substantially greater hairiness than rotor-spun yarns (Fig. 9.9b). This is a result of the differences in the principles of the two spinning methods. In ring-spinning, the phenomenon of fiber migration forces a great deal of short fibers to be positioned near the yarn surface resulting in potential hairiness.[3] In addition, the ring-traveler system in ring spinning stimulates the hairs through the rubbing effect. In rotor spinning, the back doubling effect and the low fiber migration rate result in less hairiness.[3,6]
- Compact yarn has the lowest hairiness level of all traditional spun yarns (see Fig. 9.9c). This is a direct result of the deliberate compactness of fibers in the yarn, making it a less hairy yarn (see Table 9.3).
- The difference between hairiness of yarns made for woven fabrics and that for yarns made for knit fabrics is not so large (Fig. 9.9c and 9.9d).

In addition to yarn hairiness, other fiber structural irregularities such as belt fibers in the case of rotor-spun yarns or loose wrapping fibers in the case of air-jet yarns are easily movable under the effect of rubbing against other yarns or other surfaces. This can influence the abrasion resistance and the propensity for pilling of fabrics made from these yarns.[3,6]

9.6   Conclusions

In this chapter, the emphasis has been on the design-related aspects of yarn. These were demonstrated by reviewing both conventional and specialty yarns and discussing their merits in various fibrous products and by focusing on the design-related aspects of yarn structure. Different methods of yarn spinning were mentioned here in order to illustrate the various yarn structures produced by these methods. Details of the technologies involved in these methods are outside the scope of this book as they can be found in many other literature sources including those cited in this chapter.

In closing this chapter, it is important to point out that in the design of fibrous products in which fabrics are made from yarns (spun, continuous filament, textured, compound or fancy), the choice of appropriate yarn type and yarn characteristics is critical. Commonly, design engineers focus more on fabric construction and finishing techniques to meet the performance requirements of fibrous products. Despite the benefits of such approaches, it can be very costly if not coordinated with the choice of appropriate yarn.

9.7   References

1. EL MOGAHZY Y, 'Yarn engineering', *Indian Journal of Fiber & Textile Research, Special Issue on Emerging Trends in Polymers & Textiles*, 2006, **31** (1), 150–160.
2. GOSWAMI B C, MARTINDALE J G and SCARDINO F L, *Textile Yarns, Technology, Structure & Applications*, Wiley-Interscience NY, 1977.
3. EL MOGAHZY Y E and CHEWNING C, *Fiber to Yarn Manufacturing Technology*, Cotton Incorporated, Cary, NC, USA, 2001.
4. HATCH K L, *Textile Science*, West Publishing, Minneapolis, NY, 1999.
5. BLOCK I, *Manufactured Fiber*, in AccessScience@McGraw-Hill, http://www. accessscience.com, DOI 10.1036/1097-8542.404050, last modified: May 6, 2002.
6. LORD P R, *Handbook of Yarn Production, Technology, Science and Economics*, Woodhead Publishing Limited, Cambridge, UK, 2003.
7. EL MOGAHZY Y, *Yarn types: compound and fancy yarns*, Video-lecture, Quality-Business Consulting (QBC), http://www.qualitybc.com/, 2005.
8. *Global Market for Smart Fabrics and Interactive Textiles*, report published by Textile Intelligence, http://www.mindbranch.com/products/ R674-214.html, March, 2006.
9. GONG R H and CHEN X, *Technical Yarns, Handbook of Technical Textiles*, A R Horrocks A R and Anand S C (editors), Woodhead Publishing Limited, Cambridge, UK, 2000, 42–61.
10. FERREIRA M, BOURBIGOT S, FLAMBARD X and VERMEULEN B, 'Interest of a compound yarn to improve fabric performance', *AUTEX Research Journal*, 2004, **4** (1), 1–6.
11. KLATA E, BORYSIAK S, VAN DE VELDE K, GARBARCZYK J and KRUCIŃSKA I, 'Crystallinity of polyamide-6 matrix in glass fibre/polyamide-6 composites manufactured from hybrid yarns', *Fibres and Textiles in Eastern Europe*, 2004, **12** (3), 64–69.
12. FLORY J F, MCKENNA H A and PARSEY M R, 'Fiber ropes for ocean engineering in the 21st century', paper presented at the *Civil Engineering in the Oceans Conference*, San Diego, CA, USA American Society of Civil Engineers, 1982.
13. BANFIELD S J and FLORY J F, 'Computer modelling of large, high-performance fiber rope properties', presented at *Oceans '95*, San Diego, October 9–12, 1995.
14. HEARLE J W S, PARSEY M R, OVERINGTON M S and BANFIELD S J, 'Modelling the long-term fatigue performance of fibre ropes', *Third International Offshore and Polar Engineering Conference*, Singapore, ISOPE, Golden, CO, 1993.
15. GONG R H and WRIGHT R M, *Fancy Yarns: Their manufacture and application*, Woodhead Publishing Limited, Cambridge, UK, 2002.
16. HEARLE J W S, GROSBERG P and BACKER S, *Structural Mechanics of Fibers, Yarns, and Fabrics*, Wiley-Interscience, New York, 1969.
17. ZUREK W, *The Structure of Yarn*, translated from Polish, published by USDA and the National Science Foundation, New Orleans, USA, 1975.

18. LAWRENCE C A, *Fundamentals of Spun Yarn Technology*, CRC Press, London, 2002.
19. NECKAR B, 'Yarn fineness, diameter, and twist', a paper presented in *TEXSCI '98*, Libeerec, Czech Republic, May 25–27, 1998.
20. Uster™ Technologies Technical Data, http://www.uster.com/
21. FATMA S K, *A Study of the Nature of Fabric Comfort: Design-Oriented Fabric Comfort Model*, PhD Thesis, Auburn University, USA, 2004.
22. EL MOGAHZY Y, *Quality Problems Associated with Spun Yarns during the Fabric Forming Process and during Fabric Use*, 5-hr Video lecture, http://www.qualitybc.com/, 2002.

10

Types of fabric for textile product design

Abstract: This chapter discusses the different fabric types from a product design viewpoint. Fabrics are classified into three main categories: (1) yarn-based fabrics, (2) fiber–direct fabrics and (3) compound fabrics. Yarn-based fabrics are made directly from spun or continuous-filament yarns. These include woven fabrics, knit fabrics and twisted-knotted fabrics. Fiber–direct fabrics represent a unique category of fabrics which are formed directly from fibers without the need for yarns as an intermediate component. In practice, this category is known as nonwoven fabrics. The third category consists of different fabric layers, each providing specific functions that can be made from yarn-based or fiber-based fabrics. Common examples of compound fabrics include quilted fabrics, tufted fabrics, flocked fabrics, coated fabrics, and laminated fabrics.

Key words: yarn-based fabrics; plain fabrics; wovens; knits; fiber–direct fabrics; nonwovens; mechanically bonded; media bonded; compound fabric; quilted; tufted; flocked; coated; laminated.

10.1   Introduction

In Chapter 9, the focus of the discussion was on yarn, a structure that is regarded in its simplest form as a linear, one-dimensional fiber strand. When yarns are converted into fabrics, a new structure is formed, which can be regarded, again in its simplest form, as a two-dimensional flexible flat sheet. However, just as we realized that yarn is indeed a three-dimensional structure by virtue of its structural features (volumetric density, fiber arrangement and fiber mobility), so we will find that fabric is clearly a three-dimensional structure. Indeed, the third dimension of fabric imposed by its thickness represents a key player in all performance criteria of fabric products. In this chapter, a brief review of common fabric constructions is presented. The purpose of this review is to provide a quick reference for the reader of the common types of fabrics available and their basic specifications. Again, the discussion will be focused on design-related aspects of fabric constructions.

10.2   Fabric classification

Fabrics may be classified in many different ways depending on the emphasis and the purpose of classification. The most generalized classification is based on whether the fabric is made from fibers or from other materials. In this regard, fabrics can be classified into two main categories: (a) fibrous fabrics such as wovens, knits or nonwovens and (b) non-fibrous fabrics such as plastic films, rubber sheets or metal foil. This general classification implies that unique products such as paper and leather can also be classified as fibrous fabrics. However, these products are distinguished from traditional fibrous products in that their fibers are not easily identified as separate entities with clearly defined limits; rather it is the whole aggregate which is fibrous in form. Our focus in this chapter is on common fabric structures and some of their advanced derivatives. In this regard, fabrics may be classified into three main categories: (1) yarn-based fabrics, (2) fiber–direct fabrics and (3) compound fabrics. Details of these types of fabric including their constructions, the technologies used for making them and their different specifications are discussed in numerous publications.[1-5]

10.3   Yarn-based (YB) fabrics

Yarn-based fabrics are those that are made directly from spun or continuous-filament yarns. These include woven fabrics, knit fabrics and twisted-knotted fabrics. In woven fabrics, two or more sets of yarns are interlaced, mostly at right angles to form the fabric. Knit fabrics are formed by intermeshing loops of yarn. Twisted-knotted fabrics are formed using yarns that intertwine with each other at pre-specified angles. Within each of these three categories of fabric, numerous constructions can be made to provide design engineers with unlimited options of performance criteria.[4]

10.3.1 Woven fabrics

Woven fabric is the dominant type of fabric in both traditional and function-focus products. The key reason for this dominance is superior durability and high dimensional stability compared to other types of fabric. In addition, different performance criteria can be met using many fabric parameters including thickness, cover factor, combinations of yarn types and combinations of fiber types. Most woven fabrics consist of two sets of yarns that are interlaced at right angles to each other. Yarns running along the length of the fabric are known as warp ends whilst those running across the fabric or along the width are called fillings or weft picks. The way the two sets of yarns are arranged in the woven fabric is commonly known as fabric construction. This is the most critical factor that influences fabric performance. Today's weaving machines are capable of producing hundreds of fabric constructions[1-3] that can be utilized in numerous applications. The most commonly used constructions are plain, satin, and twill weave (see Figure 10.1). These are briefly described below.

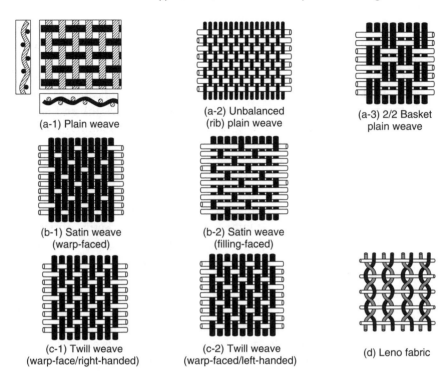

(a-1) Plain weave

(a-2) Unbalanced
(rib) plain weave

(a-3) 2/2 Basket
plain weave

(b-1) Satin weave
(warp-faced)

(b-2) Satin weave
(filling-faced)

(c-1) Twill weave
(warp-face/right-handed)

(c-2) Twill weave
(warp-faced/left-handed)

(d) Leno fabric

10.1 Basic woven structures.[4,5]

Basic weaves

The simplest woven fabric construction and the most widely used is the so-called plain weave. Indeed, one can find plain weaves used in all traditional and function-focus fibrous products. As shown in Fig. 10.1a-1, it consists of a repeat construction unit of two warp and weft yarns. It is formed by alternatively lifting and lowering one warp thread across one weft thread. The term 'plain' is commonly used to indicate smooth and flat surface of the fabric. This feature is very useful for making different printing designs and for applying uniform mechanical and chemical finishing applications. When warp and weft yarns have the same yarn count and the same number per unit length, the fabric is regarded as a balanced fabric. Typically, both the face and the back of plain-weave fabric are visually identical as a result of the consistent warp and filling interlacing. This provides ample opportunities to achieve well-controlled performance options via changes in yarn type (traditional, textured, fancy, compound, etc.), yarn count and yarn twist. Unbalanced plain weaves are commonly called 'rib fabrics' as they are characterized by the presence of ribs or ridges running crosswise on both the face and the back of the fabric (see Fig. 10.1a-2).

When more than one warp yarn interlace with more than one filling yarns (2/2 or 4/4), the plain fabric is called basket weave (Fig. 10.1a-3). Some fabrics may be made from two or three warp yarns floating over one filling yarns (2/1, or 3/1). These fabrics are commonly called half-basket weaves. Basket weaves are commonly used in many applications.[4] These include apparel fabrics such as suits, sportswear (sailcloth) and shirts. In addition, basket weaves are used in function-focus fabrics such as slipcovers for furniture, house awnings and boat covers.

Satin fabric is a smooth surface fabric that is formed by floating a warp or a filling yarn over four to 12 yarns before it passes to the back of the fabric and floats under one or two yarns (Fig. 10.1b-1 and b-2). Typically, the face of a satin fabric is more lustrous and smoother than the back, particularly when continuous-filament yarns are used. Satin fabrics are usually unbalanced and they have fewer interlacings per unit area than plain weave and twill fabrics. The smoothness and softness of the satin fabrics make them good candidates for luxurious apparels, linings and some fancy upholstery products. They are also used in some function-focus products such as protective clothing including windproof and windbreaker jackets and coats. A derivative construction of satin fabric, commonly known as sateen, is primarily made from spun yarns. This construction is typically a filling-faced satin weave in which the filling yarns are the yarns that float on the fabric face.

Twill fabric is a unique construction in which diagonal lines are running from selvedge to selvedge on the face of the fabric by the introduction of repeats on three or more warp and filling yarns. When one views the fabric along the warp direction, one can see the direction of the diagonal lines on the surface of the fabric either running upwards to the right, which is regarded as a 'Z twill or right-hand twill' or running in the opposite direction, which is regarded as a 'S twill or left-hand twill'. These two constructions are shown in Fig. 10.1c-1 and c-2, respectively. By virtue of its construction, twill fabrics will obviously have longer floats, fewer intersections and a more open construction than a plain weave fabric with the same cloth attributes.

In contrast to plain fabrics, twill fabrics are rarely printed owing to their ridged surface. They can also be very durable and, with the use of coarse yarns, their durability can significantly increase, as in the case of denim fabrics. They are also used in other applications including suits, sportswear, outdoor clothing, jackets and raincoats, work clothing, scarves and neckties.[4]

Another construction that is often used in conjunction with other weaves or independently is the so-called leno weave. This is a special construction in which warp yarns are made to cross one another, between fillings, during leno weaving.[2,3] In other words, adjoining warp ends do not remain parallel when they are interlaced with the weft but are crossed over each other. In the simplest leno, one standard end and one crossing end are passed across each other during consecutive picks as shown in Fig. 10.1d. This crossing feature can be the only feature of leno fabrics, but it can be used in combination with other weaves. The advantage of having warp yarns crossing over each other with the weft passing between them is that the warp yarns will restrict the weft movement by locking it into position. This results in an open structure (e.g. gauzes) in which virtually no thread movement or fabric distortion is observed. The same feature makes leno construction an excellent selvedge construction for binding the edge threads into position by preventing the warp threads at each side of a length of cloth from slipping out of the body of the fabric.[3,5] Leno construction can also be used in the body of the fabric at points where empty dents are created eventually to slit the fabric into narrower widths. In addition to these important applications, leno fabrics can be used for special applications including[4] sheer curtains (marquisette), mosquito netting which forms windows in tents or may be draped over a bed to provide protection from insects, also as covers for the faces of beekeepers, and as sacks for laundry, fruit and vegetables.

Specialty weaves

In addition to the basic weaves discussed above, there are other specialty woven fabrics that are constructed for particular applications. These are typically more complex structures as both the geometrical texture and the particular design must be woven in to provide the intended performance. Examples of these constructions are crepe woven, dobby, piqué, Jacquard, pile-woven, double-woven, braided and triaxial-woven. These constructions are briefly reviewed below. More details of these weaves can be found in many literature references including the ones listed here.[1–5]

Figure 10.2 illustrates the first four woven constructions, namely: crepe, dobby, piqué, and Jacquard. These are mainly fashion-oriented constructions that are primarily intended for visual esthetics. Crepe woven fabrics are characterized by random distribution of yarn floats for the purpose of producing a construction of a non-systematic repeat. This provides particular fabric textures such as crinkly, pebbly and sandy texture. Dobby woven fabrics are characterized by geometric figures woven in a set pattern. They are often compared with the more elaborative Jacquard constructions but more limited, more efficiently woven and obviously cheaper. Typically, the number of filling yarns in one pattern repeat is greater than that required for any of the basic weaves but less than that of the Jacquard fabrics. Piqué fabrics are characterized by rounded cords (ribs) in the filling direction with pronounced sunken lines between. Like the common quilts, they have

 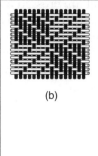 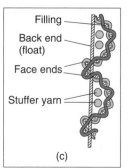 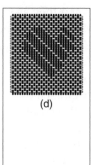

10.2 Fashion-oriented special woven constructions and product examples.[2-4] (a) Crepe, (b) dobby, (c) piqué and (d) Jacquard fabrics.

a soft, raised-surface effect. The weave on the face of the cords is often plain-woven. Jacquard fabrics are characterized by elaborate and complex woven-in patterns including some that depict paintings or complex photographs. The construction involves at least two of the basic weaves in various arrangements.

Figure 10.3 shows the basic construction of pile woven fabrics. These are obvious three-dimensional fabrics by virtue of their significant thickness (0.25–0.50 inch thick; 0.6–1.3 cm). This thickness is created by pile yarns that stand vertically from a woven ground reflecting the face of the pile fabric. The fabric ground typically consists of one set of filling and one set of warp yarns. To create the pile construction, another set of yarns called pile filling or pile warp yarns is interlaced into the fabric ground.[2-4] This type of fabric is used for a wide range of applications including[4] coats, jackets, gloves and boots. It may be used as an outer fabric lining and/or interlining. The familiar terry towels are made exclusively from pile constructions. Pile fabrics can also be used in many upholstery fabrics and some bedspreads. Function-focus products made from pile fabrics include artificial turfs with piles being cut to mimic the turf natural appearance. In Chapter 13, more discussion of fibrous artificial turfs will be presented.

Figure 10.4 illustrates the basic constructions of double-woven fabrics. These represent multiple-layer constructions in which separate layers, each consisting of interlaced warp and filling yarns, are stitched together. One obvious reason for making double-layer fabrics is to obtain heavier and thicker constructions than single-layer constructions. From a functional viewpoint, a double-layer fabric construction allows multiple options that may be reflected, for example, in double-faced fabrics with each face providing different visual effects or performance criteria. In addition, the fabric layers are separated by air, resulting in desirable thermal insulation (e.g. blanket constructions).

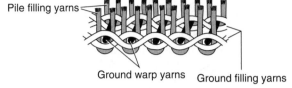

Pile filling yarns

Ground warp yarns   Ground filling yarns

10.3 Pile fabrics and product examples.[2-4]

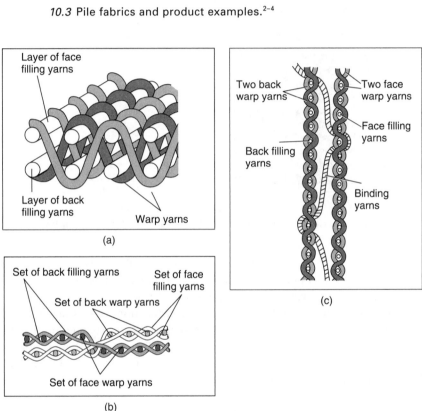

Layer of face
filling yarns

Layer of back
filling yarns

Warp yarns

(a)

Set of back filling yarns   Set of face
filling yarns

Set of back warp yarns

Set of face warp yarns

(b)

Two back
warp yarns

Two face
warp yarns

Face filling
yarns

Back filling
yarns

Binding
yarns

(c)

10.4 Constructions of double-woven fabrics.[3,4] (a) Double-faced,
(b) double-weave and (c) double-cloth constructions.

Figure 10.5 illustrates the construction of multi-axial and braided fabrics. These constructions represent another category of fabrics that are uniquely characterized by the inclination between the yarn sets forming the fabric. Triaxial fabrics are defined as cloths where three sets of threads form a multitude of equilateral triangles and two sets of warp yarn are interlaced at 60° with each other and with the weft. Another type of multi-axial is the tetra-axial fabrics. In this type, four sets of yarns are inclined at 45° to each other. One obvious advantage of triaxial fabric is dimensional stability at light weight. This makes the triaxial fabric a good candidate for many function-

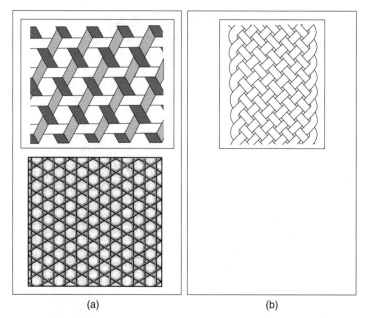

10.5 Triaxial (a) and braided (b) fabrics and product examples.[3,4]

focus applications[4,5] such as parachute gliders, sailcloths, tire fabrics, balloon fabrics, pressure receptacles and laminated structures. They are also widely used in close fit apparels and some upholstery fabrics.

Figure 10.5b shows the braided fabrics construction. In this fabric three or more yarns are interlaced in such a way that they cross one another in diagonal formation.[4] Most braided fabrics are narrow and their products are those of limited widths. These include garment trims, shoe laces and belts. Braided constructions are also used in many fiber-reinforced composite products.[6]

A combination of braid and triaxial construction can also be made to meet some unique functional performance requirements. For example, a US patent was presented in which a novel prosthesis for use in repairing or replacing soft tissue was disclosed,[7] comprising a triaxially braided fabric element with interwoven first, second and third sets of yarns, with the yarns of the second and third sets being oriented at substantially the same acute braiding angle with respect to the yarns of the first set. This patent claims that 'an elongated ligament prosthesis exhibiting the desired properties of high strength and high elasticity may be prepared by selecting high elasticity fibers for the first set, orienting first set of fibers in the longitudinal direction of the prosthesis and selecting fibers having high yield strength and high Young's modulus for the second and third sets'. The patent also claims that 'a tubular prosthesis in which high elasticity fibers are oriented in the longitudinal direction is highly suitable for use as a vascular prosthesis and that a prosthesis of the invention may also be manufactured in the form of a prosthetic heart valve leaflet.

Basic specifications of woven fabrics

Woven fabrics are characterized by many parameters that collectively reflect their performance criteria. These parameters are as follows:[1-5,8,9]

- fabric count
- fabric width
- fabric thickness
- fabric weight or area density
- fabric crimp
- fabric cover factor.

Fabric count is the number of threads per unit length (say, inch) in the warp (wpi) and filling (fpi) directions (see Fig. 10.6). These two parameters can be described independently or in a collective fashion (e.g. wpi × fpi or

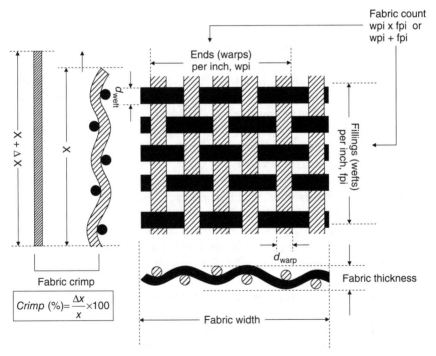

10.6 Woven fabric basic specifications.[1-4] (a) Weft and (b) warp knitting.

wpi + fpi). Standard values of fabric count are typically described for fabric off the loom or in the greige state. Obviously, the fabric width may change during dyeing and finishing and, as a result, its count may also be altered. Typical values of fabric count may range from 80×80 to 150×150. The significance of fabric count is that it reflects the tightness or the looseness of the fabric. Obviously, a tight or close fabric will require finer yarns or lower number of interlacings per unit area than a loose fabric. This point is important from a design viewpoint as it indicates two independent options for developing a tight or loose fabric, namely yarn count and number of interlacings. As indicated earlier, a balanced woven fabric is defined as the fabric which exhibits equal number of warp yarns and filling yarns per inch and contains yarns of equal size and character.

Fabric width is another characteristic that must be pre-specified prior to forming any woven fabric. With current weaving technology, woven fabrics can be made in a very wide range of width (e.g. 36–160 inch; approx 91–406 cm), but most fabrics are made in the range from 36–60 inches (approx 91–152 cm). This range does not cover the specialty narrow fabric, which is typically up to 12 inches (30.5 cm) wide (e.g. ribbons, elastics, zipper tapes, labels, carpet-edge tapes and automobile safety belts). Again, the width of the fabric is likely to change upon release from the loom and under the effects of upstream chemical and mechanical finishing.

Fabric thickness is a unique characteristic of woven fabrics as it adds a critical dimension in many applications. However, it is a difficult parameter to measure. On one hand, it is not a stable parameter as it is largely sensitive to the extent of distortion in the yarn during the weaving and finishing processes. On the other hand, even under ideal conditions, the thickness should be measured under a predetermined pressure with the higher the pressure exerted on the fabric the smaller the thickness. To make matters additionally complex, the size of the presser feet and the time elapsed before a reading is taken will also influence the accuracy of the thickness measure. These factors make it important to provide a precise description of the method of testing and the testing parameters used.

Fabric weight is a key parameter in the traditional woven fabric market with heavy weights being typically assigned for bottoms such as pants, slacks and skirts, and light weights for tops such as shirts, blouses and dresses. Commonly, fabric weight is expressed in ounces per square yard or grams per square meter. Fabrics of weights as small as 1 or less ounces per square yard are considered to be very light. A typical light fabric will have a weight, or area density, ranging from 2 to 3 ounces per square yard ($68–102 \text{ g m}^{-2}$). Heavy fabrics typically exhibit weights of 5 to 7 ounces ($173–242 \text{ g m}^{-2}$) per square yard. From a design viewpoint, fabric weight can be altered or controlled using three basic options: yarn count, fabric count and fabric construction. In addition, it can be altered using mechanical and chemical treatments of fabrics after weaving.

Fabric crimp is a direct result of yarn waviness in the fabric, which is caused by the yarns being forced to bend around each other during weaving. As shown in Fig. 10.6, crimp is determined by the relation of length of yarn in cloth to the length of the cloth in the warp or the filling direction. As a result, it is measured by the ratio between the length of the fabric sample and the corresponding length of yarn when it has been straightened after being removed from the cloth. Again, as crimp is a mode of deformation it can be difficult to measure, particularly as one has to straighten the yarn unraveled from the fabric to measure the yarn length.

Cover factor is a measure of the area of the fabric occupied by one set of yarns, warp or filling. As a result, there are two cover factors for each woven fabric: warp cover factor and weft cover factor. The total fabric cover factor, commonly known as the cloth cover factor, is determined by the sum of the warp and the weft cover factor, corrected by subtracting the overlap cover or the product of warp and weft cover factor. In practice, the cover factor may be calculated using the classic Peirce formula[8] in which the number of threads per inch is divided by the square root of the cotton yarn count. Alternatively, it may be calculated by multiplying the threads per centimeter by the square root of the linear density of the yarn, tex and dividing by 10. The latter typically reveal values that are different from those of the former by less than 5%. Typical values of cover factor may range from about 12 to approximately 16 in each direction. However, cover factors of 16 or higher may result in fabric jam particularly with plain weave constructions. Some fabrics may have even greater cover factors (e.g. duck and canvas), which requires special weaving. From a design viewpoint, cover factor can be adjusted using basic parameters such as yarn count, yarn spacing or threads per unit length, and fabric construction.

10.3.2 Knit fabrics

Knit fabrics are another type of yarn-based fabric that find significant appeal in the apparel market. Today, one can find knits in numerous apparels such as sweaters, hosiery, T-shirts, golf shirts, sweat and exercise suits, lingerie, infant and children wear, swimming suits, gloves and figure-shaping undergarments. The popularity of knit fabrics in apparel products stems from many attractive features including[4] the freedom of body movement in form-fitting garments, ease of care, resilience, soft draping quality and warmth in still air environments. Knits are also used in interior furnishing (e.g. lacy casement) and in smaller quantities in upholstery and carpets. In function-focus fibrous products, knits have exhibited

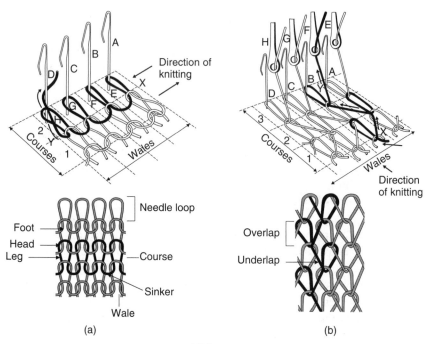

10.7 Basic knit structures.[4,10,11]

wide acceptance particularly in medical applications including[4,10] splints, antithrombosis stockings, bandages, ointment pads, flat and tubular dress-ings, dialysis filters, incontinence pads and underwear, hospital cellular blankets and stretch terry sheets, band-aid fabrics, hospital privacy cur-tains, nets for handling burn victims, fabric for artificial heart valves, and nets for blood filtration, abdominal surgery and reconstructions. Other function-focus products utilizing knit fabrics include[10] scouring pads (metallic) and fully fashioned nose cones for supersonic aircrafts.

The knit structure is basically produced by employing a continuous yarn or set of yarns to form a series of interlocking loops. The two basic types of knits are the weft and the warp knit (see Fig. 10.7). The use of warp and weft is largely analogous to the same terms used in woven structures with weft being along the fabric width and warp being along the fabric length. Specifically, weft knit fabric is characterized by a construction in which the intermeshing yarn traverses the fabric crosswise. Warp knit fabric, on the other hand, is characterized by a construction in which each warp yarn is more or less aligned along the fabric length, or the intermeshing yarn tra-verses the fabric lengthwise. Examples of weft knit constructions include plain single-jersey, rib, purl and double knits (rib and interlock). Warp knit structures include tricots and raschels. Weft knits are commonly produced in tubular form and warp knits are mostly made in flat form.

In the following two sections, some of the features of weft and warp knits will be discussed. In order to follow this discussion, it is important that the reader be familiar with some of the basic knitting terminologies that will be used in these sections. Descriptions of these terminologies are listed in Table 10.1.

Table 10.1 Basic terms used in knitting[4,9–11]

Term	Description
Course	A row of loops across the width of the fabric; courses determine the length of the fabric and are measured as courses per inch or per centimeter (see Fig. 10.7).
Wale	A column of loops along the length of the fabric; wales determine the width of the fabric and are measured as wales per inch or per centimeter (see Fig. 10.7).
Loop	This is a basic unit of knit fabric.
Weft-knit loop	This is commonly called 'needle loop'. It has a head and two legs. It also has a foot that meshes with the head of the needle loop in the course below it. The feet are usually open in weft knits; the yarn does not cross over itself. The section of yarn connecting two adjacent needle loops is called the sinker (see Fig. 10.7.a).
Warp-knit loop	This is referred to as 'an overlap' owing to the type of motion taking place to form it. The length of yarn between overlaps, or the connection between stitches in consecutive courses, is called an 'underlap'. The feet of the overlaps may be open or closed, depending on whether the yarn forming the underlaps continues in the same or opposite direction from that followed during formation of the overlap (see Fig. 10.7.b).
Stitch	Each loop in a knit fabric is a stitch. Knit constructions are commonly characterized by the stitch type. These include knit stitch, purl stitch, float stitch and tuck stitch.
Stitch density	This is the number of stitches per unit area of a knitted fabric (loops per inch^2 or per cm^2). It determines the area of the fabric.
Stitch length	The length of yarn in a knitted loop. It is the dominating factor in all knitted structures. In weft knitting, it is usually determined as the average length of yarn per needle, while in warp knitting, it is normally determined as the average length of yarn per course.
Cut or gauge	The number of knitting needles per unit length, along a needle bed or needle bar, of the knitting machine. It indicates the openness or closeness of the intermeshing loops. The greater the number of needles in the specified length, the higher the cut or gauge number and, therefore, the closer together the loops are to each other. As the cut or gauge is increased, finer-size yarn is used.

Weft knits

Weft knitting is the most common type of knitting and the one that enjoys the largest market share of traditional fibrous products. Weft knitting machines are versatile, require relatively low total capital costs and offer potential for many economical features such as small floor space, quick pattern and machine changing, short production runs and low stock-holding requirements of yarn and fabric.

The wide variety of weft knit structures provides a wide range of fabric appearance and performance characteristics. Figure 10.8 shows examples of weft knit structures. The four primary base weft knitted structures are plain, rib, interlock and purl. As explained by David Spencer in his excellent book on knitting technology,[11] each is composed of a different combination of face and reverse meshed stitches, knitted on a particular arrangement of needle beds. Each primary structure may exist alone, in a modified form with stitches other than normal cleared loops, or in combination with another primary structure in a garment-length sequence. The basic stitches that result in this diversity are knit stitch, purl stitch, float stitch and tuck stitch (see Fig. 10.8g).

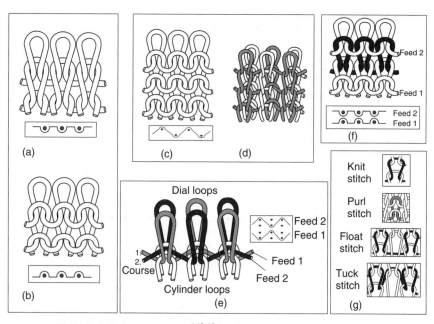

10.8 Weft-knit structures.[4,10–12] (a) Plain single jersey (technical face), (b) plain single jersey (technical back), (c) 1 × 1 rib, (d) rib double knit, (e) interlock, (f) 1 × 1 purl, (g) basic stitches.

Plain knit, shown in Fig. 10.8(a) and (b), is the simplest and most economical weft knitted structure to produce. It is typically produced by the needles knitting as a single set, drawing the loops away from the technical back and towards the technical face side of the fabric.[11] As a result, it is characterized by a difference in appearance between the face and the back, with the face reflecting v-shapes and the back reflecting arc shapes. Plain knit is the base structure of ladies' hosiery, fully fashioned knitwear and single-jersey fabrics. It has a number of important features: typically thicker and warmer than plain woven fabrics made from the same yarn, high covering power, moderate extensibility lengthwise (10–20%) and high extensibility widthwise (30–50%), curling tendency and a potential recovery of up to 40% in width after stretching.

The dominant plain knit is the so-called 'single-jersey fabric' produced on circular machines whose latch needle cylinder and sinker ring revolve through the stationary knitting cam systems which, together with their yarn feeders, are situated at regular intervals around the circumference of the cylinder. The yarn is supplied from cones mounted on a creel through tension guides, stop motions and guide eyes down to the yarn feeder guides. The fabric, in tubular form, is drawn downwards from inside the needle cylinder by tension rollers and is wound onto the fabric-batching roller of the winding-down frame. Jersey fabric is used extensively in hosiery, the foot section of socks, cotton underwear, golf and T-shirts and sweater bodies.[4]

Single-jersey knitting machines are typically simpler and more economical than rib machines. They can also have more feeders, higher running speeds and a wider range of yarn counts. The most popular diameter is 26 inches (66 cm) giving an approximate finished fabric width of 60–70 inches (152–178 cm). According to Spencer,[11] an approximately suitable count may be obtained using the formula $N_{eB} = G^2/18$ or $N_{eK} = G^2/15$, where N_{eB} = cotton spun count, N_{eK} = worsted spun count and G = gauge in needles per inch. For fine gauges, a heavier and stronger count may be necessary.

The second type of weft knits is the 1×1 rib single knit (Fig. 10.8.c). In this structure, both back and face loops (knit and purl stitches) occur along the course, but all the loops contained within any single wale are of the same sort, that is, back or face loops. In the simplest rib fabric, a 1×1, the knit and purl stitches alternate every other stitch. In other rib single knits, several knit stitches may occur in any course followed by one or more purl stitches. The rib knit is named by the number of knit and purl stitches; for example, a fabric composed of two knit stitches followed by three purl stitches is a 2×3 knit.[4,11]

A 1×1 rib appears the same on both sides, with each like the face of plain knit. It has a very high widthwise elongation (50–100%) but a moderate lengthwise elongation. It has no curling tendency as a result of the fact that it is balanced by alternate wales of face loops on each side; it therefore lies flat without curl when cut.[11] It is a more expensive fabric to produce than plain and is a heavier structure; the rib machine also requires finer yarn than a similar gauge plain machine. Rib cannot be unraveled from the end knitted first because the sinker loops are securely anchored by the cross-meshing between face and reverse loop wales. This characteristic, together with its elasticity, makes rib particularly suitable for the extremities of articles such as tops of socks, cuffs of sleeves, waist bands, collars, men's outerwear and underwear. Rib structures are elastic, form-fitting and retain warmth better than plain structures.[4,11] In circular rib machines (gauge range from 5 to 20 npi), an approximately suitable count may be obtained using the equation $N_{eB} = G^2/8.4$, where N_{eB} is the cotton count and G is the gauge in npi. For underwear fabric, a popular gauge is E 14 with a count 1/30's.

A less popular rib knit is the rib double-knit produced using two staggered layers of loops (Fig. 10.8d). This arrangement is most easily seen when a rib double-knit is viewed from a cut crosswise edge.[4] Even though there is considerable diversity in the structure and appearance of rib double-knits, they share the characteristics of not running and being rather stiff structures with low elongation. They tend to bag, snag, shrink and have high air permeability.[4]

The third primary weft-knit structure is the interlock (Fig. 10.8e). This was originally derived from rib but requires a special arrangement of needles knitting back-to-back in an alternate sequence of two sets, so that the two courses of loops show wales of face loops on each side of the fabric exactly in line with each other, thus hiding the appearance of the reverse loops. Interlock has the technical face of plain fabric on both sides, but its smooth surface cannot be stretched out to reveal the reverse meshed loop wales because the wales on each side are exactly opposite to each other and are locked together.

Interlock elongates by about 30–40% in the crosswise direction, so that a 30-inch (76 cm) diameter machine will produce a tube of 94-inch (204 m) open width which finishes at 60–66 inches (1.5–1.7 m) wide. It is a balanced, smooth, stable structure that lies flat without curl. Like 1 × 1 rib, it will not unravel from the end knitted first, but it is thicker, heavier and narrower than rib of equivalent gauge and requires a finer, better, more expensive yarn.[11] When compared to rib double-knits, interlock knits tend to have higher gauges; as fine as 40 are possible in interlock double-knits, compared to 28 for rib double-knits.[4] Interlock double-knits have better drape and are softer, lighter and thinner. They are used for a wide range of products including underwear, shirts, suits, trouser suits, sportswear and dresses.

The fourth primary weft knit is the purl knit (Fig. 10.8f). This is the only structure that has certain wales containing both face and reverse meshed loops. This is achieved with double-ended latch needles or by rib loop transfer from one bed to the other, combined with needle bed racking.[11] A 1 × 1 purl knit exhibits the same appearance on both sides (like the back of plain). It has very high extensibility both lengthwise and widthwise and no curling tendency. It is commonly used for children's clothing and for thick and heavy underwear.

In closing this section, it is important to point out that each of the above structures requires a specialized machine or machine arrangement. Thus, the diversity in weft knit structures comes at a price. For example, single-jersey machines can only produce one type of base structure. On the positive side, some machines may produce different structures provided that some rearrangements are made. For example, rib machines, particularly of the garment-making type, can produce sequences of plain knitting by using only one bed of needles. Interlock machines can sometimes be changed for rib knitting. Purl machines are capable of producing rib or plain knitting sequences by retaining certain needle arrangements during the production of a garment or other knitted article.

Warp knits

Warp knit fabric is similar to that of a woven fabric in that yarns are supplied from warp beams. The fabric is produced, however, by intermeshing loops in the knitting elements rather than interlacing warps and wefts as in a weaving machine. Warp knitted fabric is knitted at a constant continuous width. This is achieved by supplying each needle with a yarn (or yarns) and all needles knit at the same time, producing a complete course (row) at once. It is also possible to knit a large number of narrow width fabrics within a needle bed width to be separated after finishing. In comparison with weft-knit structures, warp knits are typically run-resistant and are closer, flatter and less elastic.

The two common warp-knit fabrics are tricot and raschel (Fig. 10.9). Tricot, solely composed of knit stitches, represents the largest quantity of warp knit. It is characterized by fine, vertical wales on the surface and crosswise ribs on the back. Tricot fabrics may be plain, loop-raised or corded, ribbed, cropped velour or patterned designs. It is commonly used for lingerie owing to its good drapability. It is used for underwear, nightwear, dresses, blouses and outerwear.[4] Tricot fabric is used in household products such as sheets and pillowcases. It is also be used for upholstery fabrics for car interiors.

Raschel knits have a lace-like, open construction, with a heavy, textured yarn held in place by a much finer yarn. Typically, columns of loops are

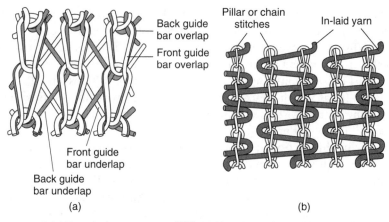

10.9 Warp-knit structures.[4,10,11] (a) Two-bar tricot, (b) raschel warpknit.

connected by in-laid yarns traversing from column to column up the fabric as shown in Fig. 10.9. They split or come apart lengthwise when the laid-in yarn is removed. Examples of products made from raschel fabrics include coarse sacking, carpets, and fine delicate laces.

Basic specifications of knit fabrics

Knit fabrics share many common specifications with woven fabrics such as fabric count, width, thickness and weight or area density. In addition, specifications unique for knit structures include stitch length, stitch density, shape factor and tightness factor (equivalent to cover factor in woven structures). Expressions and estimation approaches of knit specifications largely depend on the fabric structure under consideration.[12-14] Table 10.2 lists expressions suitable for estimating the specifications of plain single jersey fabric.

10.4   Fiber–direct (FD) fabrics: nonwovens

Fiber–direct fabrics represent a unique category of fabrics in which fabrics are formed directly from fibers without the need for yarns as an intermediate component. The early term used to describe this category of fabrics, which can be traced back to the 1930s, was 'nonwovens', or 'nonwoven fabrics'. Today, this term remains the most commonly used name despite its obvious inaccuracy, as it can imply that even knit fabrics being non-woven, also belong to this category. Attempts to correct this inaccuracy resulted in many alternative names such as bonded fabrics, formed fabrics

Table 10.2 Basic formula for single-jersey knit fabrics[10–14]

Knit fabric parameter	Formula	Description
Fabric count, stitch density and shape factor	$$\text{courses per cm (cpc)} = \frac{k_c}{l}$$ $$\text{wales per cm (wpc)} = \frac{k_w}{l}$$ $$\text{stitch density} = s = (\text{cpc} \times \text{wpc}) = \frac{k_s}{l^2}$$ $$\text{shape factor} = \frac{\text{cpc}}{\text{wpc}} = \frac{k_c}{k_w}$$	k_c, k_w and k_s are dimensionless constants; l = the stitch length
Yarn count versus machine gauge	$$\text{optimum tex} = \frac{k_{tex}}{(\text{gauge})^2} \quad \text{or}$$ $$\text{optimum } N_e = \frac{(\text{gauge})^2}{k_N}$$	For single jersey, typical k_{tex} is 1650. For double-jersey, typical k_{tex} is 1400. The gauge, G, is measured in needles per cm (npc).
Tightness factor	$$K = \sqrt{\frac{\text{tex}}{l}} = \sqrt{\frac{T_t}{l}}$$	l = stitch length (mm) For single-jersey fabrics, typical K values are: $1.29 \leq K \leq 1.64$. For most weft-knit fabrics (including single and double jersey and a wide range of yarns): $1.0 \leq K \leq 2.0$
Weight or area density	$$\text{Area density} = \frac{s \times l \times T_t}{100}\, \text{gm}^{-2} \quad \text{or}$$ $$\text{or } \frac{k_s}{l} \times \frac{T_t}{100}\, \text{gm}^{-2}$$	s = the stitch density/cm^2 k_s = constant, its value depends upon the state of relaxation, that is, dry, wet, finished or fully relaxed.
Linear density	$$\text{Linear density} = \frac{n \times l \times \text{cpc} \times T_t}{10000}\, \text{gm}^{-1}$$ $$\text{or } \frac{n \times k_c \times T_t}{10000}\, \text{gm}^{-1}$$	n = the total number of needles; k_c = a constant, its value depends upon the state of relaxation, that is dry, wet, finished or fully relaxed; T_t = yarn tex.
Fabric width	$$\text{Fabric width} = \frac{n \times l}{k_w}\, \text{cm or through}$$ $$n \times l = L \text{ (course length)}$$ $$\text{Fabric width} = \frac{L}{k_w}\, \text{cm}$$	k_w is a constant (its value depends upon the state of relaxation, that is dry, wet, finished or fully relaxed) Fabric width depends upon course length and not upon the number of needles knitting

and engineered fabrics. However, each one of these terms can also be argued on the basis that woven and knit structures are uniquely bonded, surely formed and certainly engineered. Therefore, the author of this book elected to use the term 'fiber–direct' because of its precise and accurate description of this important category of fabric. A 'fiber direct' or FD fabric is nonwoven, non-knitted and non-tufted. For the sake of the discussion in this section, both terms 'nonwoven' and FD will be used for convenience.

The term 'nonwoven' has been commonly defined by many associations as a manufactured sheet of directionally or randomly oriented fibers, bonded by means of friction, cohesion, adhesion, or combination of these, excluding paper and products which are woven, knitted, tufted, stitch bonded incorporating binding yarns or filaments or felted by wet milling, whether additionally needled or not. Fibers used for nonwoven structures may be staple or continuous filament. Fibers may also be combined with non-fibrous particles to form composite nonwoven structures. In comparison with conventional fabric forming techniques, nonwoven processes are relatively simple, productive, versatile, economic and can be very innovative.

FD or nonwoven fabrics are made using two basic operations:[15–18] fiber web preparation (or fiber laying) and fiber bonding. In most cases, these operations are performed using a continuous process from raw material to finished product. As a result, several costly and time-consuming operations are eliminated. These include traditional spinning preparation (e.g. drawing, combing and roving), yarn forming and traditional weaving preparation (e.g. winding, warping and sizing). When fiber is the direct raw material, design options to create many products and product combinations become unlimited. This is a direct result of the possibility of using many combinations of design variables such as fiber content, fiber type, fiber blend, fiber arrangement, fiber orientation, fiber properties, web preparation, fabric thickness, bonding type, bonding density and finishing techniques. In addition, FD fabrics can be combined with traditional woven and knitted fabrics to form many types of compound products, as will be discussed later in this chapter. These economical advantages have resulted in a wide range of products most of which are of an industrial nature or are function-focus products.

Despite the fact that FD fabrics are made using largely continuous processes, a review of all FD fabrics or nonwovens can be a difficult task. The main reason for this difficulty is the existence of many combinations of fiber web preparation and bonding techniques. In the discussion below, the emphasis will be on the different types of FD or nonwoven fabrics described by web preparation and bonding technologies, and common fibrous products produced from these fabrics.

FD or nonwoven fabrics may be classified by the method of laying the fiber web. In this regard, two main types of fabrics can be made: dry-laid fabrics and wet-laid fabrics. In dry-laid fabrics, the web may be prepared mechanically (using traditional carding) or aerodynamically (using a combination of mechanical fiber opening and a strong air stream). Using these methods of fiber laying, the web structure may consist of fibers that are largely oriented along the machine direction (parallel laying), largely oriented crosswise (cross laying) or largely randomly oriented. The importance of fiber orientation in the fabric web stems from the need to develop high strength in some fabric directions (anisotropic) or in all directions (isotropic) and the possibility of controling fluid or air flow through the fabric structure during the use. Dry-laid fabrics are often used in interlinings, coated-fabric backings, carpet components, diaper cover-stock, wipes and sanitary napkins.[4]

Wet-laid fabrics are made by processes similar to those used for paper making, which is a high speed process. However, since fibers are relatively longer than the wood pulp used for paper making, they are dispersed into water at a higher rate of dilution (roughly 10 times greater) to prevent fiber aggregation.[16,17] In practice, a structure containing 50% textile fibers and 50% wood pulp is considered to be a nonwoven structure; additional increase in wood pulp content will result in a structure that is better called a fiber-reinforced paper. Wet-laid fiber webs exhibit high uniformity with weight in the range from 10.4 to 554 g m^{-2} (0.3 to 16 oz yd^{-2} and thickness in the range from 0.06 to 5 mm. The dispersion of fibers results in largely random arrangement of fibers in the web. Wet-laid fabrics are often used in laminating and coating base fabrics, filters, interlining, insulation, roofing substrates, adhesive carriers, wipes, battery separators, towels, surgical gowns, diaper cover-stock and shoe components.[4,15]

Another type of FD fabric is the so called spun-laid or spun-bonded fabric. This is made by extruding filaments from thermoplastic polymers (mostly polyester and polypropylene), drawing the filaments (typically using air drawing but roller drawing is also used) and laying the filaments in the fiber web. Fiber arrangement can be controlled so that arrangement in the machine direction, some random arrangement, or some cross-direction arrangement can be achieved.[16] This type of FD fabric normally has high strength and high tear resistance. It is commonly used to make a wide range of products including[4,15] apparel interlining, carpet backing layers, bagging, packaging, filtration, wall-coverings, and charts and maps. It is also used in function-focus applications such as geotextiles, house wrap vapor-barriers and protective apparel. The use of micro-denier filaments in this type of FD fabric provides many advantages including[15] better filament distribution, smaller pores between the fibers for better filtration, lighter weight and softer feel.

Other derivatives of spun-bonded fabric are flash-spun fabric and melt-blown fabric. Flash-spun fabric is made by a specialized technique for producing very fine fibers without the need to make very fine spinneret holes.[16] In this case, the polymer is dissolved in a solvent and is extruded as a sheet at a suitable temperature so that when the pressure falls on leaving the extruder, the solvent boils suddenly. This blows the polymer sheet into a mass of bubbles with a large surface area and consequently with very low wall thickness. Subsequent drawing of this sheet, followed by mechanical fibrillation, results in a fiber network of very fine fibers joined together at intervals according to the method of production.[15,17] Melt-blown fabrics are also made by a process in which very fine fibers are produced at high production rates without the use of fine spinnerets. The polymer (polypropylene or polyester) is melted and extruded in the normal way but through relatively large holes. As the polymer leaves the extrusion holes it is hit by a high speed stream of hot air at or even above its melting point, which breaks up the flow and stretches the many filaments until they are very fine. At some point the filaments break into fine staple fibers which are collected into a batt on a permeable conveyor. Melt-blown fabrics can be layered with spun-bonded fabrics to form spun-bonded/melt-blown composites (called SMS fabrics).

As nonwovens are fiber–direct fabrics, they will rely largely on fiber-to-fiber entanglement enhanced by inter-fiber frictional forces to hold the fabric structure together. However, this cannot be the only way to achieve web integrity particularly in situations where thin webs consisting of small amount of fibers are to be produced. Thus, additional adhesives should be used to enhance the integrity of these structures. A common way to classify nonwoven fabrics is by the bonding technique used to maintain the fabric integrity. There are two major categories of nonwoven fabrics: (a) mechanically bonded fabrics and (b) media-bonded fabrics. Mechanically bonded fabrics are made using familiar methods of bonding such as needle-punching, hydro-entanglement and stitch bonding. Media-bonded fabrics are formed using external effects such as chemical adhesion and thermal exposure.

10.4.1 Mechanically bonded fabrics

Some fibers have a great ability to intermingle and entangle together by virtue of their surface morphology. The most familiar fiber that exhibits this unique characteristic is wool. Like human hair, the surface of wool fiber is characterized by natural scales that provide differential inter-fiber friction or greater friction when fibers are rubbed against scales than when rubbed with the scales. This phenomenon resulted in the development of so-called felt fabrics which consist of 100% wool or blends of wool and other fibers. To make felt fabrics, fibers are interlocked by mechanical compacting supported by heat, moisture and inter-fiber friction.[4] The web made in this way is then passed through a solution of soap or acid that causes the fabric to shrink and harden even further. Felt fabrics are known for their resiliency, sound absorption, vibration damping, shock absorption and thermal insulation. However, felt fabrics are typically stiff and often hard to manipulate (folding or bending). Felt fabrics can be found in many end products such as some interior fabrics, hats, house slippers, clothing decoration, pads for high production machinery, high-fidelity speakers, acoustic wall covering in auditoriums and public spaces, vibration-damping components for sensitive instruments, plumbing seals and marine gaskets.

Another type of mechanically bonded fabric is the needled-punched or needle-felt structure. This is formed by using special needles to induce a inter-fiber cross linking effect in the fiber web or batt.[16,17] In this case, any fiber type of staple form can be used.

Needle-punched fabrics are typically more dimensionally stable than comparable felt fabrics. In some cases, a lightweight, open-weave fabric is incorporated into the center of the web to provide additional integrity to the needle-punched fabric. End products made from needle-punched fabrics are numerous. These include[4,15] blankets and carpets (mostly acrylic or acrylic blends), outdoor carpets (mostly olefin fibers), backing in tufted carpets, supporting material in ballistics-protective vests, filter media, coated fabric backing, apparel interlinings, road underlay and auto trunk liners.

Hydro-entangled fabrics, also called spunlaced fabrics, are produced using high-pressure water streams (as high as 2200 psi; about 152 bar) that simulate the needle action to entangle the fibers in the fiber batt. The formed web (usually air-laid or wet-laid, but sometimes spun bond or melt-blown, etc.) is first compacted and pre-wetted to eliminate air pockets and then water-needled. Fabrics produced using this bonding method are typically of soft handle and have good drapability.[16,17] When hydro-entanglement is applied for dry-laid (carded or air-laid) or wet-laid webs, synthetic fibers such as PET, nylon, acrylics and Kevlar can be used. Blends of cotton fibers and sythetic fibers are also used. As cotton is hydrophilic in nature, it provides an inherent bonding ability owing to a high content of hydroxyl groups, which attract water molecules. Upon evaporation of water from the fiber batt, hydroxyl groups on the fiber surface link together by hydrogen bonds. Products made from hydro-entangled fabrics include[4,15] wipes, draperies, quilt backings, curtains, table cloths, some apparel, surgeons' gowns, disposable protective clothing and backing fabrics for coating.

Another type of bonding that can be categorized as mechanical bonding is stitch bonding. Stitch-bonded fabrics are made using cross-laid and air-laid batts bonded by a series of interlooped stitches running along the fabric length. Stitching can be applied using external yarns (sewing or warp knit stitching) or by interloping the fibers in the batt. Products made from stitch-bonded fabrics cover a wide range of applications and some of these products require special methods of stitch bonding.[18] These products include[4] interlinings for garment, shoe and upholstery fabrics, decorative fabrics, textile wallcoverings, dishcloths and backing fabrics for coatings. Most of these are made using thread stitch-bonded fabrics. Products made from threadless stitch-bonded fiber webs include backing layer for coated fabrics, textile wallcoverings, decorative felts, packing materials and insulating materials.

10.4.2 Media-bonded fabrics

Media-bonded fabrics are formed using external effects such as chemical adhesion and thermal exposure. Chemically bonded fabrics are produced using an adhesive agent (e.g. acrylic latex, styrene-butadiene latex and vinyl acetate latex) for the complete batt or only isolated portions of the batt. Good adhesion of the fibers necessitates fiber wetting, which is achieved by the surfactants that are normally contained in the lattices and in some cases using additional surfactants.[15–17] The latex is then dried by evaporating the aqueous component, leaving the polymer particles, together with any additives, on and between the fibers. During evaporation, the surface tension of the water pulls the binder particles together forming a film over the fibers and a rather thicker film over the fiber intersections.[15] Finally, the fiber batt is cured by increasing the temperature to a level higher than that used for drying (120–140°C for 2–4 min). Curing results in a significant increase in the cohesive strength by forming crosslinks both inside and between the polymer particles.

Thermally bonded fabrics are produced using melting as the key bonding technique.[16,17] In this regard, localized melting spots may be created for webs consisting of one type of thermoplastic fibers. Alternatively, a blend of fibers with different melting points, or a blend of thermoplastic and non-thermoplastic fibers can be used. Another common method is to use a bicomponent fiber with a high melting point core polymer surrounded by a sheath of a lower melting point polymer. In this case, the core of the fiber will not melt but will support the sheath in its fibrous state.

10.4.3 Nonwoven finish

In addition to web formation and bonding techniques, nonwoven structures can be treated chemically with a wide variety of finishes. Numerous methods can be used to apply chemical finish to nonwoven structures depending on the raw material used and the type of bonding. These include impregnation, printing, spraying and coating. In Chapter 11, more details of different types of finish will be presented. For nonwovens, the finish can be applied to the bulk of material or on the surface (both sides and either side). From a design viewpoint, these applications create a multiple-layer structure, with each layer performing a separate task. Chemical agents can be added to nonwoven structures to improve certain characteristics such as handle, rigidity, softness, abrasiveness, absorption, flame resistance and antistatic properties. Specialty finishes can also be added to enhance specific performance characteristics. For example, odor control and microbial attacks can be treated using special finishes such as activated carbon, zeolites (for odor control), antimicrobials agents and coatings for barrier properties. Other finishes are applied for slip control, adhesion (pressure sensitive, thermoplastic), cohesion, microspheres to increase volume, or to incorporate perfumes and other potential additives.

10.4.4 Potential for FD or nonwovens in the apparel field

In the area of human-contact products, particularly apparel goods, FD fabrics did not have the same appeal as the traditional yarn-based fabrics for the following reasons:

- The use of yarn as an intermediate component allows a great deal of structural integrity and long-term durability.
- Yarn-based fabrics enjoy relatively greater flexibility as a result of the use of self-bonding techniques such as twisting and interlacing. This is a key performance criterion particularly in relation to tactile comfort and in applications where fabrics must be manipulated (folded, bent, cut, etc.) in product assemblies.
- The use of yarn provides a great deal of surface integrity and surface conformity in applications where the fabric must come into direct contact with the human body.

Apparel products utilizing nonwoven fabrics today include interlinings, clothing and glove insulation, bra and shoulder padding, handbag components and shoe components, high-loft insulation, protective clothing, boot insulators, synthetic suede and under-collar material. In the late 1960s some effort was made to market disposable dresses made entirely from nonwoven fabrics, but with little success.

The development of complete nonwoven apparel represents an interesting design challenge. In this regard, key performance criteria include fabric drape, fabric hand, stretchability, and bulk and surface durability, particularly when washable non-disposable apparel is the target product. The development of nonwoven structures such as spun-bonded, melt-blown and stitch-bonded provides potential for moving into the apparel market. This potential is enhanced by the fact that unlike woven and knit fabrics, non-wovens do not ravel; therefore, seams do not need to be surged, making it easy to incorporate shaped hemlines into the garment design. Seams within the garments also do not require finishing. However, many challenges are still ahead.

10.5   Compound fabrics

The categories of fabric discussed in the previous sections represent the basic fabric structures and some of their derivatives. Most of these fabrics exhibit light weight, fair to excellent strength and superior flexibility compared with many non-fibrous structures. In addition, the use of different fiber types, various fiber-blend ratios and different yarn structures provides numerous design options and unlimited expansion of the use of fibrous structures even in applications that were not conceivable a few years ago. In many areas, complex and conflicting requirements may be placed and meeting these requirements using one particular fabric structure may be impossible. In these situations, compound fabrics consisting of different fabric layers, each providing specific functions, should be the focus of the design task. Fortunately, basic fabric structures can be combined in many different ways to produce compound fabrics that can serve in a wide range of traditional and function-focus applications. Common examples of compound fabrics include[4,19] quilted fabrics, tufted fabrics, flocked fabrics, coated fabrics and laminated fabrics. These different types of compound fabrics are available commercially and developments as well as investments in these exciting structures have never ceased. In the discussion below, a brief review of some compound fabrics is provided. It should be pointed out, however, that the best references dealing with accurate information about these specialty fabrics should be the technical information provided by their manufacturers.

10.5.1 Quilted compound fabrics

Quilted compound fabrics typically consist of a filler fibrous layer (e.g. nonwoven structure) sandwiched between two thin fabric layers. Quilting may be achieved using a sewing machine to apply mechanical stitch bonding or using an ultrasonic quilting machine to apply fusion bonding. The latter requires the use of high percentage of thermoplastic fibers so that the high-frequency ultrasonic vibration can cause the fabric layers to fuse together at points along the surface. Design variables associated with quilted fabrics include fabric type of the face and back layers, fiber content in the filler (volume and blend ratio), stitching rate and stitch pattern. Products made from quilted compound fabrics include cold-weather garments, upholstery fabrics, mattress pads, bedspreads and some protective vests.

10.5.2 Tufted compound fabrics

Tufted fabrics consist of tufts or loops (cut or uncut) formed by inserting yarn into a previously prepared backing fabric (woven or nonwoven). This is a form of pile fabric in which the yarn forming the pile stands vertical from the base of the fabric. In order to hold the pile yarn in place, a thin adhesive coat is often applied to the back of the fabric or a secondary backing is added. Design variables associated with tufted fabrics include the gauge length (the distance between rows of stitches), stitch length, pile height and loop condition (cut or uncut). The most commonly made product made from tufted fabric is carpet. This type of carpet is less expensive and requires lower operator skills than comparable woven carpets. Other products made from tufted fabrics include[4] upholstery, bedspreads and fur-like components used for linings of coats and jackets.

10.5.3 Flocked compound fabrics

Flocked fabric may be considered to be another derivative of pile fabric. It provides a velvet-like surface using very short or pulverized fibers attached and held to a base of fabric by an adhesive. Design variables associated with flocked fabrics include fiber type, base type, the adhesive used, pile height and pile density. Common fiber types used for flocked fabrics include rayon, nylon, polyester, olefin and acrylic. Products made from flocked fabrics include[4] velvet-like upholstery fabrics, draperies, bedspreads, wall-coverings, blankets, children's and women's clothes. They may also be used as an outer fabric for stuffed toys, air filters, non-slip patch fabrics on boat decks and swimming areas, handbags and belts.

10.5.4 Coated and laminated compound fabrics

Coated fabric is a compound fabric by virtue of the fact that the coating layer, which is typically a continuous polymeric material, is considered to be an independent layer attached to the fabric via its inherent adhesive properties. The fibrous structure can be of any of the basic structures, namely woven, knit or nonwoven. Polymeric materials used for coating are typically function oriented and they are used for many purposes including[4,16] chemical resistance, anti-stain (or low surface energy), moisture release, hydrophilicity, vapor and gas barrier, electrical conductivity and abrasion resistance. A key design challenge of coated fabrics is the need to alter the fabric surface without affecting its overall physical and mechanical properties. Numerous products can be made from coated fabrics. These include chemical-protective clothing, waterproof clothing, upholstery fabrics, wall coverings, women's shoe uppers, floor coverings, bandages, acoustical barriers, filters, soft-sided luggage, awnings, ditch liners and air-supported structures.

Laminated fabrics consist of two or more layers, at least one of which is a textile fabric, bonded closely together by means of an added adhesive, or by the adhesive properties of one or more of the component layers. A common type of laminated fabric is the so-called film laminate. In this type, one component is a film, or a thin, flexible sheet of polymeric material. This film is typically bonded using adhesive to a base fabric or between two fabrics. Film laminated fabrics can be used for waterproof applications such as shower curtains, tablecloths, shelf covers and waterproof rainwear. Another type of laminated fabric is the so-called foam laminate used primarily for insulation purposes. In this type of fabric, a thin foam layer (e.g. polyurethane foam) is present, giving small numerous air bubbles in a lofty, bulky, springy, elastic-like structure. Since foam cannot be used alone owing to its weakness, the fabric layers act as a container for the foam layer. Foam laminates are used in the manufacture of spring coats and bathrobes, where a warm and lightweight fabric is needed.

10.6　References

1. ROBINSON A T and MARKS R, *Woven Cloth Construction*, The Textile Institute, Manchester, 1973.
2. WATSON W, *Textile Design and Colour-Elementary Weaves and Figured Fabrics*, Longmans, London, 1912 (later editions revised by Grosicki and published by Butterworth).
3. WATSON W, *Advanced Textile Design*, Longmans, London, 1912 (later editions revised by Grosicki and published by Butterworth).
4. HATCH K L, *Textile Science*, West Publishing Company, Minneapolis, NY, 1999.
5. SONDHELM W S, 'Technical fabric structures-1, Woven fabrics', *Handbook of Technical Textiles*, Horrocks A R and Anand S C (editors), Woodhead Publishing Limited, Cambridge, UK, 2000, 62–94.
6. ZHENG-MING HUANG, 'Efficient approach to the structure–property relationship of woven and braided fabric-reinforced composites up to failure', *Journal of Reinforced Plastics and Composites*, 2005, **24** (12), 1289–1309.
7. SILVESTRINI T A and LAPTEWICZ JR. J E, Pfizer Hospital Products Group, http://www.freepatentsonline.com/4834755.html, *Triaxially-braided Fabric Prosthesis*, United States Patent 4834755, 1983.
8. PEIRCE F T, 'The geometry of cloth structure', *Journal of the Textile Institute*, 1937, **28**, T45.
9. THE TEXTILE INSTITUTE TERMS AND DEFINITIONS COMMITTEE, *Textile Terms and Definitions*. 10th Edition. *Table of SI Units and Conversion Factors,* The

Textile Institute, Manchester, 1995.

10. ANAND S C, 'Technical fabric structures-2. Knitted fabrics', *Handbook of Technical Textiles*, Horrocks A R, and Anand S C (editors), Woodhead Publishing Limited, Cambridge, UK, 2000, 95–129.

11. SPENCER D J, *Knitting Technology*, 3rd edition, Woodhead Publishing Limited, Cambridge, UK, 2001.

12. MUNDEN D L, The dimensional properties of plain knit fabrics, Knit. O'wr Yr. Bk., **480**, 266–71, 1968.

13. MUNDEN D L, 'The geometry and dimensional properties of plain-knit fabric', *Journal of the Textile Institute*, 1959, **50**, T448–471.

14. MUNDEN D L, 'Geometry of knitted structures', *Textile Institute, Review of Textile Progress*, 1967, **17**, 266–9.

15. SMITH P A, 'Technical fabric structures-3. Nonwoven', *Handbook of Technical Textiles*, Horrocks A R and Anand S C (editors), Woodhead Publishing Limited, Cambridge, UK, 2000, 130–51.

16. NEWTON A and FORD J E, 'Production and properties of nonwoven fabrics', *Textile Progress*, 1973, **5**(3), 1–93.

17. PURDY A T, 'Developments in non-woven fabrics', *Textile Progress*, 1980, **12**(4), 1–97.

18. COTTERILL P J, 'Production and properties of stitch-bonded fabrics', *Textile Progress*, 1975, **7**(2), 101–35.

19. HALL M E, 'Coating of technical textiles', *Handbook of Technical Textiles*, Horrocks A R and Anand S C (editors), Woodhead Publishing Limited, Cambridge, UK, 2000, 173–86.

11

Finishing processes for fibrous assemblies in textile product design

Abstract: Fabrics that have just been made are commonly labeled 'greige' fabrics. At this stage, the fabric is typically not ready for use as a garment or other end product. It may suffer some deficiencies resulting either from some adverse effect of the manufacturing operation or from inherent limitations in the technology used to make the fabric. The finishing process aims to overcome some of these deficiencies. In addition, finishing treatments can add new performance characteristics that are not inherently present in the greige fabric. These include wrinkle resistance, flame resistance, hydrophilicity, hydrophobicity and soft hand. More advanced finishing treatments can result in a smart fabric with performance characteristics that can accommodate changing environmental conditions. In this chapter, a brief overview of some of the common finish treatments is presented. Finishing treatments are divided into two main categories preparatory finishes and performance-enhancement finishes. Preparatory finishes commonly aim to produce a clean and impurity-free fabric so that further dyeing and finishing treatments can be applied uniformly and without interference. Performance-enhancement finishes are commonly applied to improve characteristics such as flame resistance, stain and water resistance, hand, durable press and soil release.

Key words: greige fabric; preparatory finish; performance-enhancement finish; flame resistance; stain resistance; water resistance; antistatic; antimicrobial; durable-press; surface roughness; nanotechnology; anti-odor; lotus effect; coating; lamination.

11.1 Introduction

Fabrics just off the loom or knitting machine are commonly labeled 'greige' fabrics. In this state, the fabric should exhibit most of the basic characteristics that were specified at the development or the design phase of the fabric. However, the fabric may suffer some characteristic deficiencies resulting from either some adverse effect of various manufacturing operations which fibers have to go through or from inherent limitations in the technology used to make the fabric. Deficiencies of the first category include excessive yarn and fabric defects and dimensional changes imposed by the mechanical stresses during weaving or knitting. Deficiencies of the second category include inherent limitations of fiber, yarn or fabric characteristics (impurities, short fibers, thick places, thin places, neps and pills). The finishing process aims to overcome these deficiencies. In addition, finishing treatments can add new performance characteristics that are not inherently presented in the greige fabric. These include wrinkle resistance, flame resistance, hydrophilicity, hydrophobicity and soft hand. More advanced finishing treatments can indeed result in a smart fabric with performance characteristics that can accommodate changing environmental conditions (e.g. thermally adapted finish and self-cleaning treatments). In this chapter, a brief overview of some of the common finish treatments is presented. This is obviously not an inclusive review as this subject is covered in numerous books.

11.2   Yarn finish

Prior to weaving, warp yarns are normally coated with a protective film in a familiar process called sizing or slashing so that they can withstand the mechanical actions applied to them during weaving. A wide variety of film materials is typically used including[1] starch, polyvinyl alcohol, carboxymethyl cellulose, gums, glues, dextrine, acrylic film, synthetic polymers and copolymers. For staple-fiber yarns (or spun yarns), starch and polyvinyl alcohol are the size films most often used. Synthetic polymers are commonly used for sizing filament yarns. Since the size film has to be removed after the fabric is formed using a desizing process, it is important to know the base size used, as each size material has to be desized using certain treatments. For example, starch can be degraded into water soluble compounds via enzymes, acid hydrolysis and oxidation. Carboxymethyl cellulose (CMC) is soluble in cold water. Polyvinyl alcohol (PVA) will also redissolve in water without initial degradation. Fabrics sized with PVA are desized through saturation with water containing a wetting agent, heating and rinsing in hot water.

The idea of sizing stems from the fact that staple-fiber yarns, as spun, encounter many problems that can adversely influence the weaving efficiency and result in many yarn breaks.[1-3] They exhibit strength that can easily fall below the applied stress during weaving (particularly the weak points in the yarn), they lack perfect uniformity and smoothness, they have many knots and slubs and they may exhibit a great deal of hairiness. These defects are virtually inevitable despite great progress in spinning technology. As a result, a protective film will mask a great deal of these defects leading to yarns that are stronger, more uniform, highly abrasion resistant and less hairy.

11.3   Fabric finish

As indicated in the introduction, prior to forming an end product, the fabric must be modified using chemical, thermal or mechanical treatments so that the desired final bulk, surface or geometrical characteristics can be obtained. In the context of product development, these treatments (collectively called finishing) sometimes represent the most critical phase of product design, as they can indeed influence many of the performance characteristics of fibrous products.

Fibrous assemblies are characterized by two basic features: (a) they are mostly polymeric based and (b) they are flexible in nature. The first feature allows manipulation and modification of the characteristics of fiber assemblies via chemical or thermal alteration and the second feature allows mechanical manipulation. Accordingly, finish applications are generally divided into three basic types: chemical, thermal and mechanical. Finish treatments may also be classified into preparatory finishes and performance-enhancement finishes. Preparatory finishes commonly aim to produce a clean and impurity-free fabric so that further dyeing and finishing treatments can be applied uniformly and without interference. Examples of preparatory finishes are as follows:[4–6]

- singeing: burning away loose projecting fibers and fuzz that are projecting from yarns or fabrics;
- desizing: removing the size film during fabric finishing;
- scouring: washing yarns or fabrics to remove natural dirt, waxes and grease;
- bleaching: applying a whitening agent to a fabric surface;
- mercerizing: treating cellulosic-based yarns or fabrics with a caustic treatment to provide luster, water absorbance, dye yield and fiber strength;
- carbonizing: treating wool material with acid to remove vegetable matter;
- heat setting: applying heat treatment to thermoplastic materials to provide dimensional stability;
- tentering: fabric drying under gripping constraints to re-balance the fabric and to reach a prespecified fabric width;
- calendering: this is analogous to ironing the fabric under high pressure (up to 1 ton per square inch; $155 \times 10^7 \, \mathrm{g \, m^{-2}}$) with the result being to flatten the fabric, making it smoother and more lustrous. It can also serve as a water-squeeze when the incoming fabric is wet or to enforce resins and other finishes added to the fabric prior to calendering.

Performance enhancement finishes are commonly applied to improve some performance characteristics such as flame resistance, stain and water resistance, hand, durable press and soil release. In the following sections, brief descriptions of many common finish treatments are presented.

11.3.1 Examples of preparatory finishes

Singeing

Singeing is essentially burning the free fiber ends that project from the fabric surface, using gas flame. It is typically used as an initial stage of fabric finishing, particularly with fabrics made from staple-fiber yarns such as cotton. The fabric is passed at high speed over rows of gas flames or between heated metal plates to avoid any burning of the yarns. Obviously, singeing cannot be used for fabrics made from thermoplastic fibers as this will result in melting the fibers. Fabrics made from continuous filaments do not require singeing. With wool fabrics, shearing or fiber clipping is used, as singeing wool fibers can result in strong sulfur odor. The main purpose of singeing is to create a smooth fabric surface and reduce pilling.

Desizing

As indicated above, this is the process of size removal from the fabric to allow application of other finishing treatments, particularly those that involve penetration through the fibers. The desizing process should be both efficient and environmentally friendly. A key aspect in this regard is to recycle size materials for repeated use through efficient separation of the water from the size liquor during desizing. Scientists at BASF AG have developed a method in which synthetic sizes can be recycled in an environmentally friendly manner.[5] This method aims to reuse the size solution as well as the water repeatedly and without any loss of size effect. The name of the synthetic size used is 'UCF-4®'. This new size material is claimed to have advantages over traditional starch in many quality attributes such as chemical consumption, energy consumption, toxicity, emission and risk potential. During desizing, UCF-4® can be removed effectively from the fabric by merely washing with hot water and without the addition of chemicals. This is achieved through an efficient ultrafiltration process during recycling to separate the water and the size liquor. During ultrafiltration, the macro-molecules of the size UCF-4® are held back by the membranes, while the water with low molecular impurities from the fabric passes through the membrane. This achieves a separation of high concentration size that can be reused for sizing new yarns and recovery of pre-heated water for the next desizing process. Using this size material, a recycling rate of over 80% is claimed without loss of size effectiveness.

As indicated earlier in Chapter 7, the issue of desizing represents an excellent opportunity for design engineers to overcome a historical problem by developing fibers that can be spun into yarns with good surface integrity so that the cost of chemical treatment can be saved. In addition, a new yarn-forming system capable of producing yarns with good surface integrity is another intriguing design idea.

Scouring

Scouring is the process of removing impurities from the fabric. In natural fibers, impurities include oils, fats, waxes, minerals, leafy matter and seed-coat fragments. These impurities are commonly removed by exposing the fabric to aqueous solutions of concentrated sodium hydroxide containing detergents and then steaming and rinsing it. In synthetic fibers, impurities can include spin finishes or oil contaminants from machinery. These are typically removed by extraction (dissolving in organic solvents), emulsion (forming stable suspensions of the impurities in water) and saponification (converting the impurities into water soluble materials). For wool and silk, carbonizing and degumming processes are used to remove impurities. In carbonizing, vegetable matter is carbonized or destroyed using sulfuric acid. Degumming is the process of boiling silk fabric with soap to remove the sericin coating the silk filament.

Bleaching

Bleaching is a whitening process used because most natural fibers exhibit off-white colors caused by the presence of colored substances that can vary within the same fabric batch or between batches. Bleaching primarily destroys these color substances by applying oxidizing and reducing chemicals to decolorize and remove off-white substances from fabrics. Bleached fabric will exhibit a consistent color after dying as a result of uniform dye uptake. Different fiber types will require different bleaching agents. For cotton fabric, the common bleaching agent is hydrogen peroxide, buffered to maintain a pH of about 10.5. Wool fabrics are typically bleached using sodium peroxide solutions.

Mercerizing

Mercerizing is the process of treating cotton fabric with caustic soda solutions at high concentration (19–26%) to complete the removal of impurities initiated by scouring or bleaching. Mercerization can substantially alter the appearance and the geometrical characteristic of fabric, making mercerized fabrics uniquely different from unmercerized fabrics. This is due to the fact that caustic soda solution results in cotton fiber swelling, breaking hydrogen bonds and weak van der Waal forces between cellulose chains, and leaving the cellulosic chains (after removal) in a reorganized state.

Mercerization can be applied in a relaxed state or under tension. The former will result in fiber swelling (coarser and shorter fibers) and fabric thickening. This leads to a stronger and more elastic fabric. The latter results in orienting the fibers along their axes making them more rounded (rod-like cross-section) and leading to a smoother and higher-luster fabric. It also results in a significant increase in strength (about 35%) by virtue of fiber orientation. In addition, the fiber also becomes more absorbent as a result of the change of the cellulose crystal unit cell from cellulose I to cellulose II and an increase in openness of the amorphous area.[6] Therefore, mercerized cotton will absorb more dye and more water than unmercerized cotton.

11.3.2 Examples of performance enhancement finishes

Moisture management: water repellency, waterproof and hydrophilicity

Moisture management is a key performance characteristic in many fibrous products such as water-repellent fabrics, waterproof fabrics, stain-resistant fabrics and comfort products. A water-repellent fabric is a fabric that simply resists water penetration through the fabric construction. This is not to be confused with waterproof fabric, which is completely water and air resistant. From a design viewpoint, water repellency can be achieved through optimization of many parameters. Fabric-related parameters include fabric construction, fabric density and cover factor. Yarn-related parameters include yarn type, yarn count and yarn twist. Fiber-related parameters include fiber type, fiber fineness, fiber absorption characteristics and fiber wicking characteristics.

The role of fabric finish is to enhance water repellency further. Chemical finishes used for this purpose are typically hydrophobic substances with a critical surface tension below that of the water because for a fabric to be water repellent, the critical surface tension of the fiber surface should be below 30 dynes cm^{-1} (pure water typically has surface tension of about 72 dynes cm^{-1}). For oil repellency, lower fiber surface tensions will be required (below 13 dynes cm^{-1}). A common substance used for water repellent treatments is paraffin wax, which is the most economical treatment. This can be applied using solvent solutions, molten coatings or wax emulsions. This is not a permanent treatment as wax can be abraded by rubbing against other objects and dissolve in dry cleaning fluids. Pyridinium compounds are also used for water repellent treatments. These are long-chain fatty amides and wax resin mixture. Silicone water repellents are also used for a variety of fabrics because of their durability as they exhibit high resistance to abrasion and they are less soluble in dry-cleaning fluids or laundry detergents.

Fluorochemical repellents have the advantage of repelling both water and oil as a result of their low surface energy. These provide a very durable finish and their application is mainly based on the reduction of the critical surface energy tension of the finished fabric surface to levels below that of the wetting liquid (in this case water). This creates a chemical barrier, which prevents penetration of the liquid. Of all textile chemicals, only fluoropolymers show this unique property of reducing surface energy to such an extent that they repel both aqueous and oily substances[7] (polar and non-polar liquids).

A waterproof fabric is very highly resistant to water penetration even under high hydrostatic pressure. As in case of water repellent fabrics, this feature will require special fabric construction of very tiny pores. The hydrophobic finish used to achieve this resistance is applied not only to the fibers but also to the fabric pores. By definition, waterproof fabrics are unbreathable, which means that just as they do not allow water drops from the rain to penetrate through the fabric, they also do not allow moisture vapor from perspiration to go through. However, unique design methods by the engineers at GORE-TEX® have resulted in fabrics that are waterproof and breathable. One GORE-TEX® idea was to use a multiple-layer fabric system through placing a membrane of PTFE fluorocarbon underneath a layer of outer fabric and adding another lining fabric layer. According to the makers of GORE-TEX®, this membrane has nine billion microscopic pores per square inch that are 20000 times smaller than a drop of water but 700 times larger than a molecule of water vapor. This allows moisture vapor from perspiration to pass through the fabric yet to prevent rainwater drops from penetrating into the fabric. Obviously, other design aspects such as fabric construction and fabric seam closeness are taken into consideration.

Another approach for moisture management is to develop hydrophilic fabrics that can absorb water. These fabrics can be made from naturally hydrophilic fibers (cotton or wool). In addition, some chemical treatments including mercerization result in further improvement in the absorption characteristics of cotton fabrics. Fabrics made from synthetic fibers such as polyester can also be made hydrophilic using special finishes such as sodium hydroxide solutions.

Oil repellent and stain resistant finishes

Oil repellent fabric is largely similar to the waterproof fabric. This requires very low surface tension of the fabric. The use of fluorochemical polymers results in fabrics that are both highly water resistant and highly oil resistant by preventing the fibrous material from being wetted and soiled, by repelling aqueous and oily soil particles and preventing adhesion of dry soil through anti-adhesive properties.[7]

Stain-resistant fabrics are treated with chemical blockers that depend on the type of potential stain. For example, waterborne stains can be resisted using silicon water repellents. On the other hand, oil-based stains can be repelled using fluorochemical polymers that have the ability to prevent oils and water from penetrating into the fabric and soils from sticking to the fiber surface. This is particularly important for fibrous products that are not routinely laundered such as upholstery fabrics and carpets. High-quality upholstery fabrics are expected to keep their pleasant appearance for a long time by being resistant to common stains in daily life such as coffee on the upholstered chair or jam on the sofa. It should be pointed out, however, that a conventional fluorochemical finish has its durability limitation, particularly in relation to abrasion resistance.[4,6,7] Abrasion effects can occur even under mild conditions such as seating and arm resting which cause abrasion of the micro-fine fluoropolymer protection film that covers the fibrous material. This can result in a constant decrease in the stain- or soil-repellent effect.

Thermal management finishes

Thermal insulation of most fibrous products is commonly made using the basic principle of trapping air to reduce heat transfer particularly by convective and conductive modes. This is a common approach, which is also used with other materials such as plastic foams and sintered refractory materials. The two key design criteria required for thermal insulation are (a) the rate of air flow (in large cells trapped air will have internal convection currents) and (b) the amount of fiber material surrounding the air (large percentages of air entrapped between fibrous materials will result in a reduction of thermal bridging between fibers). These criteria can be satisfied using a combination of parameters including fiber type, yarn bulk density, yarn fineness and fabric construction. Mechanical finishing processes that are known to influence thermal insulation are napping and brushing. In addition, special finishes can be applied using polymers that are temperature sensitive.

Unlike singeing, in which free fiber ends are removed, napping is a mechanical process used to raise fibers from the body of the fabric using wire brushes. Fibers may be raised in an upright position or oriented in some direction. The raised fibers entrap air within the fabric surface creating an insulation layer between the fabric and human skin and create less contact between the skin and the fabric so that loss of heat from the body by conduction can be reduced. Common napped products include[4] pajamas, blankets, sleepwear, sweaters and baby dresses. Napped fabrics should not be confused with pile fabrics in which a separate set of yarns, not fibers, is raised to create the pile. As indicated in Chapter 9, the napping process is facilitated by using yarns of low twist so that fibers can easily be raised from the yarns.

Brushing is a milder case of napping, which results in a lower fiber density than that witnessed with napped fabrics. The purpose of brushing is also to provide warmth but at a lower efficiency than that of napped fabrics. Yarns used for brushed fabrics could have higher twist levels than those used for napped fabrics. Another way to achieve a fuzzy surface is through a process called 'flocking' in which short fibers (rayon or nylon) are adhered to the fabric surface either on the entire surface or in pattern forms.

Thermally sensitive fabrics can also be made using chemical treatments that allow heat to be stored or released depending on the surrounding temperature. One familiar treatment is the use of polyethylene glycol (PEG) as a phase-change material (PCM). It has the ability to store excess heat at high temperature and release it at low temperature. Researchers from the USDA-ARS used PEG in combination with dimethyloldihydroxyethylene urea (DMDHEU) and some acid catalysts in a curing process to create water insoluble polymers that can be used as a phase-change finish for fabric. A phase-change fiber of PEG copolymers with polyethylene terephthalate (PET) can also be prepared by melt spinning.[8,9] The PET–PEG copolymers have solid–solid phase change characteristics at 10–60°C without any obvious liquid substance appearing, while PET/PEG blends will lose their phase-change characteristics since the PEG of the blends may melt and leak at high temperature.[8] By controlling the molecular weight and relevant proportion of PEG added, the phase-change temperature range and the enthalpy can be adjusted. These treatments have great potential for developing many products including thermal underwear, surgical gowns and socks.

Other developments in thermally sensitive fabrics include:

• the use of fibers with integral microspheres filled with phase change material or plastic crystals that can enhance thermal properties at predetermined temperatures: these fibers can be woven to form fabrics with enhanced thermal storage properties;[10]

• coating fabrics with integral and leak-resistant microcapsules filled with phase change material or plastic crystals that have specific thermal properties at predetermined temperatures;[11]

• a thermo-adhesive textile product of the type comprising a backing fabric and an adhesive layer deposited on its surface with the adhesive layer comprising a thermo-adhesive polymer and a cross linking agent which is isolated from the polymer by micro-encapsulation and freed by external action.[12]

Antistatic finishes

When two different surfaces come into contact, surface build-up of electrical charges can become a disturbing factor creating so-called static charges. This can be serious, particularly when one of the surfaces is a conductor, as this creates an unlimited supply of electrons that can be transferred to the other surface. Previous treatments of the fabric or the presence of electrolytic impurities can create these charges. Although most fibers, particularly nylon, polyester, polypropylene and acrylics are dielectric in nature (i.e. they are substantial non-conductors), static charges can be generated during the various stages of processing and during the use of fibrous products. They can result from the rapid flow of fibers or fabrics against metallic guides. Excessive drag tension can be created when surfaces develop opposite charges leading to undesirable fiber or yarn stretch. Folding fabrics can also be a problem when static charges are generated. In addition, the fabric can become a magnet for dust and other particles in the surrounding environment. When the fabric is converted into a garment, static charge often manifests itself in the garment clinging to the body and by dirt or lint attraction.

Finish treatments to minimize static charges include[6] application of static control chemicals (e.g. 3M® brand static control, aerosol sprays, sodium hydroxide solution for acetate fabrics), friction reduction using lubrication and blending fibers of potential opposite charges.

Flame-retardant finishes

For many years, the development of flame-retardant fabric has been one of the main concerns of the polymer and fiber industry. For traditional fibrous products such as children's sleepwear, carpets, drapery and upholstery, there is an obvious desire to have products that do not ignite or create a self-sustaining flames when subjected to heat sources. For specialty products such as firemen's or military uniforms, this desire is critical, as exposure to fire is inevitable. The term flame-retardant does not imply flameproof, as all fibrous products are likely to burn, melt, or char depending on the surrounding flammable situation and the period of exposure. However, some fabrics will tend to be more resistant than others. For example, a pile or napped fabric made from rayon or cotton is likely to be less flame-retardant than a tightly closed woven fabric made from highly twisted yarn.

Flame-retardant fabrics can either be developed from fibers that are inherently flame resistant or by applying a suitable finish that can slow down the burning process. Many flame-resistant traditional fibrous products are made from inherently flame-resistant fibers with the help of fabric construction factors. Specialty fibrous products are often topically treated with a flame-retardant finish. In recent years, new chemical finishes have been developed to treat fabrics of both categories for flame-resistant applications. From a design viewpoint, applications of these finishes and their add-ons have to be optimized as they may adversely affect other performance characteristics in the fabric. Examples of flame-retardant finishes include[6] a mixture of boric acid/borax (sodium borate) for cellulosic fibers, phosphorus-based flame retardants and sulfamic acid and ammonium sulfamate.

Antimicrobial finished fabrics

Antimicrobial activity can be defined as a collective term for all active principles (agents) which inhibit the growth of bacteria, prevent the formation of microbial colonies and may destroy microorganisms. In the field of antimicrobial finishes, many common terms are used including antibacterial, bactericidal, bacteriostatic, fungicidal, fungistatic or biocidal and biostatic. According to Mucha et al.[13] of the Hohenstein Institute, antimicrobial activity refers to a situation where an active agent has a negative effect on the vitality of microorganisms. If the active agent only affects bacteria or fungi, this is referred to as antibacterial or antimycotic activity, respectively. The degree of the effect is denoted by the suffix cidal (or lethal) where there is significant germicidal activity or the suffix static where the active substance serves to inhibit the growth of bacteria. Accordingly, the term bacteriostatic refers to an agent, which temporarily inhibits the growth of a specific bacterial population without destroying this population, or without being able to multiply to any significant extent. The authors add that, as yet, there is no universally accepted definition as to what constitutes significant growth or reduction in the bacterial colony. The ideal requirement of a biostasis is that a defined bacterial population remains constant in terms of its size. In practice, however, this is extremely rare as vital biological systems are in a sensitive dynamic equilibrium and react to any change in their life system by either growing or dying.

Antimicrobial finish treatments are commonly used in specialty fibrous products with a protective nature. However, the increasing awareness by consumers of hygiene-related issues has resulted in extending the use of these treatments to many consumer products. These treatments primarily aim at avoiding the adverse effects of microbial fiber degradation, limiting the incidence of bacteria, reducing the formation of odor as a result of the microbial degradation of perspiration and avoiding bacterial infection. In practice, an antimicrobial finish can be used in numerous applications, particularly to protect fibrous products from the effects of microbes especially the fungi types. Products that can be protected include uniforms, tents, geotextiles, curtains and bath mats.

One category of common finish agents are liquid additives known as sanitizers. These are used in the final laundry rinse particularly for clothing worn against the skin. In applications such as outdoor and sporting pursuits, it is important to prevent the build-up of odors on clothing. In these situations, antimicrobial agents act to prevent the microbial decomposition of perspiration on the clothing (or prevent the release of odorous substances).

Antimicrobial fibrous products can be divided into two categories: passive and active products. Passive products do not have a specific bioactive substance. They rely on fibers that are inherently antimicrobial (e.g. hydrophobic fibers with special surface morphologies such as lotus-like or micro-domain structured surfaces). In this category, bacterial cells are prevented from being adhered to the fiber surface by virtue of the surface structure and the low absorption of fibers. Active products contain an specific antimicrobial finish which can attack microorganisms. These are highly specialized agents that target specific microorganisms. Many of these finish agents have been known for years in the food and cosmetic industries. Common antimicrobial agents include oxidizing agents (aldehydes, halogens and peroxo compounds), ammonium compounds (amines, glucoprotamine), metallic compounds (silver) and natural products (chitosan). Key criteria for these agents include (1) low toxic risk and (2) resistance to washing and dry cleaning.

Durable-press finishes

Fabrics made from most synthetic fibers are inherently wrinkle free, meaning that they can be washed, dried and worn without the need for ironing, as they have good shape and appearance retention. Durable-press finishes are primarily used to maintain the dimensional stability of some fibrous products, particularly cotton and wool products, after washing. The relationship between dimensional stability and moisture is well-established, particularly when hydrophilic fibers are used.[6] Basically, dimensional instability is a result of the effect of water on the molecular structure of fibers under external stresses. In case of cotton, the mechanical stresses imposed by the washing process breaks the hydrogen bonds between adjacent molecular chains. The presence of water facilitates the slippage between the molecular chains and new bonds form (reassociation of hydroxyl groups). This new molecular configuration is difficult to recover upon drying unless external effects such as ironing are applied.

A common approach to providing wrinkle-free characteristics for cotton and wool fibers is by blending these fibers with synthetic fibers.[2,3]

When fabrics are made purely from these fibers, a durable-press finish is typically used to provide them with wrinkle-free features. For example, cotton and other cellulosic fabrics are treated with durable-press resins, which eliminate the need for starch and ironing. These fabrics should be washed carefully in cold water, as durable-press (or loosely called permanent press) chemicals may not be as durable in hot water environment. Durable-press finished products include[4] sheets, pillowcases, some shirts and blouses. Some 100% cotton trousers are also wrinkle-free as a result of the application of appropriate chemicals that create a cross-linking effect in the cellulosic chains. The merits of these chemicals often come at the expense of deterioration in critical performance characteristics such as flexibility, tenacity, elongation and abrasion resistance. This is one of the reasons that blending cotton with polyester to develop durable-press fabrics is often a better alternative, as polyester fibers, which are unaffected by the durable-press resins, can compensate for this deterioration. Another approach to reducing this deterioration is to treat cotton with liquid ammonia to strengthen the cotton fiber prior to the durable-press treatment. Mercerization can also help in conjunction with permanent-press treatments through improving the strength of cotton fibers.

For wool fabrics, the so-called permanent set, in which sharp pleats and creases are permanently inserted, is applied to the fabric to create wrinkle-free wool.[6,7] This is a result of a chemical process in which heat and steam are utilized to provide ways for the bonded protein chains to exhibit relative movement, breaking hydrogen bonds by heat and steam, breaking disulfide bonds by chemicals and allowing these bonds to reform in new positions when the fabric is set at a desired configuration. A wrinkle-free state is achieved when the bonds are allowed to reform.

Hand finishes

In general, the term 'fabric hand' describes the way a fabric feels when it is touched and manipulated by hand. It is an action noun that implies evaluation of fabric reaction to different modes of low-stress deformation imposed by the human hand. Another term that is commonly used in the industry is 'fabric handle'. This is another action noun that reflects the evaluation of fabric reaction to different modes of deformation at all levels of applied stress (low or high).

Comparisons between subjective and objective measures of fabric hand revealed that the most important parameters constituting fabric hand are surface roughness (softness), fabric drape, bending stiffness and compressibility.[14] From a design viewpoint, these parameters can be optimized through appropriate selection of raw material, yarn structure (count and twist level) and fabric construction (thickness and density). After forming the fabric, additional optimization of fabric hand can be achieved using many finish treatments. One of the common treatments for improving fabric hand is the use of softeners to give more pliability, and smoother and softer feel to the fabrics. Softeners are divided into three major chemical categories describing the ionic nature of the molecule:[4,6] anionic, cationic and non-ionic. Anionic softener molecules have a negative charge on the molecule derived from either a carboxylate group ($-COO^-$), a sulfate group ($-OSO_3^-$) or a phosphate group ($-PO_4^-$). The majority of softeners are of the sulfate and sulfonate type commonly used for cellulosic fibers and silk. These are basically negatively charged fatty acids and oils. Anionic softeners are often used for fabrics that undergo mechanical finish treatments such as napping or shearing. Cationic softeners (the common laundry fabric softeners) are ionic molecules that have a positive charge on the large part of the molecule. They also assist in reducing static charges on synthetic fibers. Non-ionic softeners can be divided into three sub-categories, ethylene oxide derivatives, silicones and hydrocarbon waxes based on paraffin or polyethylene.

Surface roughness finishes

Surface roughness can be altered either to modify the feel of fibers or to modify the surface characteristics and the visual appearance of fabric. For instance, it is commonly known that some fibers such as polyester or polypropylene can be treated with delusterant such as titanium dioxide (TiO_2) to provide different levels of surface texture or surface irregularities and, consequently, different fiber appearance (fiber dullness). Based on the weight of the fiber, the inclusion of 1% delusterant can produce a 'semidull' fiber, while doubling this amount would produce a 'dull' fiber.[15] Other studies treated nylon 66 fiber melt with TiO_2 to change the intensity of fiber luster and found that bright fibers (low TiO_2) tended to have a smooth surface whereas dull fibers have a rough surface.[16] This change in roughness was not accompanied by a chemical change in fiber surface, which remained essentially like that of the fiber material.

Titanium dioxide particles, roughly spherical and about 0.15 μm diameter (e.g. Ti-Pure R102 by DuPont), typically tend to adhere to each other forming an agglomerate. Other types of filler particles with different shapes and dimensions are also used. These include calcium carbonate (parallelepiped, produced by Fisher Chemicals), NYAD 400 wollastonite (needle-like particles by NYCO Minerals), and talc Jetfil 700C (discotic, Luzenac America). Each of these particles produces a unique surface texture[17] and, therefore, affects other surface-related properties such as friction and abrasion resistance.

11.4   Nanotechnology: finish applications

Conventional finishing in the fabric form has been effective in providing many performance features such as repellency, thermal insulation, anti-odor/anti-scent and reduced weight. However, finishing in the fabric form has also been associated with a number of challenges including manufacturing difficulties, alteration of physical and mechanical properties, poor durability, color-fastness problems, discomfort, low safety and cost. More importantly, conventional finishes have always been associated with problems related to their effectiveness, particularly in the context of add-on by weight or add-on by surface area covered by the treatment. Nanotechnology or nanoscale engineering can provide solutions to many of these problems. Before proceeding with the discussion on this important aspect, it is important to provide the reader with a brief introduction on nanotechnology.

11.4.1   What is nanotechnology?

Nanotechnology can be defined as the technology of manipulating atoms and molecules at the nanometer scale (in SI units 10^{-9} m). The word 'nano' is a Greek word, which means dwarf, or very small. It should be distinguished from so-called microtechnology, which indicates manipulation at the micrometer scale (in SI units 10^{-6} m). The full term 'nanotechnology' was coined by K. Eric Drexler in his book *Engines of Creation*.[18] In general, engineers and scientists use the term 'nano' to refer to physical quantities within the scale of a billionth of the reference unit. For example, nanometer, nanosecond, nanogram and nanofarad are used to describe a billionth of a meter (length), second (time), gram (weight) and farad (charge), respectively.

As a derivative of nano materials, some fibers are classified as 'nanofibers'. This may be defined as a one-dimensional nano-scale element, which falls under the general category of nanotubes and nanorods (very tiny diameter), but exhibits exceptional flexibility compared with the case of conventional fibers. Indeed, the flexibility nature of nanofibers often results in placing them in the category of other nano flexible elements such as globular molecules (virtually, zero-dimensional soft matter) and solid and liquid films of nano-thickness (two-dimensional elements). Some fibers that are filled with nanoparticles are also called nanofibers, while they are more like composite nanofibers. Manufacture of nanofibers can belong to one of two major categories of nanotechnology, namely: top-down (the act of removal or cutting down a bulk to the desired nano size) and bottom-up nanotechnologies (developing nanofibers using fundamental building blocks such as atoms and molecules). These technologies are now commercially available and they include drawing, template synthesis, phase separation, self-assembly and electro-spinning. Discussion of these technologies is outside the scope of this book.

In the context of finishing, nanotechnology can be used to modify conventional textile fibers at the nano level so that function-focused tasks can be performed without apparent changes in the inherent physical characteristics of the fibers. Perhaps, the key potential advantage of this technology stems from the possibility of developing a smart material that has the ability to transform or yield its properties to particular external stimuli. Technologies associated with nanochemistry and nano materials are related to material composition and interface structure, control of size between 1 and 100 nm and transformation of material to acquire special characteristic properties. Such technology is applied to optoelectronic, electronic, energy-storage and semiconductor fields, to develop new materials and key components. Related technologies also include those for nanochemical synthesis, nanostructure, performance simulation and measurement or testing of nano materials.

11.4.2 Examples of nano finishes

Anti-odor and antimicrobial additives

One of the interesting applications of nanotechnology is the use of silver at the nano scale to provide anti-odor and antimicrobial functions for the fabric. Since antiquity, silver has been known as an antimicrobial (silver dishes and chalices were used to preserve food in ancient Greece). Over the last century, silver has even been used in modern medicine to act as a disinfectant in children's eye drops and in numerous other applications, such as lining water holding tanks in boats and airplanes to ensure safe drinking for up to several months. This has resulted in this important material being considered in the development of antimicrobial finishes for fibers. In contrast to many non-permanent antimicrobial chemicals that require special handling owing to their toxicity, silver is considered to be a natural and far safer alternative. However, silver also has its own drawbacks in-cluding cost, permanence and manufacturability. Silver-coated fibers are expensive, difficult to use and shed silver flakes, reducing effectiveness and creating potential effluent problems. For example, the so-called 'nano-silver' powders applied during spinning tend to clog extruding spinnerets, do not bond well to polymers or cotton and generate airborne silver dust.[18,19]

As a result of the difficulty associated with using silver powders, a product called SmartSilver™ was developed by NanoHorizons™. This is a line of permanent anti-odor/antimicrobial additives that can be used to treat fibers such as polyester, cotton, nylon and polyurethane. This type of treatment takes place at the molecular level. According to NanoHorizons™,[19] SmartSilver-modified fibers can be used to create odor-resistant shoe linings, T-shirts, socks, gloves, carpets and more. Urethane applications include rainwear, upholstery, paints, caulks and other coatings for which SmartSilver adds odor-resistance as well as resistance to mildew and mold growth. SmartSilver is also claimed not to alter stretch, wicking, hand, or dyeability of fibers and fabrics, enabling designers to add permanent and effective anti-odor properties to fabrics and fibers.

The principle underlying the use of silver as an antimicrobial component is that silver nanoparticles take advantages of the oligodynamic effect that silver has on microbes, whereby silver ions bind to reactive groups in target cells or organisms, resulting in their precipitation and inactivation. In simple words, this means that silver ions generated by exposed silver surfaces are highly lethal to microbes in several different ways. In fact, microbes generally have a harder time developing resistance to silver than they do to antibiotics. The key to using silver effectively is to find a way to maximize the production of silver ions in a particular application. One approach is to increase the weight of silver being used. This is obviously an expensive and performance-hindering approach. Alternatively, maximization of silver ion release can be achieved by generating as much surface area as possible. This can be achieved more effectively at the nano scale. The greater the extent to which microbes are exposed to the silver ions generated by the silver surface area, the more they can interact with these ions, leading to their destruction.

Water and stain-repellent nano treatments

As indicated earlier, water and stain repellence require surface treatment with chemicals that yield a very low surface tension of the fabric, such as fluorochemicals. At the nano scale, surface wetting can be controlled through a special design in which the relationship between surface roughness and surface reaction to liquid drops is optimized. This relationship stems from the so-called 'lotus' effect; a unique natural effect which clearly demonstrates that natural phenomena can be mimicked in the development of specialty products. It will be important, therefore, to explore this subject to give the reader a greater insight into how nature could represent a major source of design ideas.

Relationship between surface wetting and surface roughness

The equilibrium shape of a liquid drop on a surface is governed by three basic forces existing at three interfaces. These are the solid/liquid, SL,

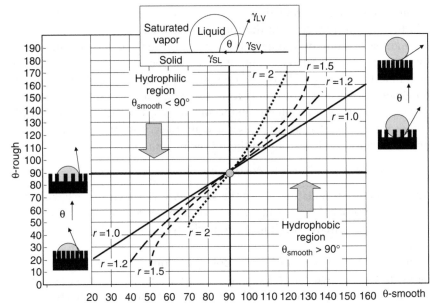

11.1 Demonstration of Young and Wenzel models.[20,21] In the hydrophilic material (left) as r increases from 1 to 2, θ decreases. In the hydrophobic material (right) as r increases from 1 to 2, θ increases.

liquid/vapor, LV, and solid/vapor, SV, surface tensions. The balance of these tensions yields the well-known Young's equation (see Fig. 11.1):

$$\gamma_{SV} - \gamma_{SL} = \gamma_{LV} \cos\theta \qquad (11.1)$$

where θ is called the intrinsic contact angle of the drop and is a parameter that reflects the degree of wettability of a surface. A surface that is wettable for a given fluid has θ less than 90° and the closer the value to zero the more rapidly wettable is the surface. The role of surface roughness in influencing wettability is given by the Wenzel equation[20,21] as follows:

$$\cos\theta_{rough} = r \cos\theta_{smooth} \qquad (11.2)$$

where r is the roughness factor defined by the ratio of rough to planar surface areas.

The above equation indicates that for a rough surface ($r > 1$) wettability is improved by the roughness of a hydrophilic surface ($\theta_{rough} < \theta_{smooth}$ for $\theta < 90°$) and hydrophobicity or dryness is improved by the roughness of a hydrophobic surface ($\theta_{rough} > \theta_{smooth}$ for $\theta > 90°$). This theoretical concept is in agreement with physical observations; a drop on a rough high-energy surface will be likely to sink into the surface. For $\theta > 90°$, the free energy of the dry surface is lower than that of the wet solid and hence the drop will be likely to recede from the roughest regions.

In an early study by Cassie and Baxter,[22] a rough surface was modeled by a heterogeneous surface composed of air pockets and the solid. They postulated that the cosine of the contact angle of a liquid drop on a heterogeneous surface corresponds to the sum of the cosines of the contact angles of the two homogeneous materials, weighted by the amount of available surface. If one of the surfaces is just air, the cosine of the contact angle on this surface is –1, leading to the following equation:

$$\cos\theta_{rough} = -1 + \varphi_S(1 + \cos\theta_{smooth}) \tag{11.3}$$

where φ_s is the surface fraction of the solid. A very rough surface will have a φ_s value approaching zero and, therefore, a rough contact angle approaching 180°, accordingly, the liquid drop will theoretically lift off the solid surface.

In summary, for the equilibrium configurations of liquids on rough surfaces, if the surface has a high interfacial energy, roughness promotes wetting, and the liquid will spread within the corrugation (surface wicking). But for a low energy surface, roughness promotes repulsion; the drop does not follow the surface corrugations, but achieves its minimum at a position on top of the corrugation. This represents the basis for self-cleaning surfaces.[23]

The question of 'how can surface texture be manipulated to enhance the cleanability of hydrophilic and hydrophobic surfaces?' has been the focus of many design activities on fibrous products, particularly those implemented in the field of biotechnology in which clean surfaces are an essential requirement. For the case of hydrophilic materials, cleaning is basically a process causing contaminant film to flow. In this case, numerous factors may influence cleanability. These include[20–23] the ability of the liquid to dissolve the contaminants (solution/contaminant chemical compatibility), the ability of the liquid to adhere to the contaminants, the pore structure (pore size and its distribution) of the hydrophilic surface, the contaminant particle size and its distribution, the extent to which the capillaries are blocked by unforeseen chemical reactions and liquid characteristics. These aspects are considered in the design of many fibrous products such as filters and re-usable absorbent medical fabrics.

In the context of surface texture, it was indicated earlier that if a material has a high surface free energy, roughness promotes wetting and the liquid will flow within the corrugation. When contaminants are present, and assuming there is no chemical interaction between liquid and contaminants, any removal of contaminant particles will depend largely on the sliding pattern of the liquid against the surface. This is because contaminants should be transported along the surface for efficient cleaning. For

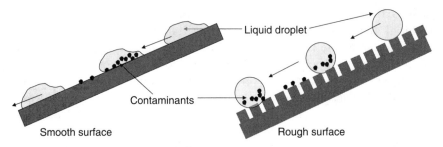

11.2 The 'lotus' effect[24,25] showing water movement against smooth and rough hydrophobic surfaces.

this, surface texture can be manipulated to accelerate the planar sliding pattern from a liquid. This can be achieved by modification of surface during spinning or by application of special finishes.

The 'lotus' effect

A fascinating phenomenon associated with hydrophobic surfaces was discovered by two botanists, Barthlott and Neinhuis from Bonn, Germany, in the course of their studies of plant leaf structures.[24,25] They noticed that the surface features of the lotus leaf, together with its waxy surface chemistry, rendered the leaf non-wettable. Indeed, the surface of the lotus leaf is one of the nature's most water-repellent surfaces. The surface has countless miniature protrusions coated with a water-repellant hydrophobic substance. As a result, water droplets form spherical globules and roll off the leaves even when they are only slightly inclined. Particles of dirt absorbed by water are removed in the process (see Fig. 11.2). Note that in comparison with human skin, another hydrophobic surface with a contact angle of about 90°, the lotus leaf exhibits a contact angle of 170°. Bird feathers, another super-hydrophobic surface, has a contact angle of about 150°.

The lotus leaf exhibits a unique porous surface texture at the micrometer scale. The air trapped in the crevices prevents water from adhering to the solid. Researchers led by H. Yildirim Erbil of Kocaeli University in Turkey[26] re-created this super-hydrophobic surface by dissolving polypropylene in a solvent and then adding a precipitating agent and applying the solution to a glass slide. After evaporating the solvent mixture in a vacuum oven, they had a highly porous gel coating at a contact angle of 160° and water-repelling capabilities comparable to those of the lotus leaf.

In a recent study, Luzinov *et al.*[27] examined a number of approaches for mimicking the behavior of the lotus leaf to create synthetic coatings with exceptional anti-wetting properties. They employed a combination of a polystyrene grafted layer (low surface energy component) and nanoparticles (roughness initiation component) and obtained a textile material that demonstrated very low wettability by water. They also attempted to create the lotus effect on a fabric surface by evaluating the deposition of both polystyrene (PS) and triblock copolymer polystyrene-b-(ethylene-co-butylene)-b-styrene (SEBS) simultaneously on a model substrate. Polystyrene was then extracted employing a selective solvent, ethyl acetate, which acted as a solvent for PS but did not dissolve SEBS. The dissolution of PS created a porous (rough) hydrophobic structure on the substrate. The controlled method of surface modification was applied to a polyester fabric which produced a practically non-wettable textile product.

On the commercial side, nanotechnology has been used to mimic the lotus effect for some fibrous products. For example, so-called Nano-Care™ for cotton, developed by Nano-Tex™ uses 'nano-whiskers' 1/1000 the size of a typical cotton fiber attached to the individual fibers. The changes to the fibers are undetectable and do not affect the natural hand and breathability of the fabric. The whiskers cause liquids to roll off the fabric. Semi-solids such as ketchup or salad dressing sit on the surface, are easily lifted off and cause minimal staining, which should be removed by laundering. In addition to the stain-resistance attributes typically provided by conventional finish treatments, Nano-Care is claimed to allow moisture to pass through the fabric (e.g. quick drying). Nano-Tex also developed Nano-Dry™ technology to provide wickability and moisture-absorption properties for nylon and polyester fabrics. This can be used for many products including high-performance outerwear.

The dynamic behavior of droplets on ultra-hydrophobic surfaces was studied in great detail by David Quéré of the Collège de France and his collaborators.[28-33] Their research indicated that the most important effect of these surfaces on liquid drops concerned the contact line of the drop, that is, the one-dimensional line of intersection of the three interfaces. Because the contact area of the drop shrinks with an increase in contact angle, the contact line can be deformed less easily, and hence the hysteresis in the contact angle between the advancing angle (θ_a, or the front angle in direction of droplet motion) and the receding value (θ_r, or the rear angle) is drastically reduced. This hysteresis is expressed by the pinning force per unit length of the drop perimeter:

$$F = \gamma_{LV}(\cos\theta_r - \cos\theta_a) \qquad (11.4)$$

This force has to be overcome by external forces (wind, gravity, etc.) to initiate droplet motion. If the hysteresis is too large and the driving force is not big enough, the liquid drop will stick or be smeared across the surface. If the contact angles are sufficiently high (>170°) viscous droplets will roll off the surfaces (not slide).

The dynamic contact angles of a drop moving down a surface are affected by the magnitude of surface roughness.[34] Surface projections of several tens of micrometers can still deform, a liquid droplet even if it is considerably larger than the projections themselves. Therefore, smaller projections are generally needed for a good ultra-hydrophobic surface.

The second most important factor affecting the dynamics of a drop on an ultra-hydrophobic surface is the velocity of impact. Typically, the impact of the liquid droplet against a surface will result in an elastic rebounding with a velocity almost equal to that of impact. This information is useful in designing repellent or drying surfaces.

The self-cleaning mechanism of ultra-hydrophobic surfaces relies on the smallness of the contact area of a drop on a surface. For the ultra-hydrophilic route to self-cleaning, the flow of the liquid film is essential. Ultra-hydrophilic surfaces are wetted easily with very low contact angle fluids; if the surface is inclined, it is the flowing liquid film that carries the contaminants along. The usefulness of this concept thus depends on the rapidity with which a liquid film runs off a surface. For sufficiently thick films (of the order of hundreds of nanometers and above), flow is hydrodynamic. For thinner films, however, the flow of the film will consist of a rapid equilibration by surface diffusion. Not all liquid will move; there will be a stagnant (solidified) layer on the microscopic scale.

11.5   Coating and lamination

The concept of fabric finish can be extended to cover a unique area that has become an independent sector of the industry owing to its critical importance particularly for function-focus fibrous products. This area is coating and lamination of fabric.[35–39] Coating is basically a surface treatment of the fabric using some form of polymeric viscous liquid, that is after drying or curing (to harden the coating) it forms a solid coat distributed evenly over the fabric surface. The idea here is not to block the inherent characteristics of the fabric, but rather to enhance them using coating materials that can provide added functions to the fibrous end product via the surface. The polymeric liquid can be a polymer melt that can be cooled for hardening and solidification, or a polymer solution, which upon cooling forms a solid film by evaporation of the solvent. Other types of coating can be applied in the liquid form and then chemically cross-linked to form a solid film.[35] Examples of coating substances and their merits are listed in Table 11.1. Note how different coating polymers provide special functions for the fibrous materials.

Lamination is basically an adhesive material combining two or more fabrics (textile laminates). Many types of adhesive can be used depending on interfacial compatibility and environmental factors. Hot-melt adhesives

Table 11.1 Examples of common coating materials and their functions[35–39]

Coating substance	Description and functions
Polyvinyl chloride (PVC)	• To be used as a coating material, it must be transformed into a soft flexible film. • The powdered PVC polymer can absorb large quantities of non-volatile organic liquids called plasticizers. It can absorb as much as its weight from cyclohexyliso-octylphthalate (a typical plasticizer). • The flexibility of the resultant film can be varied by the amount of plasticizer added. • PVC coating is resistant to acids and alkalis but organic solvents can extract the plasticizer, making the coatings more rigid and prone to cracking.[35] • Plasticized PVC forms a clear film that has good abrasion resistance and low permeability. • The film may be pigmented or filled with flame-retardant chemicals to yield low flammability. • Large dipole and high dielectric strength: this allows the coated product to be joined together by both radiofrequency and dielectric welding techniques. This factor makes it ideal for protective applications.
Polytetrafluoroethylene (PTFE)	• Expensive polymer discovered by DuPont in 1941 • Manufactured by the addition polymerization of tetrafluoroethylene • Very low surface energy (water and oil repellence) • The polymer has excellent thermal stability to up to a temperature of 250°C. • Resistant to most solvents and chemicals, but it may be etched by strong oxidizing acids (which is a feature that can be used to enhance adhesion applications).
Natural rubber	• A linear polymer of polyisoprene • Rubber emulsion can be used directly for coating, or the polymer may be coagulated and mixed at moderate temperatures with appropriate fillers.[37] • Natural rubber contains unsaturated double bonds along the polymer chain. These may be crosslinked with sulfur (i.e. vulcanization process) to give tough abrasion-resistant films or hard ebony-like structures. Cross-linking rate can also affect the rubber flexibility. • Vulcanized rubber coating is commonly used for tires and belting to provide excellent abrasion resistance.

Table 11.1 Continued

Coating substance	Description and functions
Styrene-butadiene rubber (SBR)	• Made by the emulsion polymerization of styrene and butadiene. • It is not used for tire coating because of its relatively poor resilience compared to natural rubber. • It is used for fabric coating to provide superior weather and ozone resistance. • Over 50% of all rubber used is SBR.
Polychloroprene (neoprene)	• Used as a substitute for natural rubber in many applications. • It is made during the emulsion polymerization of 2-chlorobutadiene. • It can be vulcanized and shows tensile properties similar to natural rubber. • It is famous for its excellent oil resistance, weathering and ozone resistance (belts and hoses applications).
Silicone rubber	• Silicone rubber is a unique synthetic elastomer made from a cross-linked polymer that is reinforced with silica. • It provides an excellent balance of mechanical and chemical properties required by many industrial applications. • Water resistant • Chemical resistant • Good release properties • Good adhesion • Wide operating temperature range (−50°C to +300°C possible) • Superior cold resistance (specialty silicone rubber grades can perform at temperatures as low as −85°C) • Better retention of tensile and elongation properties after heat aging • Relatively high surface friction (useful for some applications such as conveyor belts) • Resistant to weathering and UV rays • Can be made flame resistant
Polyurethanes	• Polyurethanes are made by the reaction of a diisocyanate with a diol. • Polyurethanes used for coating textiles are frequently supplied as an isocyanate-tipped prepolymer and a low molecular weight hydroxyl-tipped polyester, polyether or polyamide. • The two materials will react at room temperature although this is often accelerated by raising the temperature. • Polyurethane coatings show outstanding resistance to abrasion combined with good resistance to water and solvents, in addition they offer good flexibility.

are believed to be one of the most environmentally friendly adhesives. They are also energy efficient and produce more permanently bonded laminates at a higher speed. Polyurethane foam is still used in thin layers using so-called flame bonding. Although this process produces gaseous effluent, it typically gives a desirable bulky appearance to the final products.

11.6   References

1. GOSWAMI B C, ANANDJIWALA R D and HALL D, *Textile Sizing*, CRC Press, Boca Raton, 2004.
2. EL MOGAHZY Y and CHEWNING C, *Fiber to Yarn Manufacturing Technology*, Cotton Incorporated, Cary, NC, 2001.
3. LORD P R, *Handbook of Yarn Production, Technology, Science, and Economics*, Woodhead Publishing Limited, Cambridge, UK, 2003.
4. HEYWOOD D, *Textile Finishing*, Woodhead Publishing, Cambridge, UK, 2003.
5. STOHER K, 'Size recycling–new concept', *International Textile Bulletin*, 2002, **1**, 54–5.
6. SCHINDLER W D, *Chemical Finishing of Textiles*, Woodhead Publishing Limited, Cambridge, UK, 2004.
7. NABL W, SCHREIBER L and DIRSCHL F, 'New effects in textile finishing with innovative technologies and application of fluorochemicals', *Melliand International*, 2002, **8**, 140–3.
8. HU J, YU H, CHEN Y and ZHU M, 'Study on phase-change characteristics of PET–PEG copolymers', *Journal of Macromolecular Science, Part B*, 2006, **45** (4), 615–21.
9. MA X G and GUO H J, Studies on thermal activities of fabrics treated by polyethylene glycol, *Journal of Applied Polymer Science*, 2003, **90** (8), 2288–92.
10. BRYANT Y G, *Enhanced Thermal Energy Storage in Clothing with Impregnated Microencapsulated PCM*, SBIR Phase I Final Report, USAF Contract No. F33657-87-C-2138, February 1988.
11. BRYANT Y G, *Spacesuit Glove-Liner with Enhanced Thermal Properties for Improved Comfort*, SBIR Phase I Final Report, NASA Contract No. NAS9-18110, August 1989.
12. GROSHENS P and PAIRE C, *Thermo-adhesive Textile Product Comprising a Micro-encapsulated Cross Linking Agent*, US-Patent 4990392, Feb. 6, 1991.
13. MUCHA H, HOFER D, ABFALG S and SWEREV M, 'Antimicrobial finishes and modifications', *Mellian International*, 2002, **8**, 148–51.
14. EL MOGAHZY Y, KILINC F S and HASSAN M, 'Developments in measurements and evaluation of fabric hand, Chapter 3', *Effect of Mechanical and Physical Properties on Fabric Hand*, Behery M (editor), Woodhead Publishing Limited, 2004, 45–65.
15. SCARDINO F L, 'Surface geometry of synthetic fibers', *Surface Characteristics of Fibers and Textiles*, Schick M J (editor), Fiber Science Series, Part I, Marcel Dekker, NY, 1975, 165–91.

16. SCHICK M J, 'Friction and lubrication of synthetic fibers, Part I: Effect of guide surface roughness and speed on fiber friction', *Textile Research Journal*, 1973, **43**, 111–17.

17. GEORGE B, HUDSON S and MCCORD M G, 'Surface features of mineral-filled polypropylene filaments', *Surface Characteristics of Fibers and Textiles*, Pastore C M and Kiekens P (editors), Surfactant Science Series, Marcel Dekker, NY, 2001, Volume **94**, Chapter 6, 139–60.

18. DREXLER K E, *Engines of Creation: The Coming Era of Nanotechnology*, Anchor Books, NY, 1990.

19. SmartSilver, *NanoHorizons*, http://www.nanohorizons.com/indexNH.shtml

20. SHAFRIN E G and ZISMAN W A, *Contact Angle, Wettability and Adhesion*, Advances in Chemistry Series, Fowkes F M (editor), American Chemical Society, Washington DC, 1964, Volume **43**, 145–67.

21. WENZEL R N, 'Resistance of solid surfaces to wetting by water', *Industrial and Engineering Chemisty*, 1936, **28**, 988.

22. CASSIE A B D and BAXTER S, 'Wettability of porous surfaces', *Transactions Faraday Society*, 1944, **3**, 16.

23. BLOSSEY R, 'Self-cleaning surfaces – virtual realities', *Nature Materials*, 2003, **2**, 301–6. (www.nature.com/naturematerials)

24. BARTHLOTT W and NEINHUIS C, 'Purity of the sacred lotus, or escape from contamination in biological surfaces', *Planta*, 1997, **202**, 1.

25. VON BAEYER H C, 'The lotus effect', *The Sciences*, January/February, 2000.

26. YILDIRIM E H, LEVENT A, YONCA A and OLCAY M, 'Transformation of a simple plastic into a super-hydrophobic surface', *Science*, 2003, **299** (5611), 1377–80.

27. LUZINOV I, BROWN P, CHUMANOV G and MINKO S, *Ultrahydrophobic Fibers: Lotus Approach*, Annual Report, Project No. C04-CL06, The US National Textile Center, Delaware, USA, 2004. http://www.ntcresearch.org

28. BICO J, MARZOLIN C and QUÉRÉ D, 'Pearl drops', *Europhysics Letters* 1999, **47**, 220–26.

29. RICHARD D and QUÉRÉ D, 'Viscous drops rolling on a tilted non-wettable solid', *Europhysics Letters*, 1999, **48**, 286–91.

30. RICHARD D and QUÉRÉ D, 'Bouncing water drops', *Europhysics Letters*, 2000, **50**, 769–75.

31. AUSSILLOUS P and QUÉRÉ D, 'Liquid marbles', *Nature*, 2001, **411**, 924–27.

32. RICHARD D, CLANET C and QUÉRÉ D, 'Contact time of a bouncing drop', *Nature*, 2002, **417**, 811.

33. BICO J, TORDEUX C and QUÉRÉ D, 'Rough wetting', *Europhysics Letters*, 2001, **55**, 214–20.

34. ÖNER D and MCCARTHY T J, 'Ultra-hydrophobic surfaces. Effects of topography length scales on wettability', *Langmuir*, 2000, **16**, 7777–82.

35. HALL M E, 'Coating of technical textiles', *Handbook of Technical Textiles*, Horrocks A R and Anand S C (editors), Woodhead Publishing Limited, Cambridge, UK, 2000, 173–86.

36. BRYDSON J A, *Plastics Materials*, 7th edition, Butterworth-Heinemann, Chatswood NSW, 1999.

37. AIGBODION A I, MENON A R R and PILLAI C K S, 'Processability characteristics and physico-mechanical properties of natural rubber modified with rubber seed

oil and epoxidized rubber seed oil', *Journal of Applied Polymer Science*, 2000, **77** (7), 1413–18.

38. PELTIER G, 'Environmentally friendly dry thermoplastic adhesive films and webs for industrial bonding applications', *2nd International Conference on Textile Coating and Laminating: Assessing Environmental Concerns*, Charlotte, USA, Technomic, 1992.

39. FARRELL D, 'Reactabond – A reactive response to industry', *8th International Conference on Textile Coating and Laminating*, November 9–10, Frankfurt, Technomic, 1998.

Part III
Development and applications of
fibrous products

Part II

12

Development of traditional textile fiber products

Abstract: This chapter focuses on product development aspects associated with traditional fibrous products and discusses (a) common performance characteristics of traditional fibrous products, (b) the general relationships between performance characteristics and the attributes of fibrous components, (c) basic attributes of fabrics used for traditional fibrous products and (d) implementation of product development concepts for some traditional fibrous products. Two examples of products, denim and sportswear, that represent the largest share of the traditional market will be discussed in detail. Many aspects of the development of these two products are applicable to most traditional fibrous products.

Key words: performance-attributes relationship; structural attributes; fiber fraction; mechanical attributes; tenacity; Young's modulus; hand attributes; drape coefficient; El Mogahzy-Kilinc total hand; transfer attributes; pore size distribution; air permeability; thermal conductivity; thermal resistivity; thermal absorpitivity; thermal diffusivity; denim; comfort; durability; spandex; sportswear; hollow filaments; COOLMAX®; GORE-TEX®; thermo-physiological comfort; body-heat balance; clo units; MET; wickability.

12.1   Introduction

Throughout this book, fibrous products have been divided into two main categories: (a) traditional fibrous products (TFP) and (b) function-focus fibrous products. In this chapter, the emphasis will be on the development of traditional fibrous products. In Chapters 13 through 15, the discussion will shift to function-focus fibrous products. In order to follow the discussion in this chapter effectively, the reader should refer to Chapters 3 through 7 as the concepts presented in these chapters will be recalled in the context of the development of traditional fibrous products. Key aspects discussed in this chapter are (a) common performance characteristics of traditional fibrous products, (b) the general relationships between performance characteristics and the attributes of fibrous components, (c) basic attributes of fabrics used for traditional fibrous products and (d) implementation of product development concepts for some traditional fibrous products. In the discussion of the last aspect, two examples of products that represent the largest share of the traditional market will be discussed in great detail. These are denim products and sportswear. Concepts associated with the development of these two products are applicable to the vast majority of traditional fibrous products.

12.2 Performance characteristics of traditional fibrous products

As indicated in Chapter 3, determining and defining the performance characteristics of a product is a critical phase of product development (see Fig. 3.1, Chapter 3). Any product should be associated with performance characteristics that describe its intended function(s) and reflect the expectation(s) of the users of the product. For example, the primary performance characteristic of a raincoat is waterproof capability and that of a firefighter's uniform is flame retardance. These characteristics are greatly anticipated and naturally expected by the users of these products and any deficiency in them will be met by dissatisfaction or a total rejection of the product.

From a design viewpoint, a performance characteristic is seldom a direct attribute that can be measured and imbedded in a product in a systematic fashion to make the product perform according to its expectation. Instead, it is often a function of an appropriate product assembly leading to a combination of different attributes that collectively result in meeting the required performance. For example, suppose that the desired performance characteristic of a fibrous end product is durability. This is a characteristic that is highly recognizable, yet seldom measured by a single parameter. Similarly, comfort or esthetics represent recognized performance characteristics that require measurements of many attributes to reach their defined levels. In this regard, it is important that both the elements of the product assembly and their measurable attributes are harmonized so that their integral outcome can lead to an optimum level of the desired performance characteristics. For example, if durability is the desired performance characteristic, the selection of a fiber type exhibiting high strength will represent a key element/attribute combination. When the fibers are converted into a yarn, the new fiber assembly should still meet the same level of strength or enhance it. In this case, the new element/attribute combination to be optimized is yarn structure/yarn strength. Similarly, as the yarn is converted into a fabric, construction/strength combination of fabric should be optimized. Finally, fabric finish must be carefully selected and applied in such a way that can enhance durability, or minimize any side effects that can lead to deterioration of this critical performance characteristic.

As indicated in Chapters 8 through 10, most material attributes can easily be tested using standard techniques. These include weight, thickness, strength, thermal properties, electrical properties, and so on. Performance characteristics, on the other hand, are more difficult to test because they describe responsive behavior to specific uses or certain levels of external stimuli (environmental or mechanical). This makes it often difficult to standardize tests for performance characteristics that can reflect all applications that a product may be used for. Accordingly, they should be assessed in more simulative manner. This means that reliable simulation techniques that resemble the product performance must be incorporated into the product development cycle. Obviously, the ultimate test of a product should be based on the results of feedback from actual users about its performance. This is a critical task that engineers of fibrous products must understand and implement in order continuously to enhance the design of these products. It is important, therefore, to incorporate a 'product history' record in the product development cycle, which provides performance data derived from actual user surveys.

Most fibrous products are typically associated with two or more target performance characteristics. This makes product design more complex as the element/attribute combination suitable for one performance characteristic may not necessarily meet the desired levels of other performance characteristics. In this case, there must be a great deal of compromise. This situation is very common in many traditional and function-focus fibrous products, particularly those in which durability and comfort are required. It is for these reasons that establishing product performance characteristics and the attributes constituting this performance represents the most critical phase of product development.

In the design conceptualization phase, the performance characteristic of the intended product represents a key aspect in defining the design problem. As indicated in Chapter 4, there are two categories of design problem definition (Fig. 4.2): a broad definition and specific definitions. The broad definition of a problem should be driven by the general performance characteristics of the product. For most traditional fibrous products, general performance characteristics will include durability, comfort, esthetics, care or maintenance, and health or safety characteristics. The extent of meeting each of these characteristics will obviously depend on the type of product under consideration.

The specific definition of a design problem stems from some specifications dictated by the product user or some specialty function(s) that must be emphasized in the design process to meet the expected primary performance. As indicated in Chapter 4, a specific definition of the product can be a derivative of the broad definition that is emphasized to deal with the specific problem in hand. For example, apparel fibrous products should exhibit acceptable durability levels below which the product will not be acceptable. They should also provide comfort for the wearer. Accordingly, durability and comfort are often considered to be basic performance characteristics in the broad definition of the design problem. When an apparel product is used for specific applications (military uniform or sportswear) in which high levels of physical activity or variable environmental conditions are anticipated, specific definitions of the design problem should then be stated. These definitions should be associated with a more precise list and more specific values of performance characteristics. In this case, a more specific definition may imply the exact level of fabric strength below which the product will not be acceptable or the tolerance allowed for thermal insulation.

In light of the above discussion, both performance characteristics and their related attributes should represent the key components of the information needed to conceptualize the design. When the current state of the art warrants further efforts in design, a more specific list of the different attributes should be established in which plausible ranges of values, types of fibers leading to these values, accessibility and cost factors are clearly reported. The availability of such a list is critical prior to any design analysis as it can represent vital information in many analyses such as problem-solving, modeling and simulation, optimization, material selection and cost analysis. Since these analyses aim to reach an optimum solution to the design problem (or best performance) in terms of the effects of various attributes and their interactions, it will be important to begin any analysis by establishing a rough visual diagram of the various attributes constituting a certain performance characteristic. This diagram may be termed the 'performance–attributes diagram' as shown in Fig. 12.1. For most fibrous products, these attributes are those of the different components comprising the product, namely fibers, yarns, fabric and product assemblies. Note that material type (e.g. fiber type, yarn type and fabric type) is a common key component in the performance–attribute diagram.

With reference to Fig. 12.1, it is important to point out a fundamental difference between classic education or traditional scientific research and product development. Common education in fiber and polymer engineering typically begins by teaching students about the basic component of fibrous products, which is the fiber. Fiber courses are prerequisites for courses dealing with yarns, and both fiber and yarn courses are prerequisites for courses dealing with fabrics. This approach follows closely the arrowed solid lines in Fig. 12.1. The legitimacy of this approach stems from the fact that at this stage of learning, a pyramidal sequential approach is useful to provide students with the necessary knowledge base. In product development, the approach is quite different, as the starting point for gathering information and establishing a knowledge base is at the end of product assembly. A backward projection approach is then taken following the arrowed dotted lines in Fig. 12.1 in order to develop the end product through various manipulations of fabric construction, yarn structure and fiber parameters. In some situations, the design engineer may break the backward projection sequence by going straight to fiber parameters, as they may represent the most critical factors in determining product performance. In this regard, modeling performance characteristics in terms of

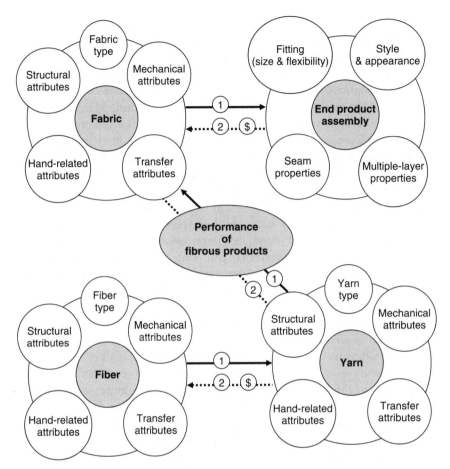

12.1 Performance–attributes diagram. (1) Classic education and scientific research, (2) product development.

these various contributing elements can greatly assist in determining the most significant factors influencing product performance. A product development backward projection should also be associated with analysis of the cost of conversion from one fiber assembly to another so that a cost-justified product can be developed.

12.3 Performance characteristics versus component attributes

From a design viewpoint, the first key components contributing to the product performance characteristic are fiber attributes. This point was clearly demonstrated in Chapters 7 and 8. In light of the discussions in these two chapters, fiber type and fiber characteristics are key design parameters in fulfilling any performance characteristic of a fibrous product. Fiber type may imply the class of fiber used or the polymer substance from which the fiber is made (e.g. natural or synthetic), the fiber form (staple or continuous filaments), fiber variety (in case of natural fibers), or a modified functional fiber using a special surface treatment or imbedded nanoparticles. Fiber characteristics are typically divided into bulk characteristics and surface characteristics, with the former reflecting the fiber integrity and durability and the latter reflecting the appearance and surface interaction between different fibers or between fibers and other objects.

When fibers are used in yarn forms, the yarn type and structural features of the yarn, discussed in Chapter 9, will represent key design components. In addition, yarn properties such as bulk characteristics (e.g. strength, yarn compactness, etc.) and surface-related parameters (e.g. abrasion resistance, surface irregularities and hairiness) can contribute greatly to the overall performance of the fabric made from the yarn. When fibers are bonded or compressed directly into nonwoven structures, the performance characteristics of the end product will be directly influenced by fiber type and fiber properties as well as by the bonding mechanism utilized to form the fabric, as discussed in Chapter 10.

Fabric should be considered to be the quasi fibrous end product, as in most applications it is the fabric construction and fabric attributes that directly reflect the performance of the fibrous end products, provided that assembling the fabric into a final product (e.g. apparel or furnishing) did not in some way prohibit the exploration of fabric performance. In Chapter 10, some of the basic fabric constructions (e.g. woven, knit and nonwovens) and their design-related features were reviewed. In the context of performance characteristics, it is important to point out that these assemblies should not be treated as merely fiber aggregates but rather as critical structures that can add significantly to the desired performance characteristic of the fibrous end product. Finally, since a fibrous product must be assembled in such a way that can directly serve the intended application of the product (e.g. apparel, furnishing, etc.), it will be important to consider the parameters associated with the final end product assembly. These include fitting or assembly flexibility, seaming integrity, appearance and style.

In the decision-making process about what elements to select and what attributes to emphasize, two key analyses should be made: (a) developing performance attributes relationships and (b) comparative analysis of different materials and attributes.

12.3.1 Performance–attributes relationships

As indicated above, the performance characteristics of any product are a function of carefully assembled elements leading to an end product, associated with a combination of different attributes that collectively result in meeting the required performance of the product assembly. Accordingly, the design of a fibrous product should begin by establishing the different elements making the product assembly and the attributes that are relevant to the desired performance characteristic(s). This point was illustrated earlier by the performance–attributes diagram. Examples of this diagram for basic performance characteristics of traditional fibrous products, namely durability, comfort, esthetic appeal, maintenance and health or safety characteristics, are shown in Fig. 12.2 through 12.6, respectively. These diagrams illustrate the common attributes related to each performance characteristic. As more specific performance characteristics are established for particular products, performance–attributes diagrams should follow closely the desired specifications through more in-depth

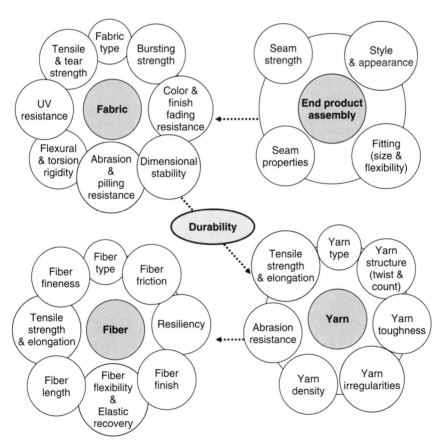

12.2 Examples of key factors contributing to the durability of fibrous products.

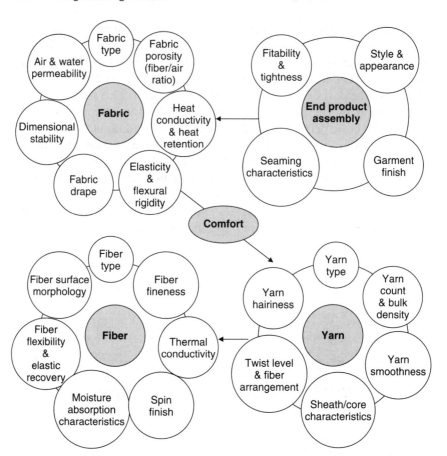

12.3 Examples of key factors contributing to fabric comfort.

analysis of the interrelationships between fiber, yarn and fabric attributes.

Ideally, the performance–attributes diagram should lead to the development of a relationship between the desired performance characteristic and the important attributes influencing this characteristic. Analytical tools that are useful in developing these relationships are as follows:

- developing simple relationships between the desired performance characteristic and one or more attributes using the one-factor-at-a-time approach in which all factors are fixed at constant levels except the factor (or attribute) under consideration;
- developing a multiple-variable relationship (multi-regression model) between the desired performance characteristic and one or more attribute;

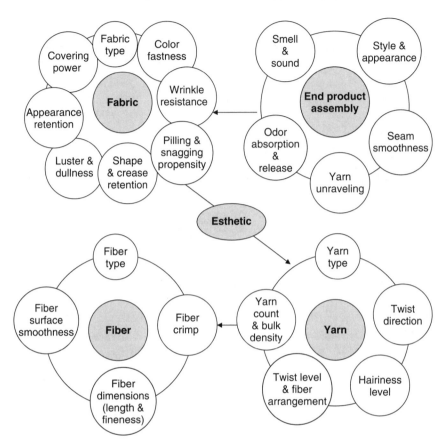

12.4 Examples of key factors contributing to fabric esthetic performance.

- performing some useful statistical techniques such as factorial design and analysis of variance to reveal the most critical attributes influencing the desired performance characteristic and the possible interaction between some attributes.

In practice, most relationships are developed using the first tool. This is due to the complexity and the high cost of performing multiple-variable relationships in a practical environment. In a fiber-to-fabric system, it is often difficult to develop performance–attribute relationships because of the discrete nature of the process of fiber to fabric conversion. For example, if one is interested in developing a relationship between a measure of fabric durability (say, fabric tear strength) and fiber attributes such as fiber strength, fiber fineness and fiber flexibility, many obstacles will be faced including the non-linear nature of the conversion process from fibers to

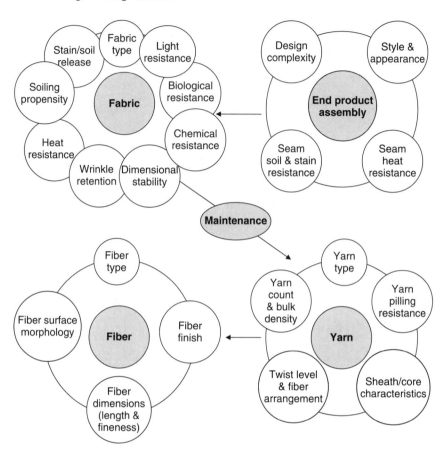

12.5 Examples of key factors contributing to fabric maintenance performance.

fabrics, the multiple operations involved in the conversion process and the multiple factors associated with this conversion. These factors make it virtually impossible to develop a reliable direct relationship between fibers and fabric characteristics. One way to overcome this difficulty is to develop incremental relationships: first, fiber-to-yarn relationships, then yarn-to-fabric relationships. Since limitations, such as the difficulty of using a wide plausible range of values and the need for many disruptive experimental trials will inevitably exist in developing these relationships, it will be important for the product developer to rely on research institutes to develop these relationships, as there are many creative ways to overcome these limitations. In recent years, many advanced techniques have been developed to allow full implementation of multiple-variable analysis in practical situations. Some of these techniques were briefly reviewed in Chapter 6 and

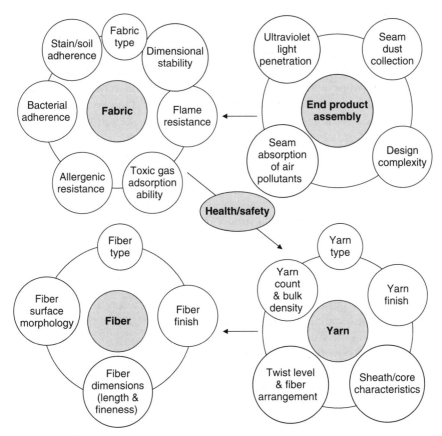

12.6 Examples of key factors contributing to fabric health or safety performance.

more details about how to implement them can be found in numerous books including one written by the present author.[1]

12.3.2 Comparative analysis of different materials and attributes

In the development of traditional fibrous products, it is often necessary to perform a comparative analysis of different materials that can potentially be used for the desired product and its attributes. When the development process aims to modify a pre-existing product, this task is normally simple because this comparison can be obtained from historical data collected in the information gathering phase of product development. In the development of new products, it will be important to perform a comparative analysis using model samples of the anticipated product elements (fibers, yarns and fabrics) to understand the relative merits and demerits of each element or of some of its attributes. This type of analysis can be time-consuming and cost-prohibitive. In this regard, cooperation between product developers and research institutes will again represent the most economical approach to conducting the analysis. However, researchers should be aware of a number of key issues regarding dealing with product development projects. These are pointed out below.

In the research environment, comparative analysis is commonly made by producing samples that are largely similar in all attributes except the one under examination. For example, a ring-spun yarn may be compared with an open-end spun yarn made of the same fiber type, yarn count and yarn twist to determine which spun yarn has better characteristics. Similarly, a plain-weave fabric may be compared with a twill-weave fabric, in order to determine the effect of fabric construction on a certain characteristic, under conditions in which fiber type, yarn type, yarn structure, fabric weight, fabric thickness and fabric density are more or less similar.

In the midst of making a fair comparison, many factors are often forced to be fixed or remain constant to such an extent that the samples made for the purpose of the study become infeasible practically or misrepresentative of the true products. In addition, some factors are often taken to levels beyond their practical limits in order to establish the extreme conditions that are useful for modeling some phenomena. For example, a researcher may use levels of twist in spun yarns that are far below or far above the optimum twist levels to compare the effect of twist on the values of strength or stiffness of two different yarns. Another researcher may produce a laboratory sample of knit fabric that has fabric count (or fabric density) exceeding the practical boundaries of the fabric, even to the extent of fabric collapse, so that the effect of fabric count on some performance characteristics can be modeled. Although these are rare research practices, the author's experience suggests that they do occur particularly among researchers with limited practical experience.

When research activities are devoted to serve product development applications, it will be important to adhere to the practical limits at which model samples can be produced or compared. In other words, any research work related to product development should be geared toward the feasibility and the practicality of the product performance as determined by actual manufacturing requirements and desired specifications and not by some research speculations that may be useful for developing a theory, but not for developing a product. In the context of comparative analysis, when two fabric samples are compared for the sake of selecting the optimum performance characteristic, each sample should be produced under its own optimum conditions and in such a way that preserves the integrity and the identity of the fabric sample. For example, some sportswear may be made from woven fabrics or knit fabrics. In this case, each fabric type should be developed and produced under practical conditions that lead to optimum levels of the desired performance characteristics of the intended product using the particular construction and the required technology suitable for the fabric sample. Comparison between the two fabrics will be justifiable from a product development viewpoint despite the inherent differences in the two fabrics.

12.4 Typical values of fabric attributes commonly considered in product development

In Chapter 10, a brief review of many of the key attributes of different fabric types was presented. In relation to product development, these attributes can be divided into four main categories: (1) structural attributes, (2) mechanical attributes, (3) hand-related attributes and (4) transfer attributes. These attributes are reviewed below with the aid of some of their typical values obtained for three woven constructions, namely plain, twill and satin, and three knit constructions, namely single jersey, pique and interlock. These values are shown in Table 12.1. The idea of reporting the values of these attributes is not to make a direct comparison between different fabrics, but rather to demonstrate the differences in their design capabilities. It should also be pointed out that the values reported in Table 12.1 are certainly non-exclusive as they only represent typical values of these fabric types. Indeed, within each fabric type a wide range of values can be found, limited only by the constructional nature of the fabric and some attributive boundaries.

12.4.1 Structural attributes

As indicated in Chapter 10, the basic structural attributes of a certain fabric construction (weave or knit pattern) are fabric thickness, fabric weight and fabric density (or fabric count). These attributes are largely determined by three main parameters: yarn count, yarn twist and the technology involved. Different combinations of values of these parameters can result in different values of fabric structural attributes as demonstrated in Table 12.1. In general, different fabric styles (the technology factor) require different combinations of values of structural attributes. For example, twill fabric used for denim is made from a heavy and thick fabric resulting from a very coarse yarn count at a medium or high level of fabric count. On the other hand, lightweight full dress satin apparel may be made from thin, lightweight and dense fabric using a fine yarn count.

Table 12.1 Typical values of different performance attributes of common woven and knit fabrics[2]

Attribute category	Attribute value	Denim	General apparel	Fashionable apparel	General dress shirts	General dress shirts	T-shirts & lightweight tops
		Twill 3/1	Plain	Satin 5	Pique	Jersey	Interlock
Fiber attributes	Fiber type	Cotton	Cotton	Cotton	Cotton	Cotton	Cotton
	Fiber length (mm)	28	32	34	32	32	34
	Fiber Fineness (mtex)	175	180	170	185	185	170
	Fiber strength (g tex^{-1})	29	32	34	30	30	32
Structural attributes	Yarn count (English count)						
	Warp	6s	35s	45s	24s	24s	40s
	Filling	20s	30s	40s			
	Yarn twist factor	4.0	3.8	3.6	3.2	3.2	3.4
	Fabric count (L + W)	140	140	220	70	80	80
	Fabric balance (L/W)	1.2	1.1	1.9	0.7	0.8	1.0
	Fabric thickness (mm)	0.7	0.3	0.35	0.8	0.6	0.9
	Fabric weight (g m^{-2})	300	110	120	200	160	180
	Fabric specific volume (cm^3 g^{-1})	2.33	2.73	2.92	4.0	3.75	5.0
	Fiber fraction	0.28	0.24	0.22	0.16	0.17	0.13
Mechanical attributes	Fabric tenacity (g tex^{-1})	11.0	6.0	9.0	3.0	3.0	3.0
	Fabric elongation (%)	6.0	12.0	10.0	60	90	120
	Young's modulus (g tex^{-1})	33.0	11.0	25.0	4.0	1.0	2.0

Hand-related attributes	Drape coefficient	0.60	0.5	0.55	0.2	0.18	0.4
	Stiffness area (N s)	3.3	0.25	0.4	0.4	0.25	0.4
	Stiffness max load (N)	5.6	0.5	0.8	0.8	0.5	0.8
	Fabric/sand friction:						
	Friction parameters $a(n)$:						
	Speed: 0.125″/min		5.0 (0.5)	4.0 (0.5)	12.0 (0.1)	13 (0.1)	15.0 (0.1)
	Speed: 0.250″/min		2.0 (0.7)	1.0 (0.9)	7.0 (0.4)	7.0 (0.5)	8.0 (0.6)
	Speed: 1.000″/min		3.0 (0.6)	3.0 (0.6)	8.0 (0.3)	8.0 (0.4)	12.0 (0.4)
	Total hand (kg_f s)	1.04	0.11	0.15	0.15	0.12	0.18
Transfer attributes	Mean pore diameter (μm)	20	65	45	30	30	22
	Air permeability	2 (0.01)	130 (0.66)	100 (0.508)	250 (1.27)	150 (0.762)	370 (1.88)
	(Air ft^3 ft^{-2} min^{-1})						
	(Air m^3 m^{-2} s^{-1})						
	Thermal resistivity (m^2 KW^{-1})	0.012	0.006	0.006	0.014	0.011	0.017

In the development of traditional fibrous products, the levels of structural attributes should be optimized in view of required levels of key performance characteristics such as durability and comfort. In general, heavier, thicker and denser fabrics (e.g. apparel bottoms, working overalls, military uniforms, curtains and furnishing products) are likely to be stronger and stiffer than lighter, thinner and open fabrics (e.g. apparel tops, underwear, some sportswear and some bed sheets). As comfort is a two-fold performance characteristic (tactile and thermal) it requires more careful evaluation of the effects of fabric structural attributes. For tactile comfort, a thinner, lighter and more open fabric may be appropriate. This is largely true for woven fabrics in which thicker fabrics are also heavier and denser than thinner fabrics (e.g. plain and satin). For knit fabrics, thickness and weight can be independent, especially considering the variety of different knit patterns. For example, the data in Table 12.1 indicate that interlock, which is thicker than piqué fabric is also lighter, despite being denser. This difficult combination was made possible by virtue of the fabric pattern and the use of finer yarns in the interlock pattern. When softness under lateral compression (hand or finger pressing against fabric surface) is considered, knit fabrics are likely to offer softer structure by virtue of their high thickness and low density combinations in comparison with those of woven fabrics. When thermal comfort is of concern, fabric thickness and fabric density will represent key structural attributes owing to their effects on thermal insulation by virtue of entrapping still air in the internal structure.

The basic structural attributes discussed above can be measured using standard techniques. Other structural attributes that can have a direct impact on the performance of traditional fibrous products include[2] cover factor, fabric specific volume and fiber fraction. These are essentially derivatives of the basic attributes and they are normally estimated, not measured. The term cover factor was introduced earlier in Chapter 10. It is an index of the area covered with fibers with respect to that covered by air in the fabric plane. Fabric specific volume is an index of bulkiness or yarn compactness in the fabric in a three-dimension geometry expressed by the equation:

$$v_{\text{fabric}} = \frac{t}{W} \, \text{m}^3 \, \text{g}^{-1} \tag{12.1}$$

where t is fabric thickness (m) and W is fabric weigh (g m^{-2}). Given the fact that fabric thickness is highly sensitive to the pressure applied on the fabric during testing, it is important to specify whether the fabric thickness was measured under a relaxed and natural state or under some level of lateral pressure. As can be seen in Table 12.1, knit structures are generally characterized by significantly higher specific volume (more fluffiness) than woven fabrics. Indeed, this attribute clearly distinguishes woven fabrics from knit fabrics.

Fiber fraction is obtained using a similar equation to that of the packing fraction of fiber in a yarn discussed in Chapter 9, after involving a correction factor for the potential still air based on yarn count and fabric count.[2] Fiber fraction provides an estimate of fiber/air weight ratio in a fabric, which is a key parameter particularly in relation to air flow and thermal insulation attributes. This requires knowledge of the values of the specific volume of both the fabric and the fibers from which the fabric is made as indicated by the:[2]

$$\text{fiber fraction, FF} = \frac{\varepsilon v_f}{v_{\text{fab}}} \qquad (12.2)$$

where v_f is fiber specific volume (e.g. about 0.65 cm^3 g^{-1} for cotton and 0.75 cm^3 g^{-1} for polyester), v_{fab} is the specific volume of fabric and ε is a correction factor accounting for yarn compactness and yarn packing fraction.[2]

The above expression yields a fraction which, if multiplied by 100, will roughly yield the percentage of fiber content in the fabric structure. The values of fiber fraction listed in Table 12.1 clearly demonstrate how fiber fraction, FF, provides a unique structural feature that clearly segregates woven fabrics from knit fabrics. Values of FF for woven fabrics ranged from 22 to 28%, while those for knit fabrics ranged from 13 to 16%. This difference represents one of the reasons for the dominant market share of knit fabrics in the apparel market today.

12.4.2 Mechanical attributes

During wear, fabric may be subjected to tension and extension, the degree of which will vary depending on the type of physical activities performed by the wearer. It is important, therefore to understand the tensile behavior of fabric. Commonly, fabric tenacity is determined by the equation:[3]

$$\text{tenacity} = \frac{\text{breaking load } (g_f)}{\text{area density}(g\,m^{-2}) \text{ or } (\text{tex}\,mm^{-1}) \times \text{width (mm)}} (g\,\text{tex}^{-1}) \quad (12.3)$$

In Table 12.1, tenacity and elongation values for the different fabrics are listed. These values largely support the common view that woven fabrics are generally stronger than knit fabrics. On the other hand, knit fabrics exhibit a substantially higher elongation or stretchability than woven fabrics. These differences have a direct impact on the levels of flexibility of these two categories of fabric, with woven fabrics being substantially stiffer under tension than knit fabrics, as indicated by the Young's modulus values in Table 12.1. The lower strength and stiffness values of knit products should not be considered as a source of inferiority of this category of fabric in comparison with woven fabrics. Indeed, they are deliberately induced, as knit fibrous products are typically used in a relaxed state and their softness as well as flexibility represent key desired characteristics in relation to comfort and hand. Most knit fabrics are stretchable and flexible under tension by virtue of their open and low dense construction. On the other hand, woven fabrics are stronger by virtue of their close and dense structure. Among the woven fabrics, twill weave yields the highest tensile stiffness and plain weave yields the lowest. Among knit fabrics, pique and interlock fabrics are generally stiffer than comparable single jersey fabrics.

Other mechanical attributes that are commonly tested for in traditional fabrics include tear strength, bursting strength, fabric abrasion, fabric pilling, fabric skew and shrinkage. These attributes will be discussed later in the context of sportswear durability.

12.4.3 Hand-related attributes

As indicated in Chapter 11, the most important parameters constituting fabric hand are fabric drape, bending stiffness and surface roughness (softness). Fabric drape is the term used to describe the way a fabric hangs under its own weight. It has an important bearing on how good a garment looks in use. In general, the draping qualities required from a fabric will differ depending on its end use; therefore a given value for drape cannot be classified as either good or bad. For example, in the context of hand and comfort, the importance of drape stems from the need for garments to follow easily the body contours. Knitted fabrics are relatively floppy and garments made from them will tend to follow the body contours. Many woven fabrics are relatively stiff when compared with knitted fabrics so that they are used in tailored clothing where the fabric hangs away from the body and disguises its contours. Accordingly, one should expect to distinguish drape behavior between knit and woven fabrics. Measurement of a fabric's drape should be capable of providing quantitative values as well as indications of the ability to hang in graceful curves.

The drape coefficients listed in Table 12.1 were measured using a circular specimen about 10 inch (0.254 m) diameter, supported on a circular disk about 5 inch (0.127 m) diameter to allow the unsupported area drapes over the edge. Since, fabrics will typically assume some folded (double-curvature) configuration, the shape of the projected area will not be circular and a drape coefficient is obtained from the equation:[2,3]

$$C_d = \frac{A_s - A_d}{A_D - A_d} \qquad (12.4)$$

The parameters presented in Equation (12.3) are illustrated in Fig. 12.7. According to the above equation, the higher the drape coefficient, the lower the fabric drapeability, or the lower the propensity to drape. The values in Table 12.1 and Fig. 12.7 reflect the expected distinguishing difference in drape behavior between woven and knit fabrics discussed above. As can be seen in this figure, knit fabrics exhibited a lower drape coefficient or a higher propensity to drape than woven fabrics. Among the knit fabrics, single jersey exhibited the highest propensity to drape (lowest drape coefficient) and interlock exhibited the lowest propensity to drape (highest drape coefficient). Among the woven fabrics, plain weave exhibited the highest propensity to drape (lowest drape coefficient) and twill exhibited the lowest propensity to drape (highest drape coefficient). Key design factors that can influence fabric drape include[2] fiber stiffness, yarn flexural rigidity, fabric thickness, fabric weight, fabric count and yarn count.

Values of fabric bending stiffness are also listed in Table 12.1. The method used to obtain these values was the D 4032–94 Standard Test Method for stiffness of fabric by the circular bend procedure. Two measures of stiffness used in this method are maximum stiffness load (Newton) and the area (N s) under the resistance force–time diagram (stiffness profile). These values reflect the ease of deformation under bending, which

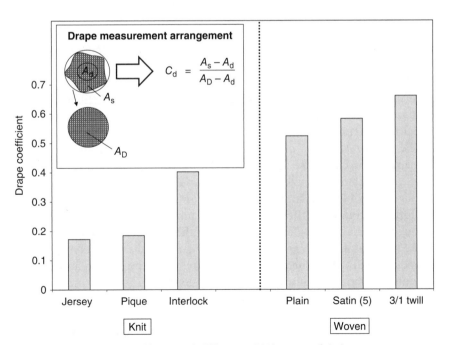

12.7 Drape coefficient of different 100% cotton fabrics.

is a critical tactile comfort characteristic. The data in Table 12.1 indicate that different levels of stiffness or flexibility can be produced within each category of fabric. In addition, a positive correlation exists between stiffness parameters and the drape coefficient values.

Surface roughness, or smoothness, is expected to vary in accordance with fabric construction, which directly influences surface texture. One common way to measure surface roughness is to examine the frictional characteristics of fabric. The most common frictional parameter is the coefficient of friction, μ, determined from the classical law of friction, $F_A = \mu P$ (where F_A is the frictional force per area and P is the lateral pressure). This law typically assumes that the coefficient of friction, μ, is constant at all levels of lateral pressure and is independent of the area of contact. This assumption has been questioned in much of the previous literature[4-6] and it was generally found to be inappropriate for materials deforming elastically or viscoelastically under lateral pressure. Fibers typically deform viscoelastically under lateral pressure. When the fibers are formed into fibrous structures or assemblies, the assumption of viscoelastic deformation continues to hold. For this behavior, an alternative formula that was empirically found in many studies is in the following form:[6]

$$F_A = a P^n \qquad (12.5)$$

where a and n are called friction parameters.[4] As can be seen in Table 12.1, knit fabrics generally exhibit higher a values and lower n values than woven fabrics at different sliding speeds and when fabrics are rubbed against a reference surface such as fine sand. In general, the parameter a largely resembles the classic coefficient of friction, μ, but it also depends on surface roughness and the true area of contact.[5] The parameter, n, on the other hand depends on the deformational behavior of the fiber assembly at the points of contact. Softer and easily deformed assemblies are likely to yield lower n values. In addition, external conditions such as sliding speed also affect the frictional parameters.

The above attributes of fabric hand were integrated in an innovative testing method called the 'Elmogahzy–Kilinc total hand method'.[2] In principle, the method is based on handling-by-pulling a circular fabric sample through a flexible light funnel (e.g. Teflon®-plastic funnels) and measuring the forces encountered through the entire handling process. Sample pulling through the contoured flexible surface of the light funnel induces anticipated hand modes such as drape, stretching, internal sample compression, lateral pressure and surface friction. During testing, these modes are encountered both simultaneously and sequentially as shown in Fig. 12.8. In addition, the funnel media allows constrained and unconstrained fabric folding or unfolding. The details of this method were described in another book.[7] The output of this method is a hand profile characterized by four distinguished areas associated with four main hand modes, as shown in Fig. 12.8. This allows characterization of hand energy values associated with different hand modes. In addition, a total hand energy index can be obtained by measuring the area under the entire hand profile. Values of this total hand index are reported in Table 12.1 and hand profiles for some of the woven and knitted fabrics listed in the table are shown in Fig. 12.9 and 12.10, respectively. These results clearly indicate the differences in hand behavior between the different fabric types as a result of the differences in their structural characteristics.

From a design viewpoint, the hand-related attributes discussed above collectively reflect the tactile performance characteristic of fabric. Design parameters that can be considered in reaching an optimum tactile comfort can be numerous, beginning with geometrical and structural parameters of fabric and extending to mechanical and surface parameters. This adds a great deal of complexity to the task of designing fibrous products for tactile comfort. The availability of a single hand index should significantly facilitate the design analysis since it can reduce the analysis from multiple-factor analysis (drape, stiffness and roughness), to a single-factor analysis (total hand). Indeed, the analysis performed by El Mogahzy and

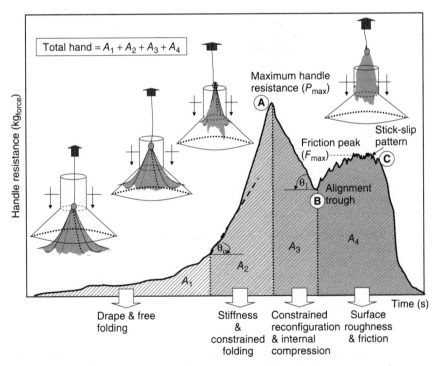

12.8 The El Mogahzy–Kilinc fabric hand profile.[2,7]

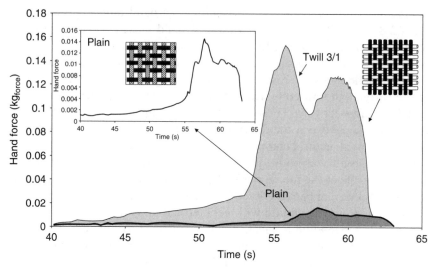

12.9 Hand profiles of twill and plain weaves.[7]

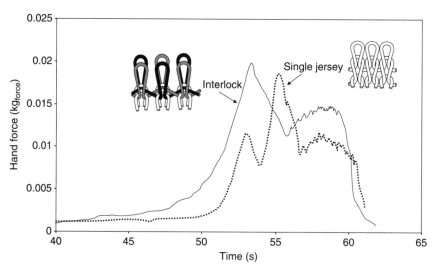

12.10 Hand profiles of jersey and interlock knits.[7]

co-workers[7] revealed significant correlations between the total hand measure and different hand-related attributes.

12.4.4 Transfer attributes

Transfer attributes are those characterizing fluid and heat transfer through fabrics. These are expected to be directly influenced by structural attributes such as fabric construction, fabric specific volume and fiber fraction. Another critical parameter that can play a vital role in relation to transfer properties is the pore size of the fabric. Fabric pores are the minute openings in the fabric structure. The existence of pores in the fabric structure is a natural consequence of the method of fabric formation. Pores can be controlled in size and number through many design options including[2] fabric structure and style, geometrical features within a given fabric structure, yarn structure, fiber properties and fabric finish.

Pore size and pore size distribution are essential parameters for determining the performance of many traditional and function-focus fibrous products. For instance, fabric wicking, one key moisture transfer phenomenon, is essentially a capillarity mechanism in which wicking can be visualized as a spontaneous displacement of a fiber–air interface with a fiber–liquid interface in a capillary system. In this regard, the height of the fluid column, h, is a critical parameter characterizing the capillary effect of fibrous assemblies.[8,9] This height is inversely related to the pore diameter. The porous structure of the fabric is also a key factor in providing flexibility, as the more porous structure is likely to produce a more flexible fabric.

The fact that the pores in fibrous assemblies are not typically uniform in size necessitates analysis of the shape, size and frequency of pores in the fabric structure.[10,11] One approach to determine pore size is by determining the relative amount of air or fluid flow through the fabric gaps.[12,13] This method is very useful in simulating the effect of pores in critical phenomena such as filtration and moisture or air transfer. However, it provides insufficient exploration of the true porous structure (size and distribution). The data on pore size reported in Table 12.1 was obtained through analysis of microscopic images of fabric. This approach required precise microscopic images (transmitted) of the fabric to be obtained and the images to be analyzed to determine pore size and pore size distribution. As can be seen in Table 12.1, woven fabrics generally exhibit larger pore size than knit fabrics. This can be attributed to the more three-dimensional nature of knit fabrics which makes it difficult to see clear pore dimensions through the fabric. It was also interesting to see that both twill weave and interlock fabrics exhibited approximately similar geometrical pore size and they represent the fabrics with the lowest pore size of both categories. This is a direct result of the closely packed structures of these two fabrics.

In addition to pore size, other transfer attributes include air permeability and thermal resistance. Values of these attributes are listed in Table 12.1. The term air permeability refers to the measured volume of air in cubic feet that flows through 1 square foot of cloth in 1 minute (0.002 m^2 s^{-1}) at a given pressure. In the context of apparel comfort, air flow through fabric is critical in two aspects: breathability and thermal insulation. A typical apparel fabric should have the capability to transfer air for ventilation and freshening purposes. In connection with thermal insulation, air is the most insulative material (thermal conductivity of air is 0.025 W mK^{-1}). This means that for fabric to exhibit good thermal insulation, it should have the ability to entrap air in its internal structure. A fabric that has low resistance to air flow (high air permeability) is likely to be a conductive fabric.

The values of air permeability shown in Table 12.1 indicate that knit fabrics generally exhibit higher air permeability (or low resistance to air flow) than woven fabrics. This expected difference can be attributed to the greater air/fiber ratio in knit fabrics in comparison with woven fabrics. Since air permeability is measured under pressure, this results in external air displacing existing air (and not hindered by solid fibers) in the internal structure of knit fabric leading to higher air permeability.

The heat flow through fabrics can be described using many measures including thermal conductivity, thermal resistivity, thermal absorpitivity and thermal diffusivity. The general heat flow equation is as follows:[14-16]

$$Q = \lambda\left(\frac{(T_1 - T_0)}{h}\right) \qquad (12.6)$$

where Q is heat flow (W m^{-2}), λ is thermal conductivity (W mK^{-1}), T_1 is heat source temperature (K), T_0 is fabric temperature (K) and h is fabric thickness (m).

As indicated in Chapter 8, thermal conductivity is a material property which expresses the heat flux Q (W m^{-2}) that will flow through the material if a certain temperature gradient ($T_1 - T_0$) exists over the material. In other words, the thermal conductivity, λ, is the quantity of heat transmitted, owing to unit temperature gradient, in unit time under steady conditions in a direction normal to a surface of unit area, when the heat transfer is dependent only on the temperature gradient. Since fabric consists of both fibers and air, the thermal conductivity of a fabric structure should be determined by this combination:[14]

$$\text{fabric thermal conductivity} = \lambda_{\text{air}}(1-f) + f \cdot \lambda_{\text{fiber}} \qquad (12.7)$$

where f is the fraction by volume of the fabric taken by fiber. This equation can clearly make use of the expression of fiber fraction given by equation (12.2).

Another measure closely related to thermal conductivity is thermal resistivity, which is a measure of the insulation of the material. It is defined as the temperature difference between the two faces of the sample divided by the heat flux:[14]

$$R_t = \frac{h_t}{\lambda} \qquad (12.8)$$

where R_t is thermal resistance (m²K W⁻¹), h_t is total thickness of the fabric (m) and λ is thermal conductivity (W mK⁻¹).

Values of thermal resistance for the different fabrics are listed in Table 12.1. These values clearly reveal the effects of fabric construction and structural factors on thermal resistivity. It should be pointed out that unlike other homogeneous solid structures, fibrous structures typically exhibit a complex thermal behavior. In the case of homogeneous solid material, and for a given material type, thermal resistance is predominantly influenced by the thickness of the material. This is because the thicker the material, the higher the material bulk density (or more material is involved). Accordingly, if the material has a low thermal conductivity, the thicker the material, the more insulative it will be. In the case of fibrous structures, thicker material does not necessarily mean higher thermal resistance. This is because a fiber structure typically consists of fibers and air and in apparel fabrics the air content can be substantially greater than the fiber content. However, the actual existence of air depends largely on the ability of the fabric structure to entrap the air inside. The more air entrapped inside the fabric structure or the lower the propensity of air to escape the fabric, the higher the thermal insulation or the lower the chance for human body to lose heat by air convection.

12.5   Implementation of product development concepts for some traditional fibrous products

In this section, product development concepts will be discussed using two major traditional fibrous products, namely denim and sportswear. These products were selected on the basis of their huge market volumes, as each belongs to a multi-billion dollar industry worldwide. They also satisfy the basic performance characteristics of most traditional fibrous products, namely durability, comfort, esthetic appeal, care or maintenance, and health or safety characteristics. In addition, some sportswear products truly represent a transition from traditional fibrous products to function-focus products, as will be discussed later in this chapter.

12.6   Denim products

Among all traditional fibrous products, no other product in history has been more widely accepted and more widely used by people of all cultures, genders, ages and origins than denim fabric, which obtains its authenticity from interweaving an indigo (blue) warp yarn and white filling threads in a twill structure. The word denim is an Americanization of the French name 'serge de Nîmes,' a fabric which originated in Nîmes, France during the Middle Ages. The brief term 'denim' was adopted by the Webster's dictionary as the English version of denim in 1864. Other related terms are jeans, dungarees or Levi's. Jeans came from cotton trousers that were worn by Italian sailors from Genoa; the term is again French, as the French call Genoa and the people who live there 'Genes.' Dungarees came from the word 'dungrí' in Hindi which means coarse cloth. This term is typically used to refer to blue denim fabric, or to trousers made from them. The term Levi refers to Levi Strauss, one of the most famous brands in the denim business.

The story of jeans or blue jeans reflects a true example of how the user of a product can indeed drive the product development process. The initial development of jeans was in two styles: indigo blue and brown cotton 'duck', which was a heavy plain weave fabric. As cotton duck was heavy and uncomfortable it was eventually terminated and replaced by denim fabric. In 1873, two American immigrants, Jacob Davis and Levi Strauss patented the so-called 'waist overalls' or work pants, the initial name for jeans. Jacob Davis, a tailor, was one of Levi's many customers who regularly purchased bolts of cloth from the wholesale house of Levi Strauss & Co. Some of Jacob's customers complained about the weakness of pockets of the pants and their continuous ripping. In an attempt to strengthen the men's trousers, Jacob came up with the idea of putting metal rivets at the pocket corners and at the base of the button fly. This development was an instant success with Jacob's customers. In order to protect the idea, Jacob patented it with his business partner Levi Strauss on May 20, 1873 (US Patent and Trademark, patent no. 139.121). This was the birthday of blue jeans. In a latter development, metal rivets were eventually replaced by reinforced stitching.

Today, denim is a multi-billion-dollar business and a multi-category apparel segment. Indeed, one can find denim products at every retail channel and at price points ranging from under US$10 to hundreds or thousands of dollars. Denim fabric can be made into a wide range of fibrous products including pants, skirts, shirts, shorts, jackets, hats, bags, upholstery, wall coverings and bed sheets. As a result, denim fabrics can be made in many weights (as determined by yarn count and fabric density, warp per inch and filling per inch). A typical range of denim fabric weight is from 5 oz yd^{-2} to 16 oz yd^{-2} (167–534 g m^{-2}). Obviously, light denim will be suitable for dresses or shirts where drape, softness and flexibility are required. Heavy denims in the range from 10 oz yd^{-2} to 15 oz yd^{-2} (333–500 g m^{-2}) are typically used for blue jeans pants and skirts. The majority of denim fabric is made from cotton fibers, but a considerable amount is made from cotton blended with other fibers at different blend ratios. Typically, most yarns used for denim fabric are open-end or rotor spun yarns, but a significant amount is made from ring-spun or compact yarns. A combination of ring-spun and open-end spun yarns may also be used in the same denim fabric.

Denim fabric is commonly treated with a variety of finishes to enhance its performance. Unlike most traditional woven fabrics, denim fabric finish begins in the yarn form. The warp yarn is dyed prior to weaving by removing individual strands of yarn from yarn packages during the warping process and prior to being gathered into a rope form suitable for dyeing. In the dye range, the yarn rope typically goes through scour/sulfur treatment, wash boxes, indigo dye vats, over a 'skying' device (to allow oxidation to occur), through additional wash boxes, over drying cans and then is coiled into tubs which are transferred to a beaming process. This process separates the dyed yarn into individual parallel strands and winds them onto a large section beam in preparation for sizing or slashing, which involves coating the yarn with a starch/wax solution and winding the yarn onto a loom beam. Weft yarns join the warp yarns in the weaving process.

In the fabric form, denim is commonly treated with a variety of finishes depending on the performance characteristics required. Some fabric may be brushed and singed prior to chemical treatment. During finishing, the fabric is normally pulled to a predetermined width, skewed, dried and rolled for the next process. Denim fabric may be categorized as dry denim or washed denim. Most denim fabric is washed after being crafted into a clothing article in order to make it softer and to eliminate any shrinkage which could cause an item not to fit after the wearer washes it. Prewashing will completely remove the size material from the warp yarns, resulting in softer fabric. Following washing, some denim is treated with abrasive or deformational applications to create a worn-in appearance for the denim garment, which is a desirable feature of most fashionable denim. Dry denim does not go through this operation. Instead, it is left to the wearer to apply repeated washing during normal use until it exhibits a natural color fade.

12.6.1 Basic tasks in denim fabric development

Figure 12.11 shows the general tasks involved in developing denim fabrics. This follows the procedures discussed earlier in Chapter 3. The first task of product development is to generate an idea for a denim product. In recent years, the market has been flooded with numerous denim products as a result of extensive developments both in the intermediate processes and in the fashion and style area. As a result the market has evolved a great deal with many transitions from old to new ideas dictated by the fashion business and customer interests. Primary transitions include:[17]

- transition from cotton-dominated denim to more utilization of cotton/ synthetic and cotton/flax denim;

- transition from open-end or rotor-spun yarn domination to more utilization of ring spun and compact yarns or combinations of open end and ring spun yarns;
- transition from conventional yarns to fancy yarns
- transition from stone-washed denim to more energy-efficient and environmentally friendly methods such as enzyme treatment, mechanical abrasion, ozone fading, water jet fading and laser treatment.

In addition to the above developments, comfort and fit have become key aspects in new denim developments. This has resulted in an increase in the use of spandex to improve fabric stretchability and in seeking ways to improve the cooling attributes of denim in hot environments.

The next task of product development is to determine and define product performance characteristics. As in most apparel products, the two common performance characteristics of denim fabric are durability and comfort. These characteristics can be translated into many attributes of fibers, yarns and fabrics as shown earlier in the performance–attributes diagrams of Figs 12.2 and 12.3. Information pertaining to denim fabric development is

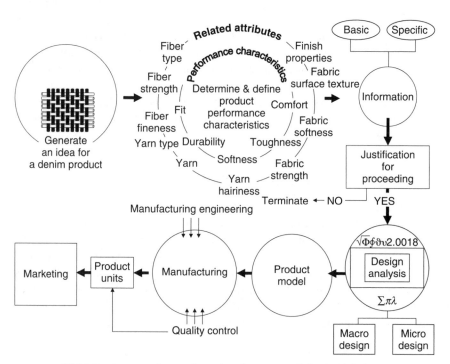

12.11 Basic steps in product development of denim fabric.

numerous. However, most of this information is of commercial type and each denim developer must establish his/her own database via general basic information obtained from various books and more specific information obtained from customers archives or directed experiments. Some of the basic information related to denim is summarized below.

Denim fibers

Most denim fabrics are made from cotton fibers; typically, a bale of cotton weighing 480 pounds (about 218 Kg) will yield over 200 pair of denim jeans. One of the fashion trends of denim fabric is to make denim from organic and naturally colored cotton (e.g. Eco jeans by Levi Strauss). The high price of this type of denim is a result of the short supply of naturally colored cotton in comparison with normal cotton and the increasing demand by consumers to use organic goods. Standards for organic cotton products have not been fully established yet, but if the demand continues in this direction, organizations such as the USDA will have to generate such standards. It should be pointed out, however, that the costs of producing organic and colored cottons and the deficiency in some of their inherent properties have made their market volume almost insignificant compared to normal cotton.

Although cotton is unanimously the common choice for denim fabric, other fibers have been used, mostly in conjunction with cotton, to provide additional performance characteristics for denim. For example, scientists and engineers in the Agricultural Research Service of the USDA have developed a cotton–flax blend to be used to make denim fabrics. The underlying concept of this development is to utilize the attributive advantages of both cotton and flax in a blend that can result in a more comfortable denim fabric, particularly in the context of moisture management.

The cotton–flax combination is a development that deserves some elaboration. Like cotton, flax fiber is a natural cellulosic fiber with bulk density equivalent to that of the cotton fiber. However, it belongs to the long-vegetable fiber category with a fiber length that could range from 4 to 40 inches (10.2–101.6 cm). With cotton fiber length of only 1 to 1.5 inches (2.5–3.8 cm), blending of the two fibers represent a challenge that must be overcome prior to making a cotton–flax yarn. The two fibers also have different diameters with cotton ranging from 10 to 14 µm, and flax ranging from 40 to 80 µm. Values of some of the basic properties of these two fibers are listed in Table 12.2. Some of the advantages of adding flax fibers to cotton are higher strength, higher elastic recovery and higher heat resistance than 100% cotton. However, flax is significantly stiffer than cotton as is evident by its higher flexural rigidity. In addition, the breaking elongation of flax is much lower than that of cotton.

Table 12.2 Values of some basic properties of cotton and flax fibers[17,18]

Fiber attribute	Cotton	Flax
Specific gravity	1.54	1.54
Moisture regain (%) at 65% RH	8.5	10
Moisture regain (%) at 100% RH	30	23
Fiber tenacity (g/denier^{-1})	3.5–5.5	>6
Breaking elongation (%)	4–5	1.8
Elastic recovery (%) at 1 g/denier^{-1} stress	50	75–80
Elastic recovery (%) at 2 g/denier^{-1} stress	<40	65–70
Flexural rigidity (g denier^{-1})	60–70	175
Abrasion resistance	Relatively low	Relatively high
Swelling % (increase in area upon water immersion)	47	21
Wet breaking tenacity (% increase w.r.t dry tenacity)	30–50	25–30
Heat resistance temperature (°C)	149	260

What makes a cotton–flax blend a good combination for denim fabric is essentially the moisture management aspect. Flax fabrics provide a very cool feeling when they are worn next to the skin. The reason for this feeling is the inherent wicking ability of flax fibers. Both cotton and flax are cellulosic hydrophilic fibers; they both absorb water and exhibit moisture regain of 8.5% and 10%, respectively, at 65% relative humidity. However, flax fiber has a greater tendency to transfer water via its surface than cotton fiber. This is because the surface structure of the flax fiber is cross-marked in the longitudinal view with multiple nodes along the length held together by a waxy film. As a result, absorption and desorption of water is more rapid for flax fiber than for cotton fiber. Note that at 100% relative humidity (soaking wet condition) flax has a moisture regain of 23% compared to 30% for cotton; this is a reverse in trend from the condition at a standard relative humidity of 65%.

In light of the above discussion, the addition of flax to cotton can provide good thermal comfort via the cooling effect caused by water transfer and water evaporation from the skin to the outside environment, particularly in hot weather. Key design factors in this regard include the type of flax fiber used, the type of cotton fiber used and the percentage of flax being added to cotton, or the cotton–flax blend ratio. These factors are critical on the grounds that a high percentage of flax fibers not only will result in processing difficulties[18] but more seriously will adversely influence the tactile comfort performance of denim owing to its high stiffness.

Other long-vegetable fibers such as kenaf, jute, ramie and hemp have also been tried in blend with cotton to make denim products.[19] However, issues such as ease of processing, blending compatibility, dye affinity, surface texture and comfort represent design challenges to the progress in this direction of product development. As a result, cost can be a major hindering factor against these developments.

Synthetic fibers are also used with cotton to enhance some of the performance characteristics of denim fabric. For example, polyester fibers are used in lightweight denim bottoms to provide better strength and higher abrasion resistance. Spandex is another fiber used in very small amounts with cotton to provide fit and tactile comfort (stretch and recovery) for denim products. As indicated in Chapter 8, spandex is an elastomeric fiber (commercially known as Lycra®) made in a filament form of deniers ranging from 20 to 5400 denier. An elastomer is a natural or synthetic polymer that can be stretched and expanded to twice its original length at room temperature. After removal of the tensile load it will immediately return to its original length.

It follows from the above discussion that cotton fiber is the dominant fiber type used in denim products. This domination should stir the curiosity of many product developers and design engineers as it does not necessarily reflect all target performance characteristics of denim products. Denim fabric is typically a durable product, yet cotton fiber has moderate durability with medium initial modulus and tenacity, and low elongation. The elastic recovery of cotton is also low (only 75% at 2% extension and less than 50% at 4% extension). Cotton fibers also act poorly under abrasive effects (e.g. breaking fiber cell walls and damaging fiber tips). However, the positive image of cotton fibers and their well-established effect on comfort and feel performance provide significant merits. The fiber will reasonably absorb water vapor emitted by the body to keep skin dry. Under dry conditions, the touch and feel of cotton is superior in most apparel products. This is directly due to its tapered and convolutionary surface structure. In addition, the structural features of cotton allow the manipulation of a variety of yarns that can indeed provide a wide range of comfort characteristics. These points provide ample opportunities for product developers in which a combination of durability and comfort features of denim are optimized based on objective and reliable measures. Figure 12.12 shows the basic steps of raw material selection following the tasks discussed earlier in Chapter 7. The reader should refer to the discussion in Chapter 7 to follow these steps.

Denim yarns

Denim fabrics are typically made from staple-fiber yarns that can be spun using rotor spinning (open-end spinning) or conventional ring spinning. As discussed in Chapter 9, these two methods of spinning produce yarns that are fundamentally different in structure. The largely truly twisted

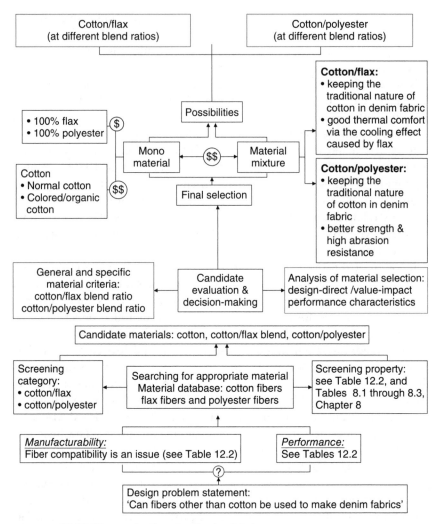

12.12 Fiber selection for denim fabric.

structure of ring spun yarn makes it stronger and denser than comparable rotor spun yarn, which typically exhibits a combination of fully and partially twisted fibers as well as many outer fibers wrapping around the yarns (see Table 9.3, Chapter 9). Ring spun yarn can also be made softer as a result of the lower levels of twist that can be used to form the yarn in comparison with rotor spun yarn. These attributes provide positive effects for both durability and comfort.

Despite the obvious merits of using ring spun yarns in making denim fabrics, rotor spun yarns have taken the largest market share for many years, particularly in the US market. This domination largely stemmed from the following reasons:[17,18,20]

- Rotor spun yarn is more economical than ring spun yarn by virtue of its higher production rate and the condensed process involved in making the yarn (i.e. no roving or winding are required for rotor spinning).
- Heavy denim fabrics are typically made from coarse yarns (e.g. Ne = 7 s to 10 s, or tex = 59 to 84) with a large number of fibers per yarn cross-section. This range of yarn count can easily be made by rotor spinning by virtue of its principle.
- Rotor spinning has a better tolerance to some of the fiber defects that cannot be easily tolerated by ring spinning (e.g. neps and short fibers). It can also tolerate blends of primary cotton and waste fibers such as noils or card waste. In addition to the cost advantage of using these fibers, many denim fabrics are insensitive to the presence of these defects in the yarns.

Studies of denim durability revealed that fabrics made from rotor (open-end) spun yarns were less durable than those made from ring spun yarns.[21] However, the difference in durability was not apparent until after 16 wash and wear cycles and was not great enough to affect consumer acceptance. When a superior combination of durability and comfort is of primary concern in denim, ring spun yarns or a combination of warp ring spun yarns and filling rotor spun yarns may be considered. When both warp and filling yarns are ring-spun, the denim yarn is commonly termed dual ring spun (also ring × ring, ring-ring, or double-ring spun). This is typically a premium and more expensive denim. Dual ring-spun denim is characterized by special texture and better softness than that of open-end or single ring-spun yarn. As indicated in Chapter 9, ring spun yarns can be twisted in two directions, Z and S twist. Almost all denim ring spun yarns are twisted in the Z-direction. In recent years, researchers at Cotton Incorporated developed combinations of yarn twist and fabric twill directions. A denim yarn of S-twist was found to add subtle textural effects to the denim fabric.

Another more recent development in denim yarns is the use of compact ring spun yarns. As indicated in Chapter 9, this is a ring spun yarn in which fibers are aerodynamically compressed to provide a denser yarn and less hairy surface.[17] The increase in cost associated with compact spinning can be justified on the basis of the lower twist required to make the yarn (softer yarn) at a comparable strength to that of conventional ring spun yarn, lower size add-on during sizing and substantially fewer stops or yarn breakage during rope beaming as a result of the reduced hairiness. The yarn is also expected to yield a fabric with greater abrasion resistance.

Denim fabrics are also made from fancy yarns, particularly slub yarns both of ring and rotor spun types. The use of slub yarns provides a wide range of appearance to the denim fabric resulting from creating slubs of identical thicknesses but varying lengths, varying thicknesses and lengths, or long sections of yarn with different counts. This direction of development opens many doors for creative fashionable denim fabrics that can provide a dynamic stimulation to this important market.

Denim fabrics

Denim fabric is predominantly made from twill fabric. As indicated in Chapter 10, twill fabrics are used for durable fibrous products that are typically used under harsh physical conditions. For this reason, they are often used in suits, sportswear, raincoats, jackets and working clothing. The reason for the high durability of twill fabric is that the few interlacings per inch in the twill structure (Fig. 10.1c, Chapter 10) allows higher fabric counts, or more yarn packing. This feature also results in low air permeability and high wind resistance. Twill fabrics with steep twill lines are typically stronger than those with reclining or with regular twill lines.[20] They also tend to be unbalanced in construction, as they have a higher proportion of warp yarns than filling yarns per inch. Another important feature of twill fabric that both designers and fabric makers should be aware of is the 'torquing' tendency resulting from its structural directional effect. Most denim is made from warp-faced twill fabric. In other words, their warp yarns lie predominantly on the face of the fabric, or warp yarns must always float over more filling yarns than they pass under (e.g. 2/1, 3/1 or 3/2 interlacing patterns). This is also one of the reasons why the blue warp is more dominant than the white filling in blue jeans.

Twill denim is typically divided into three main categories:[20,21] left-hand twill, right-hand twill and broken twill. These categories refer to the direction in which the denim is woven. Typically, left-hand twill denim is known to be softer to the touch than right hand twill. It is also easier to recognize, as the weft threads appear to move left-upward as opposed to right-upward. Broken twill contains no distinct direction of weave. It has no directional effects. In other words, it does not run to the right or left; instead it exhibits an alternating right and left with the end effect resembling a random zigzag pattern. It is interesting to know that broken twill jeans are traditionally considered to be the cowboy-preferred denim. The development of broken twill was originally inspired by the need to combat the twisting effect or fabric torque that was a characteristic of regular twill. Alternating twill directions resulted in a balance in yarn tension and minimum fabric twisting.

In addition to the unique fabric structure, denim derives its texture and familiar nature from many other sources such as the method of dyeing and the method of drying. The reason denim tends to change color by exhibiting color fading is that the blue warp yarns on the fabric face are dyed with indigo dyes in such a way that only the fibers near the yarn surface are colored, leaving the fibers in the yarn center uncolored. Abrasion effects during use (due to repeated wearing, washing and drying) typically result in removing the blue fibers partially and exposing more and more white fibers. Indigo is a poor dye by virtue of its low affinity for the cotton fiber. As a result, it does not migrate to the white filling yarns.

Denim finishes

A significant aspect of denim product development is surface finish. The impact of this aspect on consumer's acceptance often outweighs most of the other factors contributing to denim performance. Many approaches have been taken to improve denim softness and surface texture. The two prominent approaches are stone wash and enzyme treatment.[22–27] Stone wash aims to provide a fashionable stressed or deformed appearance (e.g. fuzzy texture, puckering at the seams and slight wrinkling) to denim. The process of stone washing involves drying the denim garments using pumice stones to abrade the fabric surface. This is a colorless or light gray stone or volcanic glass formed by solidification of lava, which is permeated with glass bubbles. In stone washing, factors such as garment/stone ratio, the shape and the size of the stone can make a difference to the final appearance of the garment. This is, of course, in addition to the tumbling time. The obvious drawback of stone treatment is the extent of damage that may be caused to the fabric as evidenced by the abrasive spots on some areas of the garment. This damage can be reduced using some oxidizing agents such as sodium hydrochlorite or potassium permanganate. These treatments reduce the stone damage only by reducing the tumbling time. Indeed, the oxidizing agent also results in physical damage through oxidizing the molecules of the indigo dye.

Cellulase enzyme washing is also used as a safe alternative to stone washing.[23] This is a natural protein that physically degrades the surface of the cotton fibers giving a similar appearance to that achieved by stone tumbling. Enzyme treatment also helps in conserving water, time and energy. Two classes of cellulase enzyme can be used:[23,24] acid cellulase, which works best in the pH range of 4.5–5.5 and exhibits optimum activity at 50°C, and neutral cellulase, which works best at pH 6 (can be effective at higher pH of up to 8), and provides maximum activity at 55°C. Enzyme treatments yield soft handle and attractive clean appearance of denim with minimum damage to the surface of yarn.[25] They are also inexpensive in comparison with stone washing. Indeed, they can add value to denim fabric by transforming low-grade denim to a top quality product owing to the removal of fabric hairiness and pills. Other cost benefits associated with enzyme treatments include treatment of larger quantities of garment at once, less labor-intensive and lower damage to seam edges and badges (less rejects).

Other types of finish that have been implemented more recently include[21] sand blasting, mechanical abrasion, ozone fading, water jet fading and laser treatment. The principle of sand blasting is to blast an abrasive granular soil through a nozzle at very high speed and high pressure onto specific areas of the garment surface to give a desired distressed and abraded appearance. This is a purely mechanical process in which no chemicals are used. Similarly, denim fabric can be mechanically abraded via robotic techniques to create surface effects such as sueding, raising or brushing.

Ozone fading is based on bleaching the garment in a washing machine with ozone dissolved in the water or fading denim using ozone gas in closed chambers. This method serves the fading purpose with a minimum or no loss of strength. It is also environmentally friendly since after laundering, ozonized water can easily be deozonized by UV radiation. Water jet fading is based on using hydro jets for patterning or altering the surface finish and the texture of denim garment. In this method, the extent of achieving color wash out, clarity of patterns and softness of the resulting fabric is related to the type of dye in the fabric and the amount and manner of fluid impact energy applied to the fabric.

Laser treatment is a technique used to fade denim fabric by creating patterns such as lines, dots, images, text or even pictures. Some of the laser systems use a mask to give the desired shape required on the fabric surface. The laser projects through a lens system, which expands the beam. This beam is passed through the shaped mask that comprises an aperture of the desired shape and is then deflected by a mirror to strike the fabric substrate. The duration of exposure determines the final effect on the fabric. Laser treatment can be used to create localized abrasive effects or fabric holes.

12.6.2 Denim product design

In view of the discussions above, one can imagine that the design of a denim product will involve many parameters that must be considered to meet the desired performance characteristics of this popular product. In a simple design cycle (refer to Fig. 4.1, Chapter 4), the initial task is material selection, followed by material placement or product assembly to form the final product model. Figure 12.13 shows examples of different design parameters associated with these two tasks for a denim product.

As indicated in Chapter 4, a product model is reached through appropriate analysis supported by technical knowledge and computational and/or experimental tools. The product model may take different forms including

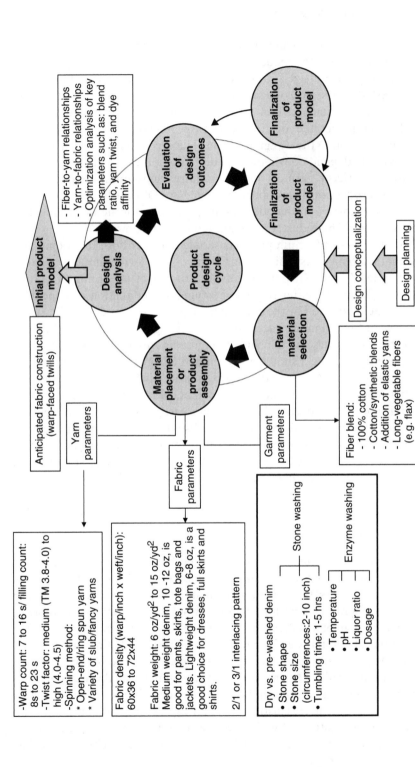

- Fiber-to-yarn relationships
- Yarn-to-fabric relationships
- Optimization analysis of key parameters such as: blend ratio, yarn twist, and dye affinity

Evaluation of design outcomes

Finalization of product model

Finalization of product model

Design analysis

Product design cycle

Material placement or product assembly

Raw material selection

Design conceptualization

Design planning

Initial product model

Anticipated fabric construction (warp-faced twills)

Yarn parameters

Fabric parameters

Garment parameters

Fiber blend:
- 100% cotton
- Cotton/synthetic blends
- Addition of elastic yarns
- Long-vegetable fibers (e.g. flax)

-Warp count: 7 to 16 s / filling count: 8s to 23 s
-Twist factor: medium (TM 3.8-4.0) to high (4.0-4.5)
-Spinning method:
* Open-end/ring spun yarn
* Variety of slub/fancy yarns

Fabric density (warp/inch x weft/inch): 60x36 to 72x44

Fabric weight: 6 oz/yd² to 15 oz/yd² Medium weight denim, 10 -12 oz, is good for pants, skirts, tote bags and jackets. Lightweight denim, 6-8 oz, is a good choice for dresses, full skirts and shirts.

2/1 or 3/1 interlacing pattern

Dry vs. pre-washed denim
• Stone shape
• Stone size (circumferences:2-10 inch)
• Tumbling time: 1-5 hrs

Stone washing

Enzyme washing

• Temperature
• pH
• Liquor ratio
• Dosage

12.13 Key tasks of a denim product design cycle.

a computer geometrical model, a mathematical model or an animated simulation model of the product assembly. As the design process approaches the final stage, a full-size prototype model may be constructed and thoroughly tested. In the case of denim products, a prototype model is typically represented by a sample of fabric woven and finished or a complete garment sample that meets the desired design specifications and provides an optimum solution to the design problem. The accessibility of such a model will obviously depend on whether the design modification can be made on a finished fabric or garment or if it requires substantial changes in all phases of production from fibers to garments. Obviously, the latter will involve more cost of product development by virtue of the involvement of many sectors of the industry in creating the product model.

Design analysis of denim products should consider both functional and style characteristics. As indicated in Chapter 4, the former implies meeting the intended functional purpose of the product at an optimum performance level and the latter implies satisfying a combination of appealing factors that are desired by the consumers. Figures 12.14 and 12.15 illustrate examples of functional characteristics and styling characteristics of denim products, respectively. Also included in these figures are examples of notes that the design engineer should make, gathered from the information phase of the design cycle. Again, and as it was indicated in Chapter 4, the distinction between functionality and styling can be largely blurred for some traditional fibrous products; denim is no exception. For example, the dimensional stability of denim fabric was listed among styling characteristics as a result of its significant direct impact on the wearer's acceptance of the product. However, it is well known that dimensional stability can indeed influence many functional characteristics including comfort, durability and fit.

Denim design conceptualization

The design cycle discussed above should be based on sound design conceptualization. As clearly indicated in Chapters 4 and 5, design conceptualization represents the foundation of any design process. This is the process of generating ideas for an optimum solution to the design problem. Figure 4.2 of Chapter 4 illustrated the basic steps towards design conceptualization. The initial step is to revisit the idea of the denim product initiated in the product development cycle and establish objective justifications for the idea in view of key criteria such as consumer-added value, potential users, producer-added value, and regulations and liability. In case of denim product development, this justification may represent a difficult task as the market is almost saturated with numerous denim products from the basic rugged denim to many high-end fashionable products. This means that only few creative approaches may be justifiable.

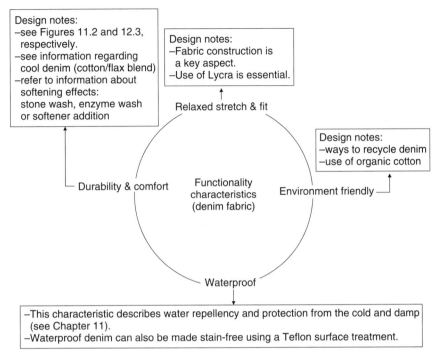

Design notes:
–see Figures 11.2 and 12.3, respectively.
–see information regarding cool denim (cotton/flax blend)
–refer to information about softening effects: stone wash, enzyme wash or softener addition

Design notes:
–Fabric construction is a key aspect.
–Use of Lycra is essential.

Relaxed stretch & fit

Design notes:
–ways to recycle denim
–use of organic cotton

Durability & comfort

Functionality characteristics (denim fabric)

Environment friendly

Waterproof

–This characteristic describes water repellency and protection from the cold and damp (see Chapter 11).
–Waterproof denim can also be made stain-free using a Teflon surface treatment.

12.14 Examples of denim functional characteristics.

Once an idea is clearly justified, the design problem should be defined. A design problem represents the primary obstacle facing the design process or hindering a desirable optimum solution. In a broad sense, a design problem may be initiated by establishing a general definition of the proposed product. Examples of this definition are as follows:

- highly durable, moderately comfortable denim
- highly comfortable, moderately durable denim
- moisture-management denim.

A broad definition should then be followed by a more specific definition of the design problem in which precise objectives and goals are stated, technical terms are defined, limitations and constraints are documented, probabilistic outcomes are discussed and evaluation criteria are perceived. For example, when moisture management is the broad definition, a specific definition may be stated in terms of the exact objectives, say 'a denim fabric that can provide cooling effects and sweat wicking in a hot environment'. This may be associated with limited fiber types options (e.g. hydrophobic and hydrophilic), yarn structure (e.g. twist levels, yarn compactness, etc) and special finish treatments. Probabilistic outcomes may involve stating issues of 'what-if' formats such as 'what-if the garment is worn in a cold

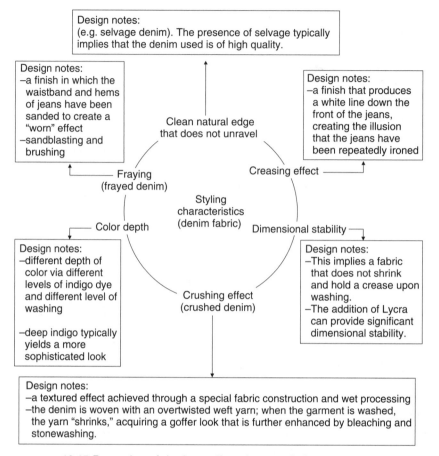

Design notes:
(e.g. selvage denim). The presence of selvage typically implies that the denim used is of high quality.

Design notes:
–a finish in which the waistband and hems of jeans have been sanded to create a "worn" effect
–sandblasting and brushing

Clean natural edge that does not unravel

Design notes:
–a finish that produces a white line down the front of the jeans, creating the illusion that the jeans have been repeatedly ironed

Fraying (frayed denim)

Creasing effect

Styling characteristics (denim fabric)

Color depth

Dimensional stability

Design notes:
–different depth of color via different levels of indigo dye and different level of washing

–deep indigo typically yields a more sophisticated look

Crushing effect (crushed denim)

Design notes:
–This implies a fabric that does not shrink and hold a crease upon washing.
–The addition of Lycra can provide significant dimensional stability.

Design notes:
–a textured effect achieved through a special fabric construction and wet processing
–the denim is woven with an overtwisted weft yarn; when the garment is washed, the yarn "shrinks," acquiring a goffer look that is further enhanced by bleaching and stonewashing.

12.15 Examples of denim styling characteristics.

environment'? Finally, any design problem should be associated with criteria that will eventually be used to judge whether the problem has been solved and whether the solution was indeed optimal.

Following the justification and problem definition tasks, information should be gathered to complete the design conceptualization. In the previous sections examples of information types were discussed. More specific information may also be needed including specific relationships between performance characteristics and potential attributes influencing them, specific scientific tools required to perform the design analysis and other information that may require some experimental trials. Ultimately, design conceptualization should lead to a reliable answer to the question of whether the current state of the art warrants further efforts with potential added success. A positive answer to this key question represents the 'go-ahead' for generating ideas for optimum solutions to the design problem and for formulating the design concept.

As discussed in Chapters 4 and 5, generating ideas for optimum solutions does not necessarily mean finding a final solution; it is rather finding an idea for a solution or a direction of thought as some design problems may be associated with many solution ideas and others may indeed have limited or no apparent solution. For example, for each one of the broad definitions of denim product listed above, design engineers may suggest some solution ideas as shown in Fig. 12.16. Some of these ideas may be feasible and valid and others may not. The key task here is to put all ideas forward, particularly in the initial stage of design conceptualization. These ideas can then be evaluated and narrowed down to the most appealing and most cost-effective ideas using the concepts and tools presented in Chapter 5, namely creativity, brainstorming and decision-making methods.

The design analysis associated with denim products can be overwhelming, particularly if a new product idea is proposed. In most situations, this analysis will focus on the following key aspects:

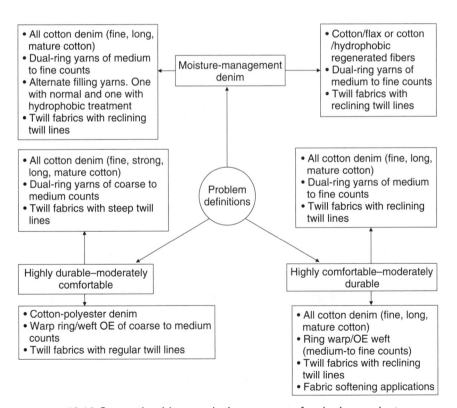

12.16 Generating ideas or design concepts for denim products.

- Determining the appropriate fiber or fiber mix that is suitable for the desired denim product. This type of analysis was covered in great detail in a book written by the present author.[17]
- Performing basic structural analysis of fibers, yarns and fabrics. These are discussed in the previous literature.[3,17,18,28]
- Establishing relationships between the desired performance characteristics and the various attributes of fibers, yarns and fabrics. The modeling analysis discussed in Chapter 6 can be very useful in this respect.
- Chemical analysis of possible finish treatments applied to the fabric.

12.6.3 Denim recycling

With millions of blue jeans being used and thrown away every year, most going into landfills, it is important to incorporate recycling or reusing of denim materials into the overall development process. In general, one can list many approaches to reusing denim fabric. The key, however, is the technological feasibility and the economics of the approach. These issues must be resolved prior to taking any approach upstream. Potential ways to recycle denim may be divided into two main categories: (a) conversion of denim garments into fibers through shredding the fabric and reusing the fibers in other products and (b) using pieces of denim fabrics obtained from waste garments in other products.

Shredding fabric into fibers is a common method of recycling of fibrous materials. In the case of denim fabrics, the recycled fibers can be used in many applications including:[17]

- reprocessing waste fibers through waste-handling spinning operations to make lower grade yarns and fabrics
- blending waste fibers with other primary fibers to produce yarns
- reprocessing waste fibers into nonwoven products such as utility fabrics, cleaning items, wadding for furniture, cushions and pillows, car wadding
- reusing waste fibers in other products such as paper and cardboard.

Technologically, the success of any of the above approaches will primarily depend on the quality of fibers obtained from the recycling process. For fibers that will be respun into yarns, the primary fiber characteristic is fiber length. Typically, fibers of 0.5 inch (1.3 cm) or shorter are considered short fibers and they cannot be respun into yarns. These fibers may be used for nonwoven or paper-making applications. The second important fiber characteristic is what may be termed as 'pre-stress history'. During the initial processing, fibers are subjected to many mechanical stresses; they are tensioned, compressed, bent and twisted. Some of these stresses may exceed the elastic limit of the fiber, leading to permanent deformation. In the weaving and knitting processes, additional mechanical stresses are applied to the fibers. During finishing, fibers are further stressed while being treated thermally or chemically. These various forms of stress certainly influence the performance of fibers during recycling by making them stiff and easily breakable. This problem is commonly handled by blending a small proportion of waste fibers with primary fibers so that the latter can act as a carrier and supportive component for waste fibers during recycling.

In the case of denim recycling, the primary challenge to reprocessing waste fibers is color. This is due to the fact that while denim filling is white, denim warp is indigo-dyed. Separation of these two yarns in the shredding process is impossible and cost prohibitive. In addition, bleaching the fibers to eliminate the indigo dye can cause a great deal of difficulties, as it is well known that indigo dye is difficult to remove or mask by bleaching. These obstacles simply mean that color in the recycled product will have to be accepted. On the other hand, the color shade can be controlled through blending with primary white fibers at different blend ratios.

As indicated earlier, cost and economical feasibility represent the primary challenge of any reclaiming effort. Cost issues are common in almost all recycling efforts. These include (a) cost differences between the waste landfill option and the reprocessing option, (b) the difference between the cost of primary fibers and waste fibers, (c) the availability of a continuous supply of the waste material to keep the recycling operation running and (d) the impact of quality of the recycled end product on its market value. These issues must be resolved in the design analysis of recycled products.

The reuse of denim through using pieces of denim fabrics obtained from waste garments in other household products has represented a campaign by many environmental advocates in recent years. This is a simple approach that requires minimum effort and minimum cost, but a great deal of awareness by different consumers of the importance of recycling or reusing waste materials. Using this approach, many creative ideas can be implemented. A quick glance at the internet can provide the reader with numerous ideas for reusing pieces of waste denim garments.

12.7 Sportswear products

Sportswear represents a critical category of traditional fibrous products that is widely used by numerous wearers from children playing soccer to professional players in different sports including boxing, wrestling, swimming, car race, mountain climbing, basketball and football. With the advance in sportswear products available in today's market, many of these products can be classified as function-focus products by virtue of their special designs associated with high costs and high prices. Another category of sport products called 'sports equipment' uses various types of polymeric-based and fibrous materials. This includes products such as tennis rackets, artificial field turfs, bicycles, boats, pool liners, and so on. This category, combined with sportswear, constitutes a multi-billion dollar market worldwide. Our focus in this section will be on the development of sportswear apparel and uniforms.

12.7.1 Performance characteristics of sportswear products

Different sportswear products may require different performance characteristics depending on a number of factors including (a) sport type, (b) the level of physical activity, (c) team or individual sport, (d) professional or amateur sport, (e) climate-indoor or outdoor sport, (c) use frequency, (d) gender, (e) age and (f) other specialty functions. In the marketplace, most sportswear products are characterized by general performance features such as fit, stretch, color and maintenance (washing and drying). In the context of product design, key performance characteristics of sportswear products can be divided into four major categories (see Fig. 12.17): durability, comfort, functionality and identity or recognition.

Durability characteristics include strength, stretchability, elastic recovery, abrasion resistance, tear resistance, color fading, body odor resistance and UV resistance. Obviously, some sports involve a great deal of physical activity that make durability an essential performance characteristic.

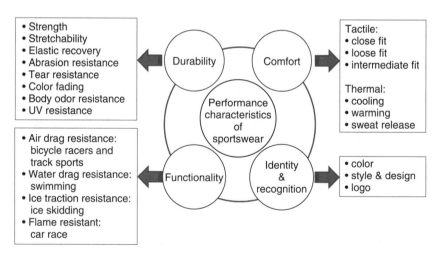

12.17 Primary categories of performance characteristics of sportswear products.

Sportswear maintenance aspects also fall under durability owing to the need to wash and dry sportswear after each use. Comfort characteristics can be divided into tactile and thermo-physiological comfort. Tactile comfort deals with the interaction between sportswear and the human body at different levels of physical activity. The parameters mentioned earlier in Section 12.4.3 fall under the tactile comfort category. The importance of tactile comfort stems from the fact that the mobility of fabric against the body at the fabric–body interface is critical in most sport applications. As a result, some sportswear applications require loose fit, some require tight fit and others require an intermediate fit. Thermo-physiological comfort is largely an issue of thermal regulation or thermal adaptability. In simple terms, this implies cooling in hot weather and warming with sweat release (transferring sweat away from the skin) in cold climates. In addition to durability and comfort, some sport applications require special performance characteristics. For example, boat or bicycle race and track sports require minimum air drag resistance, swimming and ice skidding require minimum water drag resistance and optimum ice traction, respectively, and car racing requires flame resistant sportswear. Finally, identity or recognition (the function of a uniform) is essential for team sports in which each team should have a consistent and identifiable appearance, color, style and logo.

12.7.2 Relative importance of comfort, durability and functionality in sportswear

In the context of product application, sportswear can be divided into three major categories:[29] low-physical products, high-physical products and specialty products. Both low-physical and high-physical products belong to conventional sportswear in which the primary design trade-off is between durability and comfort. It is a trade-off as a result of the fact that the enhancement in comfort characteristics often comes at the expense of durability and vice versa. Specialty sportswear products are those designed with some emphasis on special functions to enhance the player's performance in some particular sports. This category of products extends the design trade-off to accommodate functionality or function-related performance characteristics in addition to durability and comfort. Figure 12.18 illustrates the percentage relative contribution of each of these three basic performance characteristics in each product category based on a consumer's survey conducted by the present author.[2,29]

As can be seen in Fig. 12.18, for the low-physical category, comfort has the highest relative importance, followed by durability. This is a direct result of the perception associated with this category of a mildly physically demanding garment by most wearers. High-physical sportswear products

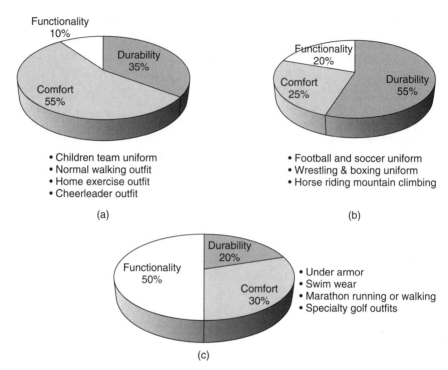

• Children team uniform
• Normal walking outfit
• Home exercise outfit
• Cheerleader outfit

(a)

• Football and soccer uniform
• Wrestling & boxing uniform
• Horse riding mountain climbing

(b)

• Under armor
• Swim wear
• Marathon running or walking
• Specialty golf outfits

(c)

12.18 Design trade-off between durability, comfort and functionality of sportswear for different product categories.[29] (a) low-physical, (b) high physical and (c) specialty sportswear.

are mainly used in professional sports that require high physical activities. Therefore, durability is of the greatest importance for this category. Comfort, on the other hand, also represents a key performance aspect, when high physical activities of long duration are expected. The increasing awareness of professional players of particular performances (e.g. moisture management and thermal stability) has resulted in a significant increase in functionality in this category over that of low-physical sportswear. Specialty sportswear products represent a relatively recent market in which the market target is mainly professional sports. The difference between this category and the other two categories stems from the specific functions provided by the product. It is also expected to be a significantly higher price by virtue of its claimed added value.

Sportswear products can be made from woven or knit fabrics. This provides wider design options, particularly in relation to durability and comfort parameters. For specialty sportswear, most design efforts have been focused on the surface characteristics of fibers and fiber assemblies. Figure 12.19 illustrates some of the important surface-related attributes. At the fiber

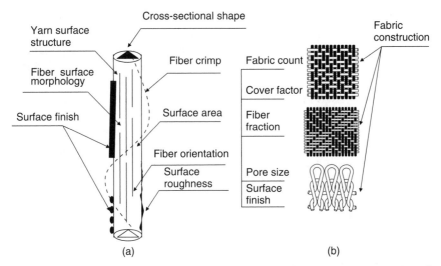

12.19 Design parameters that influence the surface characteristics of fibrous assemblies.[30] (a) Fiber and yarn parameters, (b) Fabric parameters.

level, important surface-related attributes include[30] surface area, cross-sectional shape, surface roughness, fiber crimp and surface molecular orientation. At the yarn level, yarn type and its associated structural features (e.g. fiber arrangement, fiber mobility and fiber cohesion) can influence the surface behavior of yarn. At the fabric level, fabric construction and structural features (fabric count, fiber fraction, thickness, etc.) can all contribute to the surface performance of the fabric. Obviously, many of these parameters are interrelated, requiring careful design analysis to meet optimum performance conditions.

12.7.3 Fiber types used for sportswear

Fibers used in sportswear range from conventional fibers (natural and synthetic fibers) to specialty fibers designed to meet specific performance requirements. Conventional fibers used include cotton, long-vegetable fibers, wool, nylon, polyester, viscose, polypropylene and acrylic fibers. Important attributes of these fibers were discussed in Chapter 8. Another fiber that is commonly used in numerous sportswear products is spandex (Lycra®). This fiber is used at a small percentage (10–20%) with cotton, nylon or polyester fiber to provide stretch and better fit. Recall that spandex is segmented polyurethane in which alternating rigid and flexible segments that display different stretch resistance characteristics form the fiber. The rigid segments are normally prepared from MDI (methylene diphenyl iso-cyanate) and a low-molecular-weight dialcohol such as ethylene glycol or 1,4-butanediol, while the flexible segments are made from MDI and a polyether or polyester glycol. The rigid segments have a tendency to aggre-gate and the flexible segments act as springs connecting the rigid segments which can stretch to great lengths, yet have greater stretch resistance than other rubbers and do not fail easily under repeated stretching. They also have moderate strength, high uniformity and high abrasion resistance.

In addition to conventional fibers, many specialty fibers have been developed to provide special functions for sportswear products. As indicated earlier, the functionality aspect of sportswear was introduced through the design of special surface characteristics using one or more of the parameters shown in Fig. 12.19. One of the key design parameters in this regard is fiber cross-sectional shape. This parameter can be manipulated in synthetic fibers to provide special performance features.[30] Table 12.3 shows examples of common cross-sectional shapes that are commercially available and their key design merits. In recent years, design engineers of synthetic fibers have taken a leap in realizing and utilizing fiber cross-sectional shape as a powerful functional parameter. As a result, fibers with more sophisticated cross-sectional shapes were developed. Examples of these developments are listed in Table 12.3.

Table 12.3 Examples of common fiber cross-sectional shapes[30]

Cross-sectional shape	Special features
(a) Circular	Used in most synthetic fibers, reference for other cross-sectional shapes (a circular shape factor is one) It has a low surface-to-volume ratio
(b) Hollow	Lower density at the same diameter, higher bending resistance, entraps air to provide thermal insulation, light scatter by internal surfaces leading to soil hiding and translucent characteristics
(c) Trilobal	Often used to provide higher bending stiffness and soil hiding characteristics
(d) Ribbon	Larger flat surface for sparkling appearance, directional bending characteristics

Nylon and polyester hollow filaments: These have been used for multifunctional sportswear products because of their light weight and quick drying. They can be made 20–25% lighter than conventional solid filaments. However, these fibers should be treated carefully during processing as they can be damaged by mechanical processes and during texturing or weaving. In addition to light weight, hollow fibers can provide a great deal of flexibility, better wicking and better thermal insulation than solid fibers with the same polymeric base.

C-slit cross-sectional shape: This was developed to entrap air for thermal insulation, while simultaneously improving elastic behavior.[31] The C-shaped sheath originally has an alkali-soluble polymer core reaching the external surface through a narrow longitudinal slit. After drawing, the filaments can be textured by false twisting or other means. The core is then removed using alkali finishing, giving a hollow C-shaped cross-section with a longitudinal slit. The combination of texturing and cross-sectional shape provides void fraction exceeding 30% and a springy feeling. Different fibers including polyester or polyamide based can be used. In another derivative, the sheath can be made of a blend of polyester and hydrophilic polymer.[32,36] In this case, alkali finishing introduces microcrazes and pores throughout the sheath, which allows liquid sweat absorption and transportation from the skin to the hollow core. This feature allows dryness rather than thermal insulation which is critical in many sportswear applications.

The 4DG™ fibers: This was developed by Eastman Chemical to provide several deep grooves that run along the length of the fiber.[33,34] The expanded surface area of this fiber provides about three times the amount of specific surface per denier compared to circular fibers. The geometry of the cross-section can allow transport of up to 2 liters of water per hour per gram of fiber (high capillary wicking). The grooves in a 4DG fiber are also good for trapping particles in an air or liquid stream. These grooves provide areas where eddy currents will preferentially deposit particles and where particles can collect without blocking pores in the fabric. These fibers provide increased filtration efficiency without an increase in pressure drop across the fabric. The entrapped air also provides increased thermal insulation which is suitable for sportswear products.

L-shaped fiber produced by Ciebet (Fig. 12.20): This was developed to enhance the wicking performance of sportswear. It produces close packing in conjunction with surface water wettability or wicking effect, derived from the capillary forces created in the interfiber volume that are claimed to be sufficient to wick away liquid sweat. These fibers are also claimed to eliminate the need for wettability surface treatment.[35]

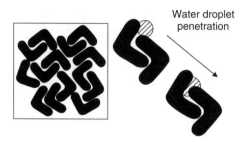

L-shaped nylon cross-section
"Ciebet"-Asahi Chemical Industry Co., Ltd

12.20 Specialty cross-sectional shapes.[31–36] L-shaped nylon cross-section.

COOLMAX® fibers developed by DuPont: This is a proprietary polyester or nylon fiber with unique engineered multiple micro-channel cross-sections. The purpose of this fiber is to provide moisture management for sportswear. It is also claimed to provide comfort through significant breathability (or air permeability).

GORE-TEX® fibers: This fiber was discussed earlier in Section 11.3.2 of Chapter 11.

12.7.4 Yarn types used for sportswear

In the context of performance characteristics of sportswear, yarn can be a key element in enhancing sportswear performance, particularly in relation to durability, stretch, elastic recovery, tactile comfort and thermal comfort. The role of yarn in enhancing these characteristics is discussed below.

As indicated earlier, durability represents one of the critical attributes of sportswear products, particularly when high physical activities are expected. In general, durability can be defined as the extent of survival and performance appropriateness under various external stresses that may be applied to the material during the lifecycle of the fibrous product. It can be measured using many parameters, each of which simulates a reaction to an external stress to which the material can potentially be subjected. These include tension, tear, bending, bursting, shear, twisting and abrasion. Figure 12.21 shows these various modes of deformation and Table 12.4 lists descriptions of these parameters and associated units.

In the context of design of sportswear, durability typically implies a combination of strength and flexibility. Strength is the ability of material to withstand maximum external stresses applied during use. Flexibility is the ease of stretching, bending, and twisting. The trade-off between strength and flexibility represents a common design challenge in most

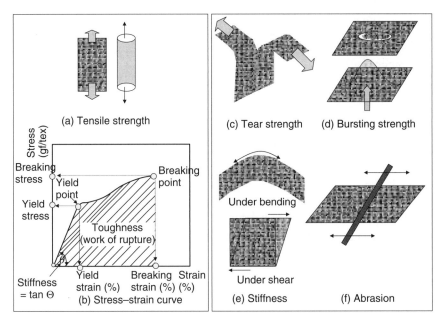

12.21 Different measures of durability.

apparel products. For sportswear, this challenge is even greater since a very stiff material can hinder the performance of players in most sportswear activities and highly flexible material may come at the expense of strength or durability. Accordingly, durability has to be great enough to allow the fibrous structure to withstand the external stresses applied during handling, yet at a reasonable level that allows the necessary flexibility that must be retained in these types of structures.

In Chapter 8 (Fig. 8.12), the main factors determining the flexibility of fibrous products were discussed. The trade-off between strength and flexibility is initially met by the selection of suitable raw material or fibers that must exhibit three basic qualifying characteristics: (1) optimum flexibility, (2) large aspect ratio (length/diameter ratio) and (3) optimum surface interaction. Fibers are then converted into yarns using twisting or some other forms of binding (wrapping or inter-filament arrangement). As indicated earlier, twist is a unique mechanism that essentially preserves the flexibility of yarns. In this regard, ring-spun yarns are preferable to other spun yarns owing to the fact that ring spinning allows yarn twist to be optimized providing the desired strength at acceptable flexibility. Twist levels that allow flexibility of sportswear products are called 'soft twist'. These levels range from 5 to 15 turns per inch (2.5 cm). At this low level, the yarn results in soft fabrics. Most knit fabrics are made from soft-twisted

Table 12.4 Measures of durability of fibers and fibrous structures[3,21]

Durability parameter	Description	Examples of units used
Tensile strength (F)	Force required to rupture a fiber, yarn, or fabric under applied tensile stress (Fig. 12.21a)	g_f, lb_f, N, or cN
Tensile stress (σ)	$\sigma = F/A$	g_f mm^{-2}, PSI, kg$_f$ m^{-2}
Specific stress (σ_s)	$\sigma_s = F/\text{tex}$ or F/denier (tex = g km^{-1}, denier = g 9 km^{-1})	g_f/tex, g_f/denier cN tex^{-1}
Strain (%)	$\varepsilon = (\Delta l/l_o)100$ = increase in length under tension/original length	
Stress–strain curve (σ–ε)	Curve describing the progressive changes in material deformation under external stress (Fig. 12.21b)	
Yield stress	Stress at which the material begins to suffer irrecoverable deformation (Fig. 12.21b)	g_f/tex, g_f/denier cN tex^{-1}
Work of rupture (toughness)	Measure of the ability of material to withstand sudden stresses, expressed by the total area under the stress–strain curve (Fig. 12.21b)	g_f/tex, g_f/denier cN tex^{-1}
Elastic recovery	Extent of recovery upon removal of external stress as expressed by the elastic recovery – stress curve	
Fabric tear strength	Force required to rupture a fabric when lateral (sideways) pulling force is applied at a cut or hole in the fabric (Fig. 12.21c)	g_f, lb_f, N, or cN
Bursting strength	Force required to rupture or create a hole in a fabric when a lateral force (perpendicular to the fabric plane) is applied to a mounted specimen (Fig. 12.21d)	lb_f or kg$_f$
Stiffness	Resistance of a fibrous structure to tension, bending, or shear (Fig. 12.21e) Under tension: Young's modulus (E) Under bending: flexural rigidity (FR) Under torsion: torsional rigidity (TR)	E = cN/tex, g_f/denier $FR = g_{wt}$ cm^2 $TR = g_{wt}$ cm^2
Abrasion resistance	Resistance to wearing away of any part of the fabric by rubbing against another surface (Fig. 12.21f)	number of cycles to rupture

g_f = grams force, kg$_f$ = kilograms force, lb_f = pounds force, g_{wt} = grams weight.

yarns. For more durable sportswear, twist levels may range from 15 to 30 turns per inch. Other types of yarn may also be used to meet the strength–flexibility trade-off. For example, compact ring-spun yarns are expected to be stronger than conventional ring-spun yarns, but also slightly stiffer by virtue of the compactness effect and the larger number of active fibers in the yarn cross-section. Alternatively, compact ring-spun yarns can be made stronger and more flexible using long and fine staple fibers. Core–sheath yarns can also be used to meet optimum strength–flexibility combinations.

Stretch and elastic recovery of sportswear fabrics can be enhanced using yarns made from fibers with high elastic recovery in staple or filament forms. Synthetic fibers such as nylon, polyester, acrylic and polypropylene fiber exhibit good elastic recovery. Textured yarns represent another approach to enhance stretch and elastic recovery. When exceptional elastic recovery is required, spandex (Lycra®) can be added as a separate yarn, or in a core–sheath structure as discussed in Chapter 9.

Yarn can play significant roles in all aspects of tactile comfort. Along the fabric length or width, yarn flexibility can enhance fabric flexibility. The hand and feel of fabric is directly influenced by yarn surface texture (flatness, twist irregularities and hairiness). In the third dimension of fabric (thickness and bulkiness), yarn plays the most critical role. In this regard, fabric compressibility and resilience are key aspects of tactile fabric. Compressibility can be defined as the proportional reduction in the thickness of a material under prescribed conditions of increased pressure or compressive loading.[37] Resilience is the degree to which a material recovers from compressive deformation. These two characteristics should be optimized to achieve good tactile comfort. In this regard, the fabric should have a moderate change in bulk or thickness at low rates of compressive loadings as substantial changes associated with poor recovery can result in significant discomfort.

The multiplicity of factors influencing fabric compressibility makes the role of yarn a complex one. However, it is generally known that yarns with high bulk will result in fabrics with high compressibility and good resilience.

Yarn bulkiness is typically a function of yarn cross-sectional shape, yarn diameter in relation to fabric tightness (loose or tight fabrics) and fabric floats (interlacings and interloping patterns). These parameters can be varied within a particular spinning system using different levels of many factors such as fiber fineness, fiber length, yarn twist and yarn count.[17] When different spinning systems are considered in the design analysis, texturized filament yarns are expected to exhibit the highest level of bulkiness, followed by spun yarns.

Yarns can also play a critical role in enhancing thermal comfort, which typically implies thermal regulation, moisture management and fabric breathability. In this regard, key design parameters of yarn include[17] yarn type (spun yarn versus continuous filament yarn, straight versus textured filament yarn, combed versus carded spun yarn and worsted versus woolen spun yarn), yarn bulkiness (or packing fraction) and yarn finish. In general, texturized and spun yarns provide better thermal comfort than continuous filament yarns.

12.7.5 Fabric types used for sportswear

As indicated earlier, fabric is the quasi-fibrous end product for any application. This is particularly true for sportswear products as they derive their performance characteristics from those of the fabrics they are made from. Sportswear products can be made from woven or knit fabrics and with various constructions, patterns and colors. Woven structures are commonly used for durable sportswear products and high physical activities. Different woven constructions have been used for making sportswear products, particularly plain and twill weave. A common example is twill weave made from polyester fibers typically used for American football shorts. This type of fabric exhibits both good softness and moderate-to-high strength. Examples of knit fabrics used for sportswear applications include a variety of swimwear knit spandex nylon/polyester tricot fabric (10–20% spandex, 90–80% nylon/polyester, 160–210 g m^{-2} weight and about 58 to 80 inch width; 147–203 cm) and polyester honey-comb knit used for sport shirts.

In addition to conventional fabrics, the demand for high-performance sportswear that can take a player's performance to higher levels and new records has driven a great deal of development of specialty sportswear products that rise to the category of function-focus fibrous products. Examples of these products include:

* *Lightweight sportswear:* These are fabrics made from fine denier nylon filament yarns (10–15 denier) to make sports jackets that weigh as low as 100 g/unit with seamless (welding) using hot melting tapes.
* *Phase-changeable clothing:* These are sportswear fabrics that contain special chemicals that can change from liquid to gel states in response to body temperature so that fabric insulation properties can be controlled in such a way as to maintain a constant body temperature even when air temperature changes.
* *Wick-away fabrics:* These are sportswear fabrics that aim to improve the wicking properties of sportswear by drawing moisture away from the skin to prevent a sweaty and clammy feeling. A common example

of this type of fabric is football shirts using the so-called sport wool yarn, which is basically a mixture of real wool and polyester, the wool being highly absorbent fiber and the polyester being essentially a wicking component.

- *Anti-odor/anti-bacterial sports socks:* These are socks that are treated with anti-odor and anti-bacterial treatments to guard against athlete foot fungus.
- *Anti-friction sports socks:* In these socks, the outer layer grips the shoe and the inner layer grips the foot so that the friction is taken up by the two layers and not by the foot skin.
- *Self-cleaning sportswear:* These are products that use nanoparticles attached to the fibers that mimic the 'lotus' concept discussed in Chapter 11.
- *High-performance swimwear:* These are fabrics that are primarily designed to reduce water drag. This can be achieved using a special chemical finish or by mimicking the way sharkskin or shark scales help sharks glide through the water.
- *Layered skiing sportswear:* These are fabrics used for activities such as skiing and mountain climbing. They consist of a layer of moisture transferring material next to the skin, an insulating layer and wind and water-resistant shell garments.
- *Waterproof/breathable sportswear:* These are fabrics coated with a polyurethane membrane via direct printing on the membrane to create a smooth surface and to protect the membrane. Simultaneously, it eliminates the sticky feeling of the polyurethane coated membrane and provides a dry smooth touch to the skin.
- *Water-repellent bonded sportswear:* These are fabrics in which a front of super water-repellent nylon stretch woven fabric is bonded to a back quick-drying polyester knitted fabric leading to a performance combination of light weight, softness, stretchability, mild waterproofing, super water repellency, wind-proofing, perspiration-absorbing and quick drying.

The examples listed above clearly reveal the significant transition toward function-focus sport products. The list also indicates that most of the new developments are geared toward comfort. As indicated earlier, the key performance characteristics of sportswear are durability, comfort, functionality and identity. Design for durability has been well achieved as a result of the availability of many fibers that can provide a great deal of durability-related attributes. The concept of making durable products through appropriate yarn structure and durable fabric construction has also been well established over many years of experience in the fiber and textile industry. In addition, the industry has a long experience with fashion, color and style. Comfort, on the other hand, has been treated for many years as a by-product of the inherent characteristics of fibrous structures (flexibility, drape, elastic recovery, fit, etc.). The list of developments mentioned above indicates that thermo-physiological comfort represents the center of the design activities of the new products. It will be important therefore to dwell on the concept of thermo-physiological comfort as this area of development still has endless potential for further innovation in the years to come. The discussion of this concept should help many engineers and product developers who are interested in designing sportswear for thermal comfort.

12.8   Thermo-physiological comfort: basic concepts

Basic concepts in thermo-physiological comfort are discussed in numerous places in the literature.[14–16,21,38–48] The most important concept is that human body must be kept at a core temperature of about 98.6°F or 37°C. A change in this temperature can be fatal. Fortunately, the human body constantly generates heat which enters the environment through metabolism and physical activities. More critically, the body exhibits a natural heat balance between the rates of heat production and heat loss. However, under severe levels of sport physical activities and/or in extreme environmental conditions, the human body has to make significant adjustments to maintain this balance.

Environmentally, thermal comfort is determined by four basic factors:[15,16,39,40] air temperature (dry bulb temperature, DBT), relative humidity, air movement (velocity in m s^{-1}) and radiation (mean radiant temperature, MRT). In response to these factors, humans use two key supporting elements: physical activity (the natural element) and shelter (the artificial element). Air temperature or the DBT is the temperature of the air surrounding human body; it is indeed the most important factor of all of the above. It is also a stochastic factor that has to be dealt with in view of a number of varying factors such as the human subjects under consideration, the type of clothing they wear and the type of activity they perform. Statistical analyses performed by the present author and others[2,38,40,41] suggest that under normal activities (office work or simple house activities), most seasonally clothed people feel comfortable at temperatures ranging from 21 to 25°C in winter and at temperatures ranging from 22 to 26°C in summer.

The humidity of the surrounding atmosphere does not have as significant effect as that of air temperature, except in extreme conditions in which the human body has to accommodate environmental changes through water evaporation for cooling at high temperature, or under high physical activities. Obviously, many sports activities fall under these conditions. Relative humidity determines the evaporation rate from the skin.[40] In a dry atmosphere, body moisture will evaporate more quickly than in a humid atmosphere. Obviously, when the surrounding atmosphere is saturated with water (very high humidity), skin evaporative cooling will be substantially hindered.

The movement of the surrounding air can result in a variety of thermal effects that can be critical in sports activities, particularly at different temperatures. When the temperature of the moving air is less than the skin temperature, air movement can increase convective heat loss. When the temperature of the moving air is higher than the skin temperature, air movement can significantly warm the skin. Air movement can also accelerate evaporation; this enhances the cooling effect. However, this particular effect will be insignificant at low humidity (<30%) since evaporation will occur without restriction in this case. It will also be insignificant at high humidity (>85%) since the surrounding air will be saturated with water in this case. Accordingly, air movement can provide cooling comfort only at medium relative humidity (40–50%).

Radiation provides effects similar to those of hot air. It is reflected in the average temperature of the media or surfaces surrounding a person. The most common radiation effect comes from sunlight through surfaces such as windows or clothes. In this case, the radiant heat is converted to long-wave electromagnetic radiation causing sensible heat (molecular movement). This heat can then be conducted through the material to the skin. Radiant heat can be decreased by simple measures such as the closing blinds and curtains or wearing appropriate clothing.

The way the unclothed human body accommodates environmental changes is truly fascinating[21,39,40] (see Fig. 12.22). When the surrounding temperature drops (cool environment), the human responds by generating body heat through food metabolism and physical activities. Simultaneously, the skin blood capillaries are constricted to slow down the circulation of the heated body blood to the skin; this provides a natural compensation for body heat loss to the environment. Typically, humans react to cold weather by shaking or shivering, which is a form of physical reaction (muscle contraction) which sends a signal to the body requesting internal heat generation. This situation is often observed when people move from a normal or warm environment to a cold environment. The key matter here is that the heat generated by the body should largely be maintained within the body and not lost to the environment so quickly. In severe conditions of temperature drop (moving into cold water or icy conditions), human body may fail to maintain the heat balance through muscle contraction and shivering, a net heat loss may result, and a hypothermia may occur.

Hypothermia is characterized by a drop in body core temperature down to less than 35°C. A further reduction to about 30–32°C can result in a

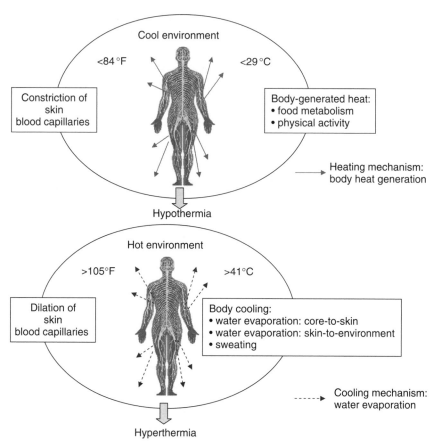

12.22 Thermo-physiological comfort: unclothed body.[21,39,40]

decreased consciousness and heart failure. In this regard, it is important to point out that the rate of heat loss to water is about 30 times greater than that to surrounding air. This should be a critical consideration for swimwear applications. Factors that can slow hypothermia include[41] body size, fat and level of activity. In general, large fat people will cool slower than small thin people.

In a hot environment, human accommodation follows a different mechanism since it must act as a cooling system rather than a heat-generating system. This means that dissipation of body heat to the environment is a positive effect in this case. In an unclothed body, the skin can easily absorb heat from the environment. As a natural response, the human body will tend to dilate blood capillaries (the opposite mechanism to blood constriction). Body cooling requires three basic evaporative actions:[21,41,42] evaporation of water from the body core to the skin, evaporation from the skin to the surrounding media and a sensible perspiration or sweating effect at high external heat (activation of sweat glands).

In contrast to hypothermia, which is the ultimate effect of body cooling, hyperthermia or heat stress is the ultimate effect of body heating. This should be a critical consideration for a wide range of sports applications from mild walking in a hot environment to marathon running. In general, signs of hyperthermia may range from an annoying but mildly disabling form of prickly heat rash to more serious forms such as heat cramps, heat exhaustion and heat stroke. Heat cramps and heat exhaustion result from dehydration and salt depletion as the body sweats to lower its internal temperature. Heat stroke occurs when the body core temperature exceeds 41°C (105°F) because its cooling mechanisms have broken down. This condition requires immediate medical attention as it can cause death.

The above phenomena are partially modeled using equations of body heat balance:[15,16]

$$M - W = Q_{sk} + Q_{res} = (C + R + E_{sk}) + (C_{res} + E_{res}) \qquad (12.9)$$

where M = rate of internal or metabolic energy production, W = rate of mechanical work, Q_{sk} = total rate of heat loss from the skin, Q_{res} = total rate of heat loss through respiration, C = rate of convective heat loss from the skin, R = rate of radiative heat loss from the skin, E_{sk} = rate of total evaporative heat loss from the skin (= $E_{sw} + E_{dif}$), E_{sw} = rate of evaporative heat loss from the skin through sweating, E_{dif} = rate of evaporative heat loss from the skin through moisture diffusion, C_{res} = rate of convective heat loss from respiration and E_{res} = rate of evaporative heat loss from respiration. All terms in the equation are in W m^{-2}.

Another way to express the above equation is through the use of the so-called rate of heat storage of human body, S in W m^{-2} expressed as follows:

$$S = M - W - (Q_{sk} + Q_{res}) = M - W - (C + R + E_{sk}) + (C_{res} + E_{res}) \quad (12.10)$$

The above equation indicates that the main components of heat balance are (i) heat production within the body ($M - W$), (ii) heat loss at the skin ($C + R + E_{sk}$) and (iii) heat loss due to respiration ($C_{res} + E_{res}$).

Metabolism is the term describing the biological processes within the body that lead to the production of heat. The metabolic rate of the human body, M, is typically a function of three basic factors:[41] the extent of physical or muscular activity, environmental conditions, body size and body fat. Human skin is a key sensorial component of thermo-physiological comfort. The thermal state of the environment is perceived by the skin surface. The temperature sensors on the skin are most sensitive around 34°C (93.2°F), where very small differences in temperature can be perceived. The heat

Resting	Btu h^{-1} ft^{-2}	MET
Sleeping	13	0.7
Seated, quiet	18	1

Walking	Btu h^{-1} ft^{-2}	MET
2.9 ft s^{-1} (2 mph)	37	2
4.4 ft s^{-1} (3 mph)	48	2.6
5.9 ft s^{-1} (4 mph)	70	3.8

Office activities	Btu h^{-1} ft^{-2}	MET
Writing	18	1
Typing	20	1.1
Filing, standing	26	1.4
Lifting/packing	39	2.1

House activities	Btu h^{-1} ft^{-2}	MET
Cooking	29–37	1.6–2.0
House cleaning	37–44	2.0–2.4
Pick and shovel work	74–88	4.0–4.8

Leisure activities	Btu h^{-1} ft^{-2}	MET
Dancing, social	44–81	2.4–4.4
Calisthenics/exercise	55–74	3.0–4.0
Tennis, singles	66–74	3.6–4.0
Basketball	90–140	5.0–7.6

12.23 Typical metabolic heat generation for various activities (MET = 58.2 W m^{-2}).[21,38,39]

produced by a resting person is about 85.83 Kcal h^{-1} (340 Btu h^{-1}). Most of this heat is transferred to the environment through the skin. It is important, therefore, to express metabolic activity in terms of heat production per unit area of skin. Using a value of about 1.82 m^2 (19.6 ft^2) as the average skin surface area, a resting adult person will have a metabolic rate of about 50 kcal h^{-1} m^{-2}) (18.4 Btu h^{-1}ft^{-2}) or about 1 MET (a common used unit). Figure 12.23 shows typical values of metabolic rates at various physical activities.

Heat from the surrounding environment can reach the body surface (the skin) by any or all of three heat transfer mechanisms: conduction, convection and radiation. Conduction is heat transfer by contact with a surface; it occurs when collisions of neighboring molecules occur in stationary matter. Convection is heat transfer by actual motion of the hot material. As we are surrounded by air, it is the main medium we exchange heat with. As mentioned earlier, when the temperature of the air is below the mean skin temperature, a net heat loss from the body by convection will occur. On the other hand, when the air temperature is above skin temperature, a net heat gain to the body by convection will occur. Radiation involves a transfer of heat by energy waves. All bodies emit radiant heat at their surface. Surrounding surfaces are radiating to the body in a similar fashion. Again, the body will lose heat if the surroundings are colder and gain heat if they are hotter. Different surfaces may have different degrees of radiant heat transfer (absorption, transmission and reflection). The heat of the sun is transferred to the earth by radiation.

Another important form of heat transfer is evaporation, or more specifically, the latent heat of evaporation. When water evaporates it extracts a quantity of heat from its surroundings. The cooling effect discussed earlier involves evaporation of water from the skin surface, which extracts with it most of the latent heat, leading to skin cooling. Indeed, it is not the sweating that results in cooling, it is the evaporation of the sweat. This is a very effective mechanism as the evaporation of 1 g per minute will be equivalent to 41 W.

In the context of the heat balance equation discussed earlier, thermal balance exists when the heat produced by the body is fully dissipated to the environment. The heat produced in the body core is continuously transported to the skin surface (and to the lungs). It is then emitted to the environment by convection, radiation, some conduction and evaporative heat transfer.

Despite the great deal of knowledge about the heat balance and the different factors influencing thermo-physiological comfort, when the analysis is applied to human subjects, other factors come into the picture that make it nearly impossible to establish universal conclusions. As a result, thermal comfort is commonly defined as the state of mind in which human feels and expresses satisfaction with the thermal environment.[15,16] No matter how controlled the experimental factors are, it is virtually impossible to control the state of mind. Although skin thermal sensors send their signals to the brain, it is the brain that actually controls the unconscious thermoregulatory actions. To make matters additionally complex, skin thermal signals on their way to the brain are internally integrated with core temperature signals in an unknown non-linear fashion. Now the brain has to analyze these signals and in doing so the brain must continuously consult the individual robustness of accommodating the heat or cold (the psycho-physical status). Irrespective of how severe the conditions may be, it is the individual feeling of being pleased or displeased that counts at the end. More critically, it is the individual's ability to express how he/she actually feels, that seldom makes it a one-to-one relationship between comfort and the physical conditions underlying it.

12.8.1 Development of sportswear fabric for thermo-physiological comfort

The role of clothing in providing thermo-physiological comfort is well recognized. A clothed body represents a system in which the fabric acts as an intermediate environment between the surrounding media and human skin. In sport activities performed in a cold climate, the fabric can act as an insulator to prevent the dissipation of the heat generated by the body. In a hot environment, the fabric can act as a barrier against heat absorption by the skin and as a regulator of the evaporation or wetness processes. The benefit of using clothing for thermo-physiological comfort can be appreciated from that fact that clothes can allow humans to survive in conditions ranging from $-20°C$ in the snow to more than $40°C$ in a desert. Obviously, different fibrous structures will exhibit different heat and moisture balance capabilities, which is a key aspect of design for comfort. However, this aspect can be further complicated by other external factors such as the level of physical activity performed by the body, the residence time in a cold or hot environment, body size, body fat, human physical or health status and the rate of change in temperature or temperature gradient.

The total insulation of the clothing may be expressed as the sum of the contributions from the individual layers of clothing being worn. Obviously, these layers are separated by air layers entrapped between them which should be counted in the final analysis. One of the most commonly used measures of clothing insulation are the so-called clo units, defined by the insulation necessary to keep a person comfortable at $21°C$.[42,43] The underlying concept of the clo is that an individual can exert a considerable degree of control over most forms of heat exchange between his/her body surface and the environment by choosing appropriate clothes. However, calculation of heat transmission through clothing can be difficult.

The clo unit (the more technical unit is $m^2K\ W^{-1}$ with $1\ clo = 0.155\ m^2K\ W^{-1}$) was devised to simplify this analysis. A 1 clo unit will maintain indefinite comfort in a sedentary man at 1 MET in an environment of $21°C$ ($69.8°F$), 50% RH and $0.01\ m\ s^{-1}$ ($20\ ft\ min^{-1}$) air movement. Assuming no wind penetration and no body movements to pump air around, clothing insulation is simply calculated by 0.15 times the weight of clothes in lb. (i.e. 0.15 clo per lb of clothes). Accordingly, 10 lb of clothes is equivalent to

1.5 clo. The clo scale is designed so that a naked person has a clo value of 0.0 and someone wearing a typical business suit has a clo value of 1.0. An overall clo value can be calculated for a person's dress by simply taking the clo value for each individual garment worn and adding them together. Another insulation unit used more commonly in Europe is the 'tog', which is 0.645 clo. A fabric has a thermal resistance of 1 tog when a temperature difference of 0.1°C between its two faces produces a heat flow equal to $1\ W\ m^{-2}$.

The relationship between the level of physical activity (MET) and the clothing insulation (clo) is typically governed by the surrounding temperature. Under normal levels of temperatures of say 20°C, people performing high levels of physical activity (MET > 2.0) require clothing with a low clo values (<0.5). As the temperature increases or decreases, the relationship between clo and MET remains essentially the same; the higher the physical activity level, the lower the clo value. However, the desired values of clo and MET will vary significantly with the variation in temperature. For example, at low temperatures, say <10°C, high levels of both clo and MET are desired and at high temperatures, say >25°C, low levels of both clo and MET are desired. The relationship between MET and clo is also influenced by other factors such as sex and age; women tend to need a slightly higher temperature to be comfortable than men perhaps owing to fewer blood vessels near their skin and older people also prefer higher temperature as a result of circulation problems.

The insulation capability of sportswear clothing will depend a great deal on key design factors such as fiber type, yarn structure and fabric structure. By comparison with many other materials, fibers exhibit good resistance to heat flow (low conductivity). Different fiber types exhibit different thermal insulation properties as shown in Fig. 8.9 in Chapter 8. As the fibers are converted into a yarn, the packing density of the yarn becomes a key aspect in governing the heat insulation capability of a fabric since it determines the air/fiber ratio of the yarn, with air being an excellent heat insulator. These factors are clearly reflected in the performance of fabric. More importantly, they can be enhanced or controlled using particular fabric constructions. All these factors can result in fabrics with different insulation levels. A critical point in this regard is the level of air movement within the fabric structure; the more still the air is, the better the insulation capability of fabric. The large surface area of fabric typically allows air to adhere to the fabric structure and exhibit slow movement. In addition, a larger fabric thickness is likely to provide greater thermal resistance. Finally, garment design (collars, cuffs and waists) and garment fit represent key aspects of thermo-physiological comfort.

Another way by which clothing can provide thermo-physiological comfort is by acting as an evaporation regulator. As an intermediate medium between the skin and the surroundings, clothing can effectively influence evaporative cooling through slowing down the diffusion of water vapor from the skin. It can also absorb excess moisture next to the skin. In other words, it can act as an intermediate moisture reservoir from which evaporation and extraction of latent heat occur, giving a supportive cooling effect.

The human body continuously loses water as insensible perspiration. The majority of this water is lost through the skin (more than 60%) and the remainder is discharged through the lungs to the air. In a clothed body, the fabric in contact with the skin handles this moisture through the vapor pressure gradient mechanism, when a difference exists between the vapor pressure on one side of the fabric and the other. In this regard, three media are involved: the moisture on the skin, the moisture in the fabric and the moisture in the surrounding atmosphere. The water will always flow from the moist zone to the dryer zone. Accordingly, insensible perspiration loss requires wetter media from the skin to the fabric and finally to the surrounding air. In a dry environment, this is typically the case. When the surrounding temperature increases (also when the level of physical activity increases), sensible perspiration (sweating) occurs. As indicated earlier, evaporation of the sweat is critical to cool the body. This requires considerable heat to be generated by the body. Thus, the skin must first be heated before it is cooled. At this stage, a critical trade-off must be achieved; evaporation must continue for cooling, yet the sweat must not run out so that evaporation can continue. This is where clothing, acting as an intermediate moisture reservoir becomes critical.

The concept of fabric being an intermediate regulating moisture reservoir involves a great deal of engineering design. Ideally, the fabric should act as a rapid pathway for insensible perspiration from the skin to the environment; this is to keep the skin reasonably dry under normal conditions. In this case, the fabric structure should provide a good wicking effect or capillary action. In other words, the fabric should contain fibers that do not highly absorb the water and it should not interfere or resist water vapor flow or become moisture laden. Instead it should allow the water vapor to flow smoothly through its structure to the environment. In this regard, one of the common terms used is 'breathability'. A breathable fabric is a fibrous structure that allows water vapor to be transmitted to the environment without the structural clogging that can prevent the basic insulation capability of fabric (through removal of inter-structure air).

In a hot environment, fabric wickability is highly desired as it provides a continuous but regulated removal of the sweat from the skin. The regulation aspect here stems from the fact that for wicking to occur, the water level has to be high to create a continuous vapor flow. Thus, the wicking effect will be largely proportional to the amount of sweat generated by the human body. It is important to point out here that the presence of a still water layer between the skin and the fabric is one of the most irritating factors particularly for sports requiring a high level of physical activity.

From a design viewpoint, synthetic fibers are often more suitable for wicking effects than natural fibers. Most synthetic fibers do not absorb a great deal of water. Natural fibers on the other hand are highly absorptive. In other words, they allow water molecules to penetrate into their internal structures. In this regard, two common terms should be distinguished: absorption and adsorption. Absorption is the penetration of water into the internal fiber structures. Adsorption on the other hand is the attachment of water molecules to the fiber surface by some intermolecular forces. Normally, the adsorption characteristics are discussed in the context of the behavior of fabric assemblies. In this regard, a geometrical parameter, such as the surface area of the fiber assembly, is a key design parameter; the larger the surface area, the larger the amount of adsorbed water (cotton towels with looped surfaces are far more adsorbent than cotton towels with cut pile surfaces).

12.8.2 Examples of commercial sportswear fabrics developed for thermo-physiological comfort

On the basis of the above concepts, much effort has been made to develop sportswear fabrics that can provide thermo-physiological comfort. The three primary focuses of this effort has been thermal regulation, moisture management and wind resistance. Most sportswear garments developed for thermal regulation aim to keep body heat in the thermo-neutral zone (within an optimum temperature of $37° \pm 1°C$). Some sportswear garments used in cold weather are designed with good realization of radiant heat loss (via the use of infrared reflective materials and insulative materials) and convective heat loss (via a wind barrier). A commercial example of this category of garments is the so-called 'X-Static®', developed by InSport International, which has a pure silver surface coating that aims to reflect all radiative thermal energy produced by the body in cold weather. In a hot environment, the makers of X-Static® claim that silver, being an excellent conductive element, can quickly distribute the conductive energy produced in hot weather to the environment, keeping the wearer cooler.

Moisture regulation sportswear fabrics primarily aim to remove perspiration quickly so that the fabric remains dry without undergoing dimensional instability (shrinkage and shape loss). The concept underlying the development of most moisture-management sportswear fabrics is illustrated in Fig. 12.24. As can be seen in Figure 12.24a, the presence of moisture absorbent fabric against the skin can create discomfort as it will store the sweat of a wet body via absorption and create a high moisture

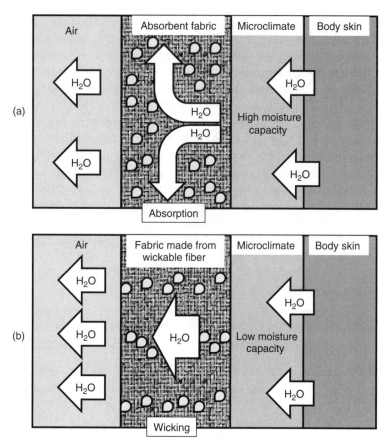

12.24 The concept underlying moisture-management sportswear.
(a) Absorbent fabric against the skin, (b) wickable fabric against the
skin.

capacity between the skin and the fabric. Replacing the absorbent fabric
by a wickable fabric (Figure 12.24b) can result in rapid flow of moisture
through the fabric leading to a cooling evaporative effect and low moisture
capacity at the skin/fabric interface. A familiar example of a commercial
moisture-management fabric is 'Dri-release®', developed by Optimer®.
This fabric is developed with a copolymer polyester acting as the wicking
medium. The yarn is also treated with anti-odor finish (Freshguard®),
which is a permanent protection technology that is claimed virtually to
eliminate body odor. Fabrics made from this yarn are used for shirts, tanks,
singlets, shorts and socks.

Another example of a moisture-management fabric is 'Ever Dry®', developed by Perfectex®, which is a special fabric with a wicking function provided by the shape of the filaments of the yarn, not added by a chemical

wicking material. This fabric is completely washable and the wicking is permanently retained. The filaments of Ever Dry® allow a ditch to exist between the filaments. This ditch provides an excellent conduit for moisture. Perspiration can flow quickly through the fabric during exercise and diffuse quickly and vaporize, allowing the surface of the skin to stay dry, breathable and comfortable. Ever Dry® is available in either nylon or polyester for T-shirts and all exercise wear. The same company also produces fabrics with special coatings or inserted membranes to provide windproof, waterproof or water repellent, anti-UV, anti-bacterial and breathable properties for skiwear, raincoats, jackets and backpacking products.

12.9   References

1. EL MOGAHZY Y, *Statistics and Quality Control for Engineers and Manufacturers: from Basic to Advanced Topics*, 2nd edition, Quality Press, Atlanta, USA, 2002.
2. EL MOGAHZY Y, *Developing a Design-oriented Fabric Comfort Model*, Project Final Report, National Textile Center, Project S01-AE32, 2003.
3. HEARLE J W S, GROSBERG P and BACKER S, *Structural Mechanics of Fibers, Yarns, and Fabrics*, Wiley-Interscience, New York, 1969.
4. EL MOGAHZY Y, *A Study of the Nature of Fiber Friction*, PhD dissertation, North Carolina State University, Raleigh, NC, 1987.
5. GUPTA B S and EL MOGAHZY Y, 'Friction in fibrous materials. Part I: structural model', *Textile Research Journal*, 1993, **61**(9), 547–55.
6. EL MOGAHZY Y and GUPTA B S, 'Friction in fibrous materials, Part II: Experimental study of the effects of structural and morphological factors', *Textile Research Journal*, 1993, **63**(4), 219–30.
7. EL MOGAHZY Y, KILINC F S and HASSAN M, 'Developments in measurements and evaluation of fabric hand, Chapter 3', *Effect of Mechanical and Physical Properties on Fabric Hand*, Behery H, (editor), Woodhead Publishing Limited, Cambridge, UK, 2004, 45–65.
8. HSIEH Y-L and YU B, 'Liquid wetting, transport and retention properties of fibrous assemblies', *Textile Research Journal*, 1992, **62**(11), 677–85.
9. HSIEH Y-L, 'Liquid transport in fabric structure', *Textile Research Journal* 1995, **65**(5), 299–307.
10. GUPTA B S, 'The effect of structural factors on absorbent characteristics of nonwovens', *TAPPI Journal*, 1988, 147–52.
11. NECKAR B and IBRAHIM S, 'Theoretical approach for determining pore characteristics between fibers', *Textile Research Journal*, 2003, **73**(7), 611–19.
12. AHN K J and SEFERIS J C, 'Simultaneous measurements of permeability and capillary pressure of thermosetting matrices in woven fabric reinforcements', *Polymer Composites*, 1991, **12**(3), 146–52.
13. MILLER B and TYOMKIN I, 'Liquid porosimetry: new methodology and applications', *Journal Colloid Interface Science*, 1994, **162**, 163–70.
14. HES L, *Recent Development in the Field of Testing Mechanical and Comfort Properties of Textile Fabrics and Garments*, paper presented at the Institute of Textiles and Clothing, Dresden University, Germany, 1997.

15. ASHRAE Standards 55–1992: *Thermal Environmental Conditions for Human Occupancy – ANSI Approved*, American Society of Heating, Refrigerating and Air-Conditioning Engineers, NY, USA.
16. GAGGE A P, BURTON A C and BAZETT H C, 'A practical system of units for the description of the heat exchange of man with his environment', 1941, *Science*, **94**, 428–30.
17. EL MOGAHZY Y and CHEWNING C, *Fiber To Yarn Manufacturing Technology*, Cotton Incorporated, Cary, NC, USA, 2001.
18. LORD P, *Handbook of Yarn Production, Technology, Science and Economics*, Woodhead Publishing Limited, Cambridge, UK, 2003.
19. BEL-BERGER P, VON HOVEN T, RAMASWAMY G N, KIMMEL L and BOYLSTON E, 'Cotton/kenaf fabrics: a viable natural fabric', *Journal of Cotton Science*, 1999, **3**, 60–70.
20. HATCH K L, *Textile Science*, West Publishing Company, Minneapolis, NY, 1999.
21. MORRIS M A and PRATO H H, 'End-use performance and consumer acceptance of denim fabrics woven from open-end and ring-spun yarns', *Textile Research Journal*, 1978, **48**(3), 177–83.
22. CARD A, MOORE M A and ANKENY M, 'Performance of garment washed denim blue jeans', *AATCC Review*, 2005, **5**(16), 23–7.
23. CAVACO-PAULO A and ALMEIDA L, 'Cellulase activities and finishing effects', *AATCC International Conference and Exhibition*, Green Ville, SC, Book of Papers, 545–54, 1995.
24. HOFFER J M, 'Identifying acid wash, stone wash pumice', *Textile Chemist and Colorist*, 1993, **25**(2), 13–15.
25. BUSCHLE-DILLER G, INGLESBY M K, EL MOGAHZY Y and ZERONIAN S H, 'The effect of scouring using enzymes, organic solvents and caustic soda on the properties of hydrogen peroxide bleached cotton yarn, *Textile Research Journal*, 1998, **68**, 920–9.
26. KOCHAVI D, VIDEBAEK T and CEDRONI D, 'Optimizing processing conditions in enzymatic stonewashing', *American Dyestuff Reporter*, 1990, **79**(9), 24, 26, 28.
27. LANTTO R, MIETTINEN-OINONEN A and SUOMINEN P, 'Backstaining in denim wash with different cellulases', *American Dyestuff Reporter*, 1996, **85**(8), 64, 65, 72.
28. ZUREK W, *The Structure of Yarn* (translated from Polish), USDA and the National Science Foundation, USA, 1975.
29. EL MOGAHZY Y, 'Apparel products used in sportswear: categories and performance', *QBC News letter*, http://www.qualitybc.com/, No. 8, 2006.
30. EL MOGAHZY Y, 'Friction and surface characteristics of synthetic fibers, Chapter 8', *Friction in Textile Materials*, Gupta B S (editor), Woodhead Publishing Limited, Cambridge, UK, 2008.
31. BERKOWITCH J E, 'New hollow filaments from Kanebo', *Trends in Japanese Technology*, Industrial Report published by US Department of Commerce, Office of Technology Policy, Asia-Pacific Technology Program, 92–8, December 1996.
32. BERKOWITCH J E, 'Highly perspiration-absorbing quick drying polyester from Mitsubishi Rayon', *Trends in Japanese Technology*, Industrial Report pub-

lished by US Department of Commerce, Office of Technology Policy, Asia-Pacific Technology Program, 100–5, December 1996.

33. PRAMANICK A K and CROUSE B W, 'Application of deep grooved polyester fiber in composite high absorbent paper', *TAPPI Nonwovens Conference*, St Petersburg, Florida, IPST Technical Paper Series Number 693, 1998.

34. HAILE W A, 'Deep grooved polyester fiber for wet lay applications', *Tappi Journal*, 1995, **78**(8), 139.

35. BERKOWITCH J and YOSHIDA P G, *Trends in Japanese Textile Technology*, Department of Commerce, Office of Technology Policy, December 1996.

36. HONGU T and PHILIPS G O, *New Fibers*, Ellis Horwood, West Sussex, England: 1990.

37. GOSWAMI B C, MARTINDALE J G and SCARDINO F L, *Textile Yarns, Technology, Structure & Applications*, Wiley-Interscience, John Wiley & Sons, London, 1977.

38. EL MOGAHZY Y E, *Understanding Fabric Comfort, Human Survey, Textile Science 93 International Conference Proceedings*, Vol. 1, Technical University of Liberec, Czech Republic, 1993.

39. UMBACH K H, 'Protective clothing against cold with a wide range of thermo-physiological control', *Melliand Texilber*, 1981, **3**, 360–4 (English edition).

40. VOKAC Z, KOPKE V and KEUL P, 'Physiological responses and thermal, humidity, and comfort sensations in wear trials with cotton and polypropylene vests'. *Textile Research Journal*, 1976, **46**, 30–8.

41. UMBACH K H, 'Protective clothing against cold with a wide range of thermo-physiological control', *Melliand Texilber*, 1981, **4**, 456–62 (English edition).

42. HOLMER I, 'Heat exchange and thermal insulation compared in woolen and nylon garments during wear trials', *Textile Research Journal*, 1985, **55**, 511–8.

43. SPENCER-SMITH J L, 'Physical basis of clothing comfort, Part V: The behavior of clothing in transient conditions'. *Clothing Research Journal*, 1978, 21–30.

44. WEHNER J A, MILLER B and REBENFELD L, 'Dynamics of water vapor transmission through fabric barriers', *Textile Research Journal*, 1988, **58**.

45. BEHMANN F W, 'Influence of the sorption properties of clothing on sweating loss and the subjective feeling of sweating', *Applied Polymer Symposium*, 1971, **18**, 1477–82.

46. CASSIE A B D, ATKINS B E and KING G, 'Thermo-static action of textile fibers', *Nature*, 1939, **143**, 162.

47. CASSIE A B D, 'Fibers and fluids', *Journal of the Textile Institute*, 1962, **53**, P739–P745.

48. HOCK C W, SOOKNE A M and HARRIS M, 'Thermal properties of moist fabrics', *Journal of Research of the National Bureau of Standards*, 1944, **32**, 229–52.

13

Development of technical textile products: materials and applications

Abstract: The term 'function-focus fibrous product' (FFFP) has been used collectively to describe all nontraditional fibrous products that make up a wide range of technical, industrial and other products in numerous applications. The superiority of fibrous structures in these products is the result of a broad range of key attributes including light weight, durability, heat resistance, fireproof, and so on. Function-focus fibrous products touch the lives of billions of people under medical care by providing comfort, safety and protection against contamination and infection. Common performance characteristics are discussed and this chapter reviews different application categories of function-focus fibrous products. Function-focus fibrous products are categorized by their application.

Key words: durability; dimensional stability; biodegradation; soil compatibility; fatigue resistance; chemical resistance; environmental effects; radiation effects; bacterial effects; temperature effects; bullet-proof vests; medical textiles; high-modulus reinforcements; superalloys; composites; semiconductor materials; photonic material; e-textiles; biomaterials; load-bearing; weather resistant agro-fiber.

13.1 Introduction

The term function-focus fibrous product (FFFP) has been used in this book collectively to describe all non-traditional fibrous products such as technical and industrial products, in addition to a wide range of products that are used in many areas and numerous applications. Indeed, these products touch upon every aspect of modern living and are felt by all people of all ages at home, at work and everywhere they go. FFFPs can be recognized on the move every day, whether on the road, in the sea or through the air. Indeed, the use of fibrous products in transportation applications has provided billions of people with safety, comfort and many useful functions. FFFPs can be found in critical products such as car and airplane seats, safety air bags, seat belts, upholstery (seat covers), carpets (tufted, needled), filters (non-woven cabin filters engine filters and fuel filters), head liners, hood liners, luggage racks (nets), luggage covers, blankets, soft tops for convertibles and vehicle body parts. The superiority of fibrous structures in these products is a result of key attributes such as light weight (fuel saving and reduction in CO_2 emissions), durability, heat resistance, fireproof, dimensional stability, light fastness, color fastness, abrasion resistance, pilling resistance, liquid and stain repellence (cleanability), hydrolysis resistance and molding resistance. Chapter 14 is entirely devoted to the discussion of the development of fibrous products for transportation applications, particularly safety air bags, seat belts and car seats.

FFFPs also touch upon the lives of billions of people under medical care by providing comfort, safety and protection against contamination and infection. Indeed, when the wound is open, there is no other material that can prevent many microorganisms from contacting and infecting the wound better than a fibrous material, assembled in products such as gauzes, bandages, sutures, hydro gels, hydro colloids, surgical drapes, gowns and clean air suits. All hospital environments rely totally on fibrous products to prevent microbial infection and contamination, to resist penetration by microorganisms in dry and wet conditions, to protect against contamination resulting from liquid penetration and to stay durable against all forms of tear and bursting deformation in operating rooms. Chapter 15 is largely devoted to the discussion of the development of fibrous products for health-related applications such as non-implantable products, implantable products, extracorporeal devices and healthcare/hygiene products.

FFFPs also save the lives of millions of people who work in many risky occupations such as in extreme climate conditions, military and law enforcement battles and flame or fire exposed environments. In these applications, fibrous products largely meet all functionality requirements for protection. In addition, they fulfill the most challenging aspect associated with these applications, which is the trade-off between functionality and comfort. In Chapter 15, developments in protective clothing will be discussed in detail.

In addition to the above major categories of FFFPs, there are many other applications in which fibers and fibrous products are utilized, or can be potentially utilized. In this chapter, a review of different application categories will be presented. Before proceeding with this review, it will be helpful to discuss some of the common performance characteristics of function-focus products.

13.2 Performance characteristics of function-focus fibrous products

Unlike traditional fibrous products, where common performance characteristics can be found that are applicable to almost all products, FFFPs will have different types of performance characteristics and different levels of these characteristics depending on the specific application under consideration and the extent to which they meet the requirements of this application. It is difficult, therefore, to establish performance characteristics and attributes or design parameters that are common to all products, or even to a particular application category without a high degree of specificity.

As will be seen shortly, a survey of most FFFPs will reveal many types of performance characteristics depending on the intended function. However, in most situations a number of characteristics are likely to be useful to various degrees. These include durability, dimensional stability, heat resistance, biodegradation, soil compatibility, solar radiation resistance, fatigue resistance, chemical resistance, frost resistance, wind breaking, thermal screening, hail resistance, gas-release and chemical penetration. For FFFPs used on, or against, the human body (e.g. protective clothing, medical textiles, car seats, seat belts, etc.), light weight and comfort represent critical performance characteristics.

The concept of durability was discussed in Chapters 9 through 12 in the context of traditional fibrous products. Durability of FFFPs is not only related to strength and mechanical resistance but also to many other sources that can influence product durability including environmental effects, radiation effects, bacterial effects and temperature effects. In addition, durability over time (or reliability) is critical for a number of applications. Some products may be stored for months or even years prior to use (e.g. air bags and some medical products), others may be used in changing conditions from hot to cold over time (e.g. protection systems and agricultural-related products), and others may be used for a very short period of time in which intense or severe external effects are applied (e.g. flame-resistant and bullet-proof vests). It is important therefore to extend the concept of durability in the design analysis of function-focus products to accommodate these various effects.

As indicated above, for FFFPs used on, or against, the human body (e.g. protective clothing, medical textiles, car seats, seat belts, etc.), light weight and comfort become necessary. In this regard, the key design challenge is the trade-off between meeting the functional characteristics of the product and the human comfort requirements. This particular issue still represents a challenging task in developing many products particularly protective clothing. In Chapter 15, more discussion on this aspect is presented in the context of protective clothing.

FFFPs are associated with a long supply chain that must be taken into consideration in the product development phase. This chain begins with polymers or fibers suitable for the intended function, then yarns and fabrics revealing the characteristics of the fibers and finally an end product that can be 100% fibrous-based, fibrous-based structure coated with functional substances, or a fiber-other material mixture. In most function-focus products, finishing, coating, lamination and membranes are key components of product performance. Accordingly, the performance characteristics added by these treatments should also be taken into consideration.

The concepts of product development discussed in Chapter 3 can be applied directly to FFFPs. In contrast with the development of most traditional fibrous products, where ideas often stem from conventional wisdom, fashion or style change and long experience with existing products, the development of function-focus products stems from a full realization of the specific function of the fibrous structure in relation to the integrated assembly of the final product. This is a direct result of the fact that function-focus fibrous components are seldom stand-alone products as they must be used in conjunction with or in adherence to other components of the final product assembly. It is important, therefore, to evaluate generated ideas not only in view of the performance characteristics of the fibrous assembly but also in view of the performance of the final product assembly or subassembly. This is a key aspect that should be highly emphasized in the design conceptualization process.

In view of the above discussion, design engineers of FFFPs should be highly knowledgeable of the applications in which fibrous assemblies are used. For example, the design of fabric for automobile airbags should be based on good knowledge of the airbag system components and of the deployment mechanism, as will be discussed in Chapter 14. Similarly, the design of a particular suture for medical application requires knowledge of the factors influencing and influenced by the use of such a suture, as will be discussed in Chapter 15. For these reasons, the development of FFFPs involves examination and analysis of performance characteristics that are extended beyond the conventional characteristics discussed in the previous chapter.

It is also critical to point out that the development of FFFPs should partially stem from a good understanding of the basic characteristics of traditional fibrous products and the basic concepts associated with producing fibers, yarns and fabrics. This point is very important particularly in view of the transition that many schools around the world have decided to make in recent years from a focus on traditional fibrous products to a total emphasis on function-focus products. In this transition, there is often a tendency to eliminate ties with traditional equipment such as spinning and weaving machinery in the research laboratories. It is the author's opinion that a great deal of what we know today about the special functionality of fibers stems from our conventional knowledge of the inherent capabilities of fibers and fibrous assemblies. It is important, therefore, not to break the bridges between traditional and function-focus products.

In many situations, meeting the performance functions of a product should also be associated with understanding of the basic performance characteristics of traditional fibrous products such as comfort and maintenance. This is particularly true for medical and protective fibrous systems as well as for many transportation applications. Furthermore, a great deal of the functional applications used in specialty products can be implemented in traditional fibrous products in order to add value to these products. These include antimicrobial, antibacterial and fire-resistance applications.

13.3   Different categories of function-focus products

One of the key issues mentioned in the above section is the need for design engineers to learn a great deal about the applications in which function-focus products will be either fully or partially used. It is also important to learn about competing materials in various applications so that ways to improve fibrous materials can be established and ideas to substitute competing materials with fibrous materials can be generated. These are essential requirements in the development of any FFFP warranted by the significant expansion of fibrous material utilization in numerous applications. Indeed, the network of FFFPs has been expanded to virtually every aspect of modern life. Figure 13.1 shows the major categories of applications in which FFFPs are used. Within each of these categories are literally

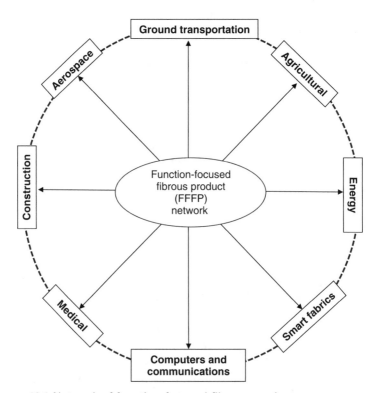

13.1 Network of function-focused fibrous products.

hundreds of fibrous products many of which are replacing products that have been well-established over the years and many others have been newly developed taking advantage of the unique characteristics of fiber materials.

Many companies in the USA, Europe and Japan which have been in the traditional fibrous products business for many years have shifted toward the development of FFFPs either by adding functionality features to traditional fibrous products or by switching over to complete lines of FFFPs. Many new investors in these markets have entered the business of FFFPs with great optimism that their added value will bring about sound profits. One of the assumptions underlying these trends is that FFFPs are less vulnerable to import and price change. While this assumption may be valid on a short-term basis, the reality learned from history is that the key to business robustness and long-term survival stems solely from the dynamic strive to develop new products, add new features and create new markets or expand existing ones.[1-4]

From a product development viewpoint, the success in the market of FFFPs will largely depend on the extent of change in the industry from a commodity-based industry to a value-added industry. This change requires high levels of investment in product innovations. In addition, design conceptualization must become a key element in the industry structure. Furthermore, developers of FFFPs must invest in market research to identify opportunities and explore new ideas that serve consumers in various sectors. Competition will always be a fact of life and it is likely to get tougher. This means that innovation and creative ideas should become a continuous streak and not a static establishment. Creative design can indeed yield products that are unique and robust against duplication or imitation. In this regard, information about market needs and the dynamic changes of these needs is essential for a growing market.

In view of the above discussion, design engineers of FFFPs should be highly knowledgeable of the applications in which fibrous assemblies are used. In this section, a review of some of these applications will be presented. First, an introduction to the application category will be made, then existing or potential uses of fibrous products in the category will be discussed. Since material selection is typically the key to any development, the discussion will emphasize the role of material in different applications. In addition, performance characteristics associated with different product categories will be addressed.

13.4   Materials for energy

In all phases of energy handling including generation, distribution and conversion, material plays a vital role, as it can contribute to consuming energy, saving energy, or becoming a source of energy. As a result, there is a category of materials that is commonly called 'energy materials'. These materials are evaluated and selected on the basis of some classifications.[5,6] For example, they may be divided into passive and active materials, or classified by their association with energy systems such as conventional, advanced and future systems.

Passive energy materials are those that do not contribute directly to the actual energy-conversion process but act as supporting systems (e.g. containers and tools), or energy media (reactor vessels, pipelines, turbine blades or oil drills). Active energy materials are those that play a direct role in energy conversion. These include solar cells, batteries, catalysts and superconducting magnets. Materials associated with conventional energy systems (e.g. fossil fuels, hydroelectric generation and nuclear reactors) are well studied.[6] These materials typically face common performance criteria such as structural mechanical properties and corrosion resistance. Advanced energy systems are in the development stage and are in actual use in limited markets. These include oil from shale and tar sands, coal gasification and liquefaction, photovoltaics, geothermal energy and wind power.[7,8] These will also be associated with material performance criteria including mechanical performance and environmental effects.

Future energy systems are those that are not yet commercially deployed to any significant extent. They are currently under ongoing research that may change the world in years to come, provided that cost obstacles are resolved. Examples of future energy systems include hydrogen fuel and fast-breeder reactors, biomass conversion and superconducting magnets for storing electricity. In recent years, the energy crisis has reached a substantial peak with the price of oil jumping almost 300% in the period from 2006 through 2008. If this trend continues, efforts toward discovering new energy resources will be likely to accelerate.

Both material and energy represent essential resources for product development. A classic example of the relationship between material and energy is engine efficiency. A heat engine converts heat energy to mechanical work by exploiting the temperature gradient between a hot source (e.g. steam) and a cold sink. Heat is transferred from the source through the working body of the engine, to the sink. In this process, some of the heat is converted into work by utilizing the properties of a working substance (e.g. a gas or liquid). The efficiency of a heat engine relates how much useful power is output for a given amount of heat energy input. A simple expression of efficiency is as follows:[9]

$$\eta_{max} = 1 - (T_c / T_h)$$

where T_h is the absolute temperature of the hot source and T_c is that of the cold sink. The first generation of heat engines (steam engines) had an efficiency lower than 1%. Now, heat engines have very high efficiencies (e.g. 3% for ocean power, 25% for most automotive engines, 35% for supercritical coal plant and 60% for a steam-cooled combined cycle gas turbine). These processes gain their efficiency from the temperature drop across them.[9,10]

The evolution of heat engines was a result of continuous improvement in engineering design with material selection being a key component of design. In general, the larger the difference in temperature between the hot source and the cold sink, the larger is the potential efficiency of the conversion of heat to work. The cold side of any heat engine is typically close to the ambient temperature of the environment (26.85°C or 80.33°F). Accordingly, any effort to improve the thermodynamic efficiencies of various heat engines must focus on selecting appropriate materials suitable for combustion chambers, pistons, valves, rotors and turbine blades that can function at very high temperatures.

The first material used for engine design was cast iron and then ordinary steel. With the development of metal alloys, high-temperature alloys containing nickel, molybdenum, chromium and silicon were used. These materials were capable of handling high temperatures of more than 540°C (1000°F). At higher temperatures, metals can fail owing to serious structural dislocations imposed by the planes of atoms that do not extend through the crystal, and free outer electrons, which are good for metal ductility (ease of reshaping) but result in excessive plastic flow under stress, particularly at high temperatures.[5,9] These limitations have resulted in significant efforts to develop materials that can accommodate very high temperatures at the highest efficiency possible (minimum waste of heat). These include[9,10] intermetallic compounds (e.g. nickel aluminide) and ceramics (silicon nitride or silicon carbide). These materials have highly localized electrons in the form of valence or ionic bonds that hold them together. As a result, they exhibit higher melting points than metals and high resistance to chemical attacks. Unfortunately, these materials are also brittle (fail to flow under high stresses) and are subject to crack propagation. These are challenges that call for creative design ideas and new materials or material composites.

Fibrous materials can play a major role in energy applications. They can be a major source of energy and they can certainly save energy. The relationship between fibers and energy goes back thousands of years, as natural organic fibers have been burned to produce energy or converted into gas and used for fuel. This is what is commonly known as 'biomass' and the process of generating energy from these materials is called 'bio-energy'. In principle, heat can be used chemically to convert biomass into a fuel oil, which can be burned like petroleum to generate electricity. Biomass materials can also be burned directly to produce steam for electricity production or manufacturing processes. In a power plant, a turbine usually captures the steam and a generator then converts it into electricity.[11]

Today, the point where the cost of producing energy from fossil fuels exceeds the cost of biomass fuels has been reached. Indeed, with a few exceptions, energy from fossil fuels will cost more money than the same amount of energy supplied through biomass conversion. World production of biomass is estimated at more than 150 billion metric tonnes a year. Although this is mostly wild plant growth, one can imagine how the world will look in the future if all waste produced from fibrous and polymeric products were utilized for energy purposes. One study[12] suggests that nearly 68% of the energy in raw biomass is contained in the charcoal and fuel oils made at the production facility under study. The charcoal has the same heating value in Btu as coal, with virtually no sulfur to pollute the atmosphere. The pyrolytic fuel oil has similar properties to no. 2 and no. 6 fuel oil. The remaining energy is in non-condensable gases that are used to co-generate steam and electricity. Every tonne of biomass converted to fuels in this manner produces approximately 27% charcoal, 14% pyrolytic fuel oil and 59% intermediate-Btu gas.

Fibrous products can also result in substantial energy saving by virtue of their light weight. This is reflected in many applications including machining, automobiles and aircraft. The ease of material manipulation during fabrication and the ease of handling afterwards also results in major energy saving. The need for less frequent maintenance (as in the case of the carpet pile example in Chapter 7) also results in energy saving.

As indicated earlier, fibers have great potential for supporting critical energy applications such as biotechnology. Another dimension of this support can be realized from understanding the functional performance of bioreactors. Typically, bioreactors utilize the specific function of bio-catalysts to produce new and useful fine chemical material effectively, to generate energy and to remove pollutants. A key performance criterion of bioreactors is the immobilization of biocatalysts using many enzymes. In addition, there is a great need for immobilizing organelles, microorganisms or cells to boost the efficiency of bioreactors. In this regard, fibrous structures can be used as carriers for immobilization processes instead of the traditional bead-like, spherical or membrane-type carriers.[13] Positive features that support this trend include an infinitely extended surface area that can be enhanced via innovative fiber cross-sectional shapes and hollow fibers.

13.5   Materials for ground transportation

The term 'ground transportation' can be used to imply transportation vehicles such as automobiles, buses, trucks and trains, or to cover materials used for building highways and railroads. In this section, the focus will be on transportation vehicles. Materials used for highway applications will be covered later under the category of materials for construction products. The most prevalent ground transportation vehicles are automobiles and trucks, with world production exceeding 40 million cars annually. Since fuel efficiency has been and will always remain a critical performance criterion, engineering design has focused for many years on designing fuel-efficient vehicles. This has been a classic design problem in which material selection has played a central role by the development of new auto-body materials, the discovery of new energy sources, or both. The selection of appropriate materials for automobiles has been an evolutionary process of great interest. For many years, competition between steel and aluminum has been well documented.[14] Now, more materials have appeared on the scene in a continuous striving largely to displace metals and replace them with lighter metallic alloys and non-metals. The well-known battle between aluminum and steel has opened engineers' eyes to more innovative materials.[15,16]

Aluminum is typically one-third less dense than steel and this makes it a serious competitor in many automobile parts including doors, hoods, trunk decks and roofs (which collectively make up more than 60% of a vehicle's weight). Using aluminum alloys, yield strength equal to that of moderately strong steel can be reached (similar resistance in fender dent). The problem, however, is that alloying aluminum does not significantly affect its elastic modulus, which is one-third that of steel.[14] A good elastic modulus will prevent automotive door panels or the hood from being easily and largely deflected by external stresses. Early thoughts for resolving this performance issue involved increasing the thickness of aluminum alloys to three times that of steel. Obviously, this would have defeated the whole purpose of the substitution as it would have resulted in a substantial weight increase equal to that of steel. This design problem resulted in extensive research by two British material scientists Michael Ashby and David Jones, in the 1980s, in which the way that components actually deflect stresses was taken into consideration and the result was to increase the thickness of the aluminum of panel doors only slightly to reach equivalent performance.[15] Accordingly, through understanding the relationship between material properties and structural design, a compromise was made in which the net result was a 33% weight saving owing to the substitution of aluminum for steel in such body components.

Another more recent evolution in automobile design is the use of polymeric materials, driven by the need for further reduction in automobile weight. In general, plastics are one-sixth the weight of steel and one-half that of aluminum per unit volume.[16] As expected, polymeric materials had a tougher task in competing against metals in this very profitable application. In general, the strength of most plastics (e.g. epoxies and polyesters) is roughly one-fifth that of steel or aluminum and their elastic modulus is one-sixtieth that of steel and one-twentieth that of aluminum.[16] As a result, the only way for polymeric material to be a part of the automobile body was through composite structures. In this regard, the key factor was to combine relatively weak low-stiffness resins with high-strength, high-modulus reinforcements. As is always the case, cost consideration was an issue in using composite structures and compromises had to be made. For example, the use of carbon fiber (five times the modulus of steel) generater a relatively high cost which has limited its use in automobiles and moved it to the aerospace industry where the high cost can be justified. Other options that have been entertained include glass fibers (1.5 the modulus of aluminum) and mixtures of glass and carbon fibers. Now, the use of composite structures in automobiles has become an accepted reality, a long road that began in 1953 with the introduction of fibreglass-reinforced plastic skins in General Motors' 1953 Corvette sports car and has continued until today. In 1984, General Motors introduced into the market the Fiero, whose entire body is made from composites. Now, composites represent the dominant material in automobiles.

As indicated earlier, striving for ever increasing fuel efficiency has never ceased. Now, Hypercar® vehicles are available with weights less than half those of conventional cars. This is certainly creating a radical 'dematerialization' in the automobile industry. Some Hypercar® vehicles use about 92% less iron and steel than conventional cars. Indeed, the only type of metal that will increase in consumption as a result of this type of car will be the copper material used for electric drives. In 2007, the US President signed a bill, approved overwhelmingly by the House of Representatives, raising automotive fuel economy standards for the first time in more than three decades, requiring a corporate average of 35 miles per gallon by 2020. This bill also boosts federal support for alternative fuel research and energy conservation efforts. This represents new opportunities for further development in energy saving and alternative energy sources which fiber and polymer engineers should give great consideration.

In addition to fibers being a part of auto-composite structures, many other fibrous products are also used in typical vehicles. Indeed, fibrous products effectively contribute both to the constructional features and to the safety features of a car. In a typical vehicle, fibrous products represent many components including[17,18] seat belts, carpets, air bags and tires. Fibrous products can also be found in smaller amounts in other automobile components including[17] cabin air filters, battery separators, hood liners, wheel-arch liners, hoses and belts. In Chapter 14, a great deal of discussion will be presented on the development of seatbelts, safety airbags and car seats. A brief discussion of automobile carpets and tires is presented below.

Estimates suggest that carpets cover about 4 m² on average in every car.[17] These carpets may be made by tufting or by needle-punching.[19] Materials used for car carpets can be made from tufted continuous filament nylons or needle-punched polyesters or polypropylene. Tufted carpets typically have better wear and tear resistance than needle-punched carpets by virtue of potential fiber extraction in the latter, but needle-punched carpets have better mouldability.[17] The relatively poor to moderate compression resilience of polyester prevents it from being widely used in tufted carpets. It should also be pointed out that carpet material can only contribute to fuel efficiency through the use of lighter weights (e.g. 12 oz yd⁻² or 400 g m⁻² for tufted and lower than 8 oz yd⁻² or 267 g m⁻² for needle-punched). Many of the carpet-related performance aspects discussed earlier in Chapter 7 can be applied to the carpets used in transportation vehicles.

An automobile tire may be considered to be a multi-material product in which different materials such as rubber, metal, fibers and chemicals are combined together to form one of the most critical products in human use. Indeed, the construction of a single tire may contain up to 200 different material components.[5,20] A typical tire may consist of nearly 40% rubber (natural and synthetic), 30% carbon black, silica and carbon chalk, nearly 20% of reinforcing materials (steel and fibers) 10% plasticizer (oils and resins) and other vulcanization chemicals (sulfur, zinc oxide) and antioxidants.

The main performance criterion for tires is to provide a cushioning effect between the vehicle and the road. Since the road can be very pumpy, the tire must be robust, flexible and resilient to accommodate road irregularities and provide safe and controlled driving maneuvering. From a design viewpoint, one of the main tire performance limitations is the difference between the elastic properties of rubbers and fibrous materials. This results in poor fiber-rubber bonding, heat build-up and hence poor durability.[17] Shrinkage is also a critical criterion because yarn movement when the tire is subject to heat during the vulcanization process or during use, could lead to distortions and reduced performance and durability. The friction between the tires and the road determines the maximum acceleration and, more importantly, the minimum stopping distance.

In order to make this incredible product, a number of fundamental steps must be taken.[17,20] Rubber is first mixed with chemicals (rubber mixing) to enhance its use performance and create a uniform compound. Prior to tire building, fibers and steel plies are sandwiched in rubber compound using calendering machines. Treads and sidewalls are formed by an extrusion process followed by cooling and cutting processes. Steel wires are covered with the rubber compound and wound to the desired diameter using so-called bead processing. Finally, the tire is assembled, cured (for strengthening), welded and tread patterns are engraved. The tread is the grooved outer layer that is in direct contact with the road. The rubber compound is designed to grip the road, resist general wear and tear and cope with high temperatures generated by tire/road friction. The grooves and tread sipes on the shoulder of the tire are specially designed to channel water away from the surface of the tire and maintain maximum wet grip. Different rubber compounds are suitable for different grip and driving conditions.[17] For example, racing cars have tires that can work at very high temperature ranges with optimum grip, enabling prolonged usage at high speeds on the track. These tires wear more rapidly than typical road tires which are balanced to provide optimum steering, braking, road holding and wear capabilities.

In the context of which fibers to use for making tires, it will be useful to give the reader a glance at the evolution of fiber use in tire applications.[5,8,17,21] Fibers were used for the first time as a reinforcement of tires in 1888 by Dunlop who used a canvas fabric as rubber tire reinforcement. Woven fabric used earlier in making tires was later replaced by a unidirectional arrangement of cords, sometimes with a small number of weft threads across them. Today, fibrous structures represent about 6% of the total weight of a radial tire and 21% of the total weight of a cross-ply tire. In a radial tire, a steel cord, or a breaker, is typically used between the rubber and the fibrous structure for added shock resistance.

Tires have also made use of fiber developments with the first fiber used being cotton, replaced later by rayon owing to its thermal stability. However, the fact that rayon has a smooth surface created bonding problems with rubber, which had to be overcome using resorcinol–formaldehyde latex bonding. High tenacity rayon yarns soon took over. After the discovery of nylon, it was used for tire applications because of its toughness and light weight. However, both nylon 6 and nylon 66 had the problem of 'flatspotting', which restricted their growth, particularly in private cars. This resulted in further developments and more competition between rayon and nylon fibers in the field. Polyester fibers were later used and again presented bonding problems with rubber. The poor dynamic performance of polyester at elevated temperatures has restricted its use to passenger car tires and not to aircraft tires. Yet, the relative low cost of polyester has made it the most popular fiber in tire applications.

In recent years, more advanced materials such as aramids and carbon fibers have been incorporated in tire design.[17] As indicated in Chapter 8, aramids offer the highest strength-to-weight ratio coupled with high temperature resistance. This has made them well suited to specialty cars and aircraft. The obvious limitation here was cost. DuPont developed a fiber–rubber combination using very intimate Kevlar short fiber/rubber dispersions which allowed the special properties of Kevlar (high tensile strength, high modulus and thermal and flexural performance) to be transferred to the rubber. The result was a tire with better resistance to tears and cuts, punctures and actual wear. The high strength–weight ratio of carbon fiber made it a good candidate for tire applications, particularly when vibration dampening and abrasion resistance are of primary interest. As a woven composite fabric, it was increasingly used in high-tech racing cars and jet planes, and even in rockets and satellites. In some advanced tire products, the outboard sidewall of the tire is reinforced with a high-tech carbon fiber insert that helps to provide stiffness for responsive handling and steering precision.

13.6   Materials for aerospace

Products for aerospace applications represent top-end structures in which performance, value and cost are all at their highest levels as a result of the critical sensitivity and precision of these products. In general, key performance criteria for aerospace products include[5,22] fuel efficiency and light weight to increase the distance traveled, high resistance to crack and structural failure for safety and low maintenance requirements and high resistance to elevated temperatures. This three-performance combination largely makes metal alloys and advanced composites dominant in this critical application.

With regard to fuel efficiency, the issues discussed earlier under materials for energy are still valid for aerospace applications but at the higher constraints imposed by the very high speeds and the severe requirements of advanced propulsion systems that operate at temperatures exceeding 1000°C (1800°F). For moderate temperature levels, polymer–matrix composites can be used to take advantage of their light weight. At extremely high temperatures, appropriate material candidates will include metal alloys, metal–matrix composite and ceramic–matrix composite. The high resistance to crack and structural failure makes metal alloys and composite structures the appropriate choices of material.

Keeping up with the historical evolution of materials, metal alloys have been used as the primary material for aerospace applications for many years. As indicated in Chapter 7, metal alloys are composed of two or more metals, or metals and non-metal substances combined together by dissolving in melted conditions. For aerospace applications, it is often desirable to achieve some directional effects using materials that can be tailored to accommodate stresses applied in certain directions (e.g. along aircraft length or in wings). This cannot be achieved using grainy metals that are mostly isotropic (i.e. similar properties in all directions). With metal alloys, a familiar technique called directional solidification, which follows the melting process, is used to provide a great deal of directional effects. In this technique, the temperature of the mold is precisely controlled to promote the formation of aligned stiff crystals as the molten metal cools.[22,23]

Another category of metal alloys called superalloys has contributed largely to the enhancement of aerospace products performances.[5,22] These materials exhibit high strength, high-temperature resistance and high surface stability. One of the common superalloys used in aerospace products is a nickel-based superalloy. This is typically used for the turbine section of jet engines. The inherent weak resistance of this metal alloy to oxidation at high temperatures makes it important to add other superalloys to it such as cobalt, chromium, tungsten, molybdenum, titanium, aluminum and niobium. For engine components subject to intermediate to high temperatures, aluminum–lithium alloys can be used. These are stiffer and less dense than conventional aluminum alloys. The fine grain size of aluminum–lithium alloys makes them 'super plastic'. Some aircraft wings and body skins are partially made from these types of metal alloys. For turbine engines, the high temperatures require titanium alloys. These are also used for military aircraft.

The second candidate materials for aerospace applications are composite structures. The need for composites in aircraft stemmed from the desire to have light-weight structures, an area in which metals and metal alloys cannot compete very well. Another driving force for the use of composites is thermal stability. As indicated in Chapter 7, composite structures can be polymeric, metallic or ceramic. Since polymeric materials tend to degrade at elevated temperatures, polymer–matrix composites (PMCs) are restricted to secondary structures in which operating temperatures are lower than 300°C (570°F). For higher temperatures, metal–matrix and ceramic–matrix composites are required. In polymer–matrix composites, a variety of reinforcements can be used with both thermoset and thermoplastic PMCs, including[22,23] particles, whiskers (very fine single crystals), discontinuous (short) fibers, continuous filaments and textile preforms (made by braiding, weaving or knitting yarns together in prespecified designs).

In Chapter 7, it was clearly indicated that composite structures can be fabricated to provide different directional performances depending on the application. This can be achieved through a number of fiber arrangements including: (a) random (or multi-direction) arrangement for stresses that are homogeneously applied to the product in many directions, (b) unidirectional arrangement for applications in which uniaxial loads are applied to the structure or (c) bi-directional in which stresses are concentrated on two specific directions. It is also important to realize that continuous filaments are more efficient at resisting loads than are short ones, but it is more difficult to fabricate complex shapes and arrangements from materials containing continuous filaments than from short-fiber or particle-reinforced materials.

Metal–matrix and ceramic–matrix composites are justified for aerospace products on the ground that some finished parts may perform at temperatures high enough to melt or degrade a polymer matrix. Metal matrices offer strength, ductility (for toughness) and high-temperature resistance. The problem, however, is that they are relatively heavier than PMCs. In addition, they are more difficult (costly) to process than PMCs. Yet, their durability makes them a superior candidate for sensitive product elements such as the skin of a hypersonic aircraft. For product elements that operate under temperatures high enough to melt metals (e.g. wing edges and engines), the choice left will eventually be ceramic–matrix composites, supported by metallic elements. What works in favor of ceramic–matrix composites for aerospace applications (particularly, aircraft engines) is two-fold:[22] superior heat resistance (or greater combustion efficiency) and low abrasive/corrosive properties. However, what works against them is brittleness (unless fibers or whiskers are added).

Another type of advanced material used for aerospace applications are carbon–carbon composites. These consist of semi-crystalline carbon fibers embedded in a matrix of amorphous carbon fabricated using highly precise techniques.[23] The structures made from this type of composite can retain their strength at a stunning 2500°C (4500°F). For this reason, they are used for the nose cones of re-entry vehicles. Typically, they are protected by a thin layer of ceramic to minimize potential oxidation at such high temperatures. Another category of advanced materials that is being developed is the near-zero coefficient of thermal expansion materials, or perfectly thermally stable materials.[22,23] These will have great potentials for high-speed civilian aircraft.

13.7   Materials for information, computers and communications

In today's information era, we witness the computer revolution. Typically, witnessing a revolutionary development in progress makes it difficult to report it fully. Indeed, historians are now busy recording notes and adding points so that when this revolutionary development stabilizes, they can report it accurately and precisely. The question is when it will stabilize? Obviously, this will not be soon. Interestingly, this revolution is all about using various forms of materials for processing and transmitting information in the form of signals representing data, patterns, sounds and documents.[24,25] These are the basic functions of computers and communications systems. Since signals are created, transmitted and processed as moving electrons or photons, the basic material groups involved are classified as electronic materials and photonic materials.[6] In order to imagine today's revolution of computers and communication simply it should be recalled that in a period of only 35 years (1955–1990), improvements and innovations in semiconductor technology increased the performance and decreased the cost of electronic materials and devices by a factor of one million.[13] This incredible achievement has come as a result of the introduction of new materials and an exponentially upward spiral of capital investment that has provided a drastic reduction in manufacturing cost and created many cost-effective and flexible products.

Electronic material in its natural state can be recognized as[26–28] silicon, Si (the most important material of electronics), silicon dioxide, SiO_2, aluminum, Al, and slightly exotic electronic substances such as titanium nitride, TiN, and tungsten, W. In a more integrated form, electronic materials include various crystalline semiconductors, metalized film conductors, dielectric films, solders, ceramics and polymers (formed into substrates on which circuits are assembled or printed) and gold or copper wiring and cabling. Indeed, in this era of computer and information technology, the term 'semiconductors' has become a household word. This is best defined by an element, such as silicon or germanium that is intermediate in electrical conductivity between conductors and insulators, through which conduction takes place by means of holes and electrons. In other words, a semiconductor is neither a good conductor of electricity (like copper) nor a good insulator (like rubber). These materials can be doped to create an excess or lack of electrons. Computer chips, both for the central processing unit (CPU) and memory, are composed of semiconductor materials. Semiconductors make it possible to miniaturize electronic components, such as transistors.

Photonic material is an emerging leader in optical crystal manufacturing for several fast growth applications including[5,6,27] telecommunications, detector technologies and lasers. Photonic materials include[27] compound semiconductors (designed for light emission or detection), elemental dopants (to serve as photonic performance-control agents), metallic-film or diamond-film heat sinks, metalized films for contacts, physical barriers, bonding, silica glass, ceramics and optical fibers. The best way to imagine and realize the nature of photonic materials is through a biological view. For so many years, technology has not caught up with our desire to create fully three dimensional, micrometer scale, periodic structures. Drawing patterns on a surface presents few problems to the integrated circuit industry, but that third dimension represents the ultimate complexity that has not been fully overcome yet. On the other hand nearly all of biology is engineered on the micrometer scale and with the help of DNA, very complex structures are manufactured. This is why optical properties are frequently exploited, nowhere with more spectacular effect than in the butterfly.

In the context of fibers and fibrous materials, conventional fiber materials do not represent a direct component in electronics or photonic structures. However, fibrous products can support these revolutionary materials in many ways. One logical way is to make fibrous products that can act as a platform to deploy electronics. In this regard, an emerging line of fibrous products categorized as 'E-textiles' has been commercially introduced to consumer markets. This market was driven by the abundance of fabrics in our life, which offer unlimited possibilities for integrating electronics into wearable fibrous products and household fibrous components.[1-4]

From a design viewpoint, the integration of electronics into fibrous products requires a great deal of precise design, particularly in housing and protecting electronic or photonic components in the fibrous structures while achieving durable and precise performance. This combination creates a new way of thinking, as durability has been a traditional design aspect of fibrous products, but precision has not been a common aspect. When one combines these criteria with other critical aspects such as simulation environments and precepts, the design of these products must be supported by defining the electronic fibrous or textile architecture, creating a simulation environment, defining a networking scheme and implementing hardware prototypes.[29] This means that innovative software/hardware architecture should be developed. The software environment acts as a functional modeling and testing platform, providing estimates of design metrics such as power consumption.

In addition to the simulation environments, developing E-textile architecture to serve as software/hardware architecture for a specific class of applications will require the establishment of a set of precepts. These will assist the wearer or user in controling the systems and making decisions governing the application in question. These precepts are likely to be based on past experience and newly developed concepts.[29]

Applications of E-textiles can be extended to inexpensive, flexible, large-area systems that can be draped over a vehicle or a tent or inserted easily in many components, a difficult task to achieve using conventional technologies. In addition, fibrous products exhibit the ability to conform dynamically to new requirements of most applications. Consequently, components such as sensors, actuators and processing elements can be changed and their relative positions can be altered.[30-34]

Many models of fibrous products used in association with computers and communication have been presented under the category of E-textiles. Some are still under development and others have been commercially introduced. The extent of success of these models and the volume of their potential markets are difficult to predict. However, one can only observe how humans are interacting with the surrounding environment today in comparison with 20 or 30 years ago, and then imagine the forms these interactive modes may take in the next 20 to 30 years to realize the incredible potential of these markets. Surely, some models will fail, others will be met with strict regulations or resistance, but many will find their ways to future users. One potential model may aim to develop a 'location awareness' system that determines a user's location within a building or perhaps somewhere in the mountain or the battlefield. Research in this area[35] revealed some promising results using a moderate number of ultrasonic range transceivers as the sensing elements. Given a set of range readings from these sensors, the system attempts to match those actual readings to expected readings associated with a set of candidate locations for the wearer. These expected readings are calculated using a simulation model of the propagation of ultrasonic signals within a building. An additional algorithm is given to determine the wearer's movement between locations, allowing for the uncertainty associated with sensor readings in complex, multi-location environments.

Another model is a hybrid fabric system with a temperature sensing capability.[34,35] Commonly, measurement accuracy comparable to commercial products such as thermistors and thermocouples is used as a good target performance. This is typically about 0.5K over a temperature range of 10°C to 60°C. Some of these hybrid fabric models aim not only to sense temperatures but also to transmit signals.[36] This type of model can also be expanded to accommodate other sensing capabilities such as tracking body postures and blood pressure monitoring.[37–39] The basic design challenge of these models is to incorporate electronics into fabric structures in such a way that allows good performance of these electronics, easy maintenance of the fibrous product and minimum sacrifice of the inherent flexible, drape and the soft nature of the fibrous products. Some of these photonic textiles are developed to offer interactive and communication features using orientation and pressure sensors and Bluetooth or GSM components.[40,41]

In closing this section, one cannot overlook the role of fiberglass or optical fibers in communication and information technology. These fibers can transmit digitized data as an electromagnetic wave and clad optical fibers allow transmission over many miles without the need for signal boosting.[42] The high frequency of optical signals allows a bandwidth of 10^5 MHz, permitting significantly higher data capacity than conventional copper cables, while appropriate fiber design allows signal dispersion to be minimized. Most commercial optical fibers consist of silica glass core-clad structure, made as a 'bulk preform' and drawn down into an optical fiber of diameter 125 μm.

13.8   Materials for medical applications

Ideally, the human body is a self-supporting system that can function on its own to perform many incredible tasks; this is how humans lived for thousands of years. It was a human choice to seek shelter and clothing to create environments that support and enhance human performance. This fundamental concept has been expanded over the years and now we are at a stage in which more support is pursued for the human body in various forms such as replacement of diseased organ implants using artificial organs, insertion of artificial body-support components such as contact lenses and even body-alteration plastic surgery. In addition, the human body occasionally interacts with other objects during surgery, which requires body-friendly components, contamination-free elements and certainly body-compatible materials.

In order to meet the above criteria, a special category of materials commonly called 'biomaterials' has been on the market for many years.[43] The dominant type of material in this category is synthetic polymers such as the polyurethanes or Dacron (polyethylene terephthalate). Other types of materials, used to a lesser extent, include biological polymers (such as proteins or polysaccharides), metals and ceramics. The main reason that some of these materials were reclassified as biomaterials stems from the unique characteristics that they must exhibit to provide safe and effective interaction with the human body. These characteristics can vary widely depending on the application in question. Common characteristics may range from soft and delicate water-absorbing hydrogels made into contact lenses to resilient elastomers found in short- and long-term cardiovascular devices or high-strength acrylics used in orthopedics and dentistry.[43,44] Almost all biomaterials should have excellent interfacial behavior, particularly if they come in contact with blood or host tissues.

In today's medical field, biomaterials are being used in a wide range of applications. These include[44,45] (a) common or high-volume products, such as blood bags, syringes and needles and (b) advanced or complex products, such as implantable devices designed to augment or replace a diseased human organ. The latter devices are used in cardiovascular applications (e.g. heart valves, heart pacemakers and large-diameter vascular grafts), orthopedic (e.g. hip-joint replacements) and dental applications as well as in a wide range of invasive treatment and diagnostic systems.

From a design viewpoint, material selection is the most critical aspect in determining the performance of biomaterial products. Earlier, it was mentioned that synthetic polymers are the dominant category of biomaterials. This is due to the many performance-related characteristics that these materials exhibit including:[33-45]

- mechanical properties: hardness, tensile strength, modulus and elongation; fatigue strength, which is determined by a material's response to cyclic loads or strains; impact properties; resistance to abrasion and

wear; long-term dimensional stability, which is described by the
material's viscoelastic properties
- liquid and gas interactions: swelling in aqueous media; and permeability to gases, water and small bio-molecules
- biocompatibility: human tissues and fluids interaction with biomaterials.

Among the above characteristics, biocompatibility is the most critical one. Most analyses of biocompatibility reveal that it is mainly a surface interaction phenomenon. This is based on the common use of most bio-materials in which the bulk of the material interacts with the human body via the material surface. This surface often exhibits an interface with body materials such as human tissues and fluids (adsorbed layers of water, ions and proteins). These in turn, represent the surface layers of human components filled with cells dispersed in biological fluid.[44,45] As a result, the two key issues associated with biocompatibility are thrombosis, which involves blood coagulation and the adhesion of blood platelets to bio-material surfaces, and the fibrous-tissue encapsulation of biomaterials that are implanted in soft tissues. When implants are the primary products, property requirements should include load-bearing capacity, corrosion resistance, low weight, very low allergic risk, X-ray transparency, durability and easy handling.

History tells us about many problems that have resulted from the use of incompatible biomaterials with body organs. These are lessons that design engineers of biomaterials must be aware of and can learn from. For example, Teflon, a material that has been adopted in biomaterial applications owing to its seemingly biocompatible features such as low friction with human tissues and chemical inertness, was proved to be a major failure in heart valve applications. Teflon has a relatively poor abrasion resistance and this did not set well in heart valve applications. Thus, as an occluder in a heart valve or as an acetabular cup in hip-joint prosthesis, Teflon may eventually wear to such an extent that the device fails. From a design viewpoint, this means that a material that is biocompatible in one application may not necessary meet the requirements of other applications.

Another example that has received worldwide attention is silicone breast implants. In April 1992 the US Food and Drug Administration (FDA) established a ban on the use of this product. It did allow continued use of the implants for women who had undergone mastectomies. It also allowed a small number of women who wanted implants for cosmetic purposes to enroll in long-term studies. The problem was essentially an inner body-exposure to silicone and its potential adverse impact on human life. Studies suggest that there are two ways that silicone can get into the body from an implant.[46-48] One is when microscopic droplets of silicone fluid 'bleed' through the envelope of a gel-filled implant. The other way is through rupture. According to the FDA, about 4–6% of silicone gel implants have ruptured.

In light of the above discussion, it is clear that material selection is an essential aspect of biomaterial design. Indeed, and unlike many of the applications discussed earlier, developing biomaterial products necessitates significant interdisciplinary efforts, not only among engineers of different fields (materials, polymer and biomedical engineers) but also with experts in medical fields (pathologists and clinicians).

The role of fibrous materials in medical applications is truly immense. Indeed, this is one of the fastest growing areas of function-focus products. In Chapter 15, many of these applications will be discussed in detail. A glance at some of the commercial medical fibrous products is presented here to give the reader a brief idea about this growing field:[49–56]

- cotton, nonwoven fabrics and gauze used for traditional medical products
- operation room gowns and drapes, wound dressings, bandages, feminine pads, sanitary products, incontinence aids and hospital bedding
- hollow fibers that are used for applications such as artificial kidney, artificial lung, condensation of alcohol and blood purification
- antithrombotic or hemolytic fibers that are used for blood cell exchange
- carrier fibers that are used for biocatalysts
- resolving, slow-release and biodegradable fibers that are used in numerous applications including sutures and artificial blood vessels
- novel disposable nonwovens that combine the functional properties with good esthetics and comfort
- fiber-reinforced polymer composites for surgical implants.

As can be seen from the above list, developments in fibrous biomaterials have been phenomenal in recent years. This has been a result of major improvements in the quality of commercial absorbent products, major progress in fiber and nonwoven technologies and the use of super absorbent polymers (incorporated in the absorbent core of biomaterial products). For more details on these developments the reader is encouraged to refer to a book on new fibers by Hongu and Phillips,[13] a book on wound closure biomaterials and devices edited by Chu et al.[48] and the *Proceedings of Medical Textiles for Implantation*.[49]

13.9   Materials for construction

In this section, the term 'construction' is used to indicate materials that are used in civil applications from house building to highway structures.

This represents another huge market that continually attracts a great deal of new ideas and major investments owing to its direct impact on human welfare and daily living. Different types of material can be used in these applications, mostly in assembly forms and not in isolation, with factors such as the types of assembled components, shapes and forms representing key design aspects.

In addition to the many material categories used to support construction projects, soil materials should not be overlooked as they are a key partner in most of these applications. The importance of soil stems from its major role in numerous areas including establishing building foundations, highway construction, reinforcing road structures and maintaining roads integrity through controlled drainage and environmental effects. Soil is an enormously complex structure of organic and inorganic components that can generally be characterized by its basic properties such as texture structure and color.[57] Soil texture is represented by the relative proportion of sand, silt and clay size particles in a sample of soil. Clay size particles are the smallest, being less than 0.002 mm in size. Silt is a medium size particle ranging from 0.002 to 0.05 mm in size. The largest particle is sand with diameters ranging between 0.05 mm for fine sand to 2.0 mm for very coarse sand. When clay is the dominant component of soil, the material is described as fine textured. When larger particles are the dominant ones, the material is described as coarse textured.

Soil texture directly influences all other soil properties, particularly those that have a direct impact on soil performance criteria. Different soil textures can result in different structures, compounds, porosity and permeability. Like many fibrous structures, soil exhibits unique natural porosity, the open spaces between its particles. These are created by the immense contact between different irregularly shaped soil particles. Obviously, fine textured soil will have smaller porosity than coarse textured soil as a result of the high packing density of the many small particles. Consequently, fine textured clay soils will have better integrity and can hold more water than coarse textured soils (or sandy soils). Coarse textured soils, on the other hand, have higher water permeability than fine textured soils.

Soil structure is represented by the aggregative manner of soil particles into the so-called peds.[57] The shape of these peds can vary widely depending on the texture, composition and environment. In this regard, two basic soil structures are identified:[57–60] granular soils and platy soils. Granular soils are basically crumb structures with open structures that allow water and air to penetrate through soil. Platy soils are basically closed structures that impede water penetration through soil. In this regard, soil compactness determined by the bulk density of soil is a key structural feature. Logically, bulk density will increase with the increase of clay content.

Soil material is typically very strong under compression, particularly when it is appropriately compacted. Yet, it cannot carry tensile loading. This deficiency has resulted in the use of many materials for soil reinforcement. In this regard, fibrous assemblies represent a logical candidate material for soil reinforcement as they can easily accommodate tensile stresses. Fibrous assemblies used for this purpose are generally called 'geotextiles'. These can be used in many construction applications including[58–61] road reinforcement, filtration, embankment reinforcement, drainage, separation and erosion control. Many fibers have been used in geotextile applications, but the most common fiber types used are polypropylene, polyethylene and polyester. The key performance characteristics of geotexiles are mechanical properties, filtration ability and chemical resistance.

Soil or soil mixtures can be converted into harder structures such as concretes or stones for use in many construction projects such as buildings, roads, bridges and dams.[61] Concrete is a composite material which is made up of a filler and a binder. The binder (cement paste) binds or glues the filler together to form a synthetic conglomerate. The constituents used for the binder are cement and water, while the filler can be fine or coarse aggregates. Cement is a mixture of compounds made by burning limestone and clay together at very high temperatures ranging from 1400 to 1600°C.

The performance of concrete structures is largely determined by factors such as the type of aggregate or cement (shape, texture and size), aggregate specific properties and conversion techniques. The strength and flexibility of an ordinary structural concrete is largely determined by the water-to-cement ratio; the lower the water content, all else being equal, the stronger the concrete. However, the concrete mixture should have just enough water to ensure that each aggregate particle is completely surrounded by the cement paste, that the spaces between the aggregate are filled and that the concrete is liquid enough to be poured and spread effectively.[62,63] The strength of the concrete is also determined by the amount of cement in relation to the aggregate (expressed as a three-part ratio – cement to fine aggregate to coarse aggregate). For especially strong concrete, relatively less aggregate will be needed.

Concrete integrity and strength are largely sensitive to changes in environmental effects such as temperature and moisture.[62] Typically, the strength of concrete is measured by the force per area (e.g. pounds per square inch or Newtons per square meter) that is needed to crush a sample of a given age or hardness. If concrete is allowed to dry prematurely, it can experience unequal tensile stresses which, in an imperfectly hardened state, cannot be resisted. For this reason, the concrete is kept damp for some time after pouring to slow the shrinkage that occurs as it hardens (curing process). Low temperatures can also adversely affect concrete strength. To compensate for this, an additive such as calcium chloride is mixed in with the cement.[62] This accelerates the setting process, which in turn generates heat sufficient to counteract moderately low temperatures.

Despite efforts to form strong concretes, ordinary concrete is brittle and is not typically very strong, particularly under bending, making it vulnerable to external factors such as wind action, earthquakes and excessive vibrations. As a result, it is not suitable for many structural applications. The common way to overcome this deficiency is by reinforcing concrete with other materials such as metals (mainly steel) to form *the* so-called ferroconcrete.[62] This results in a unique performance combination of high tensile strength (steel) and high compression strength (concrete). This also provides various formation possibilities such as rods, bars or mesh. The fluidity of the concrete mix makes it possible to position the steel at or near the point where the greatest stress is anticipated.

The benefits obtained from reinforcing concrete with steel are often limited by the relatively poor corrosion resistance of the steel by salt, which results in the failure of these structures. As a result, constant maintenance and repair is needed to prolong the lifecycle of steel-reinforced concrete used in civil structures. Alternatively, fiber-reinforced concrete (FRC), in which the steel is partially or completely replaced by fibers, can provide a better option.[62,63] Typically, Portland cement concrete is reinforced with more or less randomly distributed fibers by dispersing thousands of small fibers randomly in the concrete during mixing. Fibers can greatly assist in improving post peak ductility performance, pre-crack tensile strength, fatigue strength, impact strength and eliminate temperature and shrinkage cracks.

The key design criterion in making FRC is the fiber type used. Before the idea of FRC was fully developed, many natural fibers had already been used to reinforce concretes. Over the years, the concept of concrete reinforcing by fibers has evolved as a result of the increasing capabilities of testing and the introduction of new fibers. These include synthetic organic fibers such as polypropylene or carbon, synthetic inorganics such as steel or glass fibers and natural organics such as cellulose or sisal fibers.

When concretes are reinforced by fibers, design engineers should carefully evaluate key fiber performance characteristics such as fiber dimensions (diameter, length and specific gravity), mechanical properties (tensile strength, elastic modulus and toughness) and fiber–cement compatibility or reactivity with the surrounding environment (fiber chemical properties in terms of their inertness). In general, a plain concrete structure cracks into two pieces when it is subjected to excessive tensile loading. A FRC may also suffer crack at the same peak tensile load, but it will not completely fail even at higher levels of deformation. Thus, the difference will primarily be in the enhancement of toughness, as the area under the stress–strain curve of a fiber-reinforced concrete will be larger than that of an ordinary concrete. Thus, the real advantage of mixing fibers with cement will be in creating fiber bridges to minimize cracking under a pullout process.[62]

13.10 Materials for industrial applications

The numerous industrial products available in today's market clearly imply that all types of materials from metals to foams have been used to make these products. A review of these materials will be a review of all industrial products, which certainly deserves a separate book, if not books. In this section, we will focus on two examples of industrial products for the purpose of demonstration: (1) filters and (2) load-bearing products. These two types of material were chosen primarily for their wide applications in numerous fields.

13.10.1 Materials for filtration

Fibers and fibrous materials are the primary candidates for filtration applications used by many industries such as textile, power, cement, chemical, mining, food, biotechnology and petroleum. In principle, the fibrous structure with its porous nature can be used as a filter to separate or retain substances (e.g. pollutant particles) from a fluid environment (e.g. air or other gaseous flow) that can easily penetrate the fibrous structure. This results in a clean fluid that can be used several times. Filters may take different shapes including cartridges, panels, bags, sleeves and rolls.

Unlike many of the markets of FFFPs, the filtration market is typically scattered as a result of the diverse fields in which filters can be used.[64–67] All types of fabric can be used to develop filters. However, the most commonly used type is nonwovens, particularly in the hydro-entangled form.[64] These are typically 60% lighter than needle-punched nonwovens and exhibit equivalent resistance against external stresses. Nonwoven structures are also used for developing dual-density filters. These are compound filters in which each component increases the filtration capacity of the other one. They can be made from a combination of a spun-bonded nonwoven with a melt-blown cloth, which protects and reinforces the material used as filter, or a combination of a melt-blown non-woven cloth and a fabric sheet as a protector element.

Many fibers can also be used for filtration applications including[65,66] polyester, polypropylene and acrylic fibers. In some special situations where durability and high temperature resistance are key criteria, high-performance fibers such as meta-aramides, polyamides and polytetrafluoroethylene are used. The temperature resistances of these fibers are considerably higher than the traditional fibers, exceeding 200°C. For

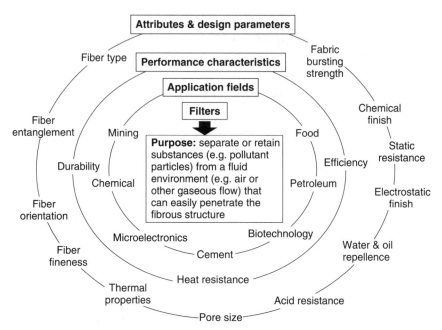

13.2 Performance characteristics and related attributes of fibrous products used for filters.

temperatures exceeding 300°C, fibers can be covered with an organic gel of zirconium acid, which is turned into an inorganic covering by thermal treatment.[64]

To improve filter efficiency, key factors include[564,65] contact surface, porous size and distribution, water and oil repellence, static charges and acid resistance. The use of hydro-entangled nonwovens and micro-denier or nano fibers can assist in increasing the contact surface and controlling the pore size, allowing a higher retaining rate of particulates.[64] Different types of chemical finish can also be used to enhance the other factors. It is also possible to charge the fibers electrostatically in the filters by applying a stable electrostatic corona discharge that allows a fast configuration of the dipole, where each fiber is made of electric polarization cells or 'electrets', similar to a set of magnets, so that particles are attracted by fibers. Different performance characteristics and attributes associated with filters are shown in Fig. 13.2.

13.10.2 Materials used for load handling

Load handling is a common application in which loads may be lifted, suspended or dropped using flexible, yet very strong materials. It has traditionally been performed using metal chains or metallic slings (cables) particularly in applications where medium to very heavy loads are utilized. Fibrous components have also been used, particularly for light to medium load applications. In recent years, fibrous materials have been used in many super heavy-load applications particularly in deep ocean lifting and suspension applications.[68,69] In Chapter 9, fibrous ropes were discussed in the context of yarn types. In today's heavy-load applications, fiber ropes are used in handling extremely heavy loads owing to their strength, fatigue resistance, heat resistance and flexibility. In this regard, many fatigue models were developed to assist in optimizing the design of ropes used for heavy-load handling. Some of these ropes may be reinforced with metallic components either as a part of their structure or as terminals to facilitate the union and suspension of loads. Another fibrous product is flexible slings with a flat tape used to join loads with the hook of a crane or elevation equipment.[70] The tapes are typically made from woven polyamide, polyester or polypropylene multifilaments. They can also be made of two or more layers of identical tape superimposed lengthwise and joined by seams. Tubular slings represent another important fibrous product in which high tenacity polyamide, polyester or polypropylene multifilaments covered with fabric are used. Figure 13.3 shows common performance characteristics for fibrous products used in load handling and their associated attributes and design parameters.

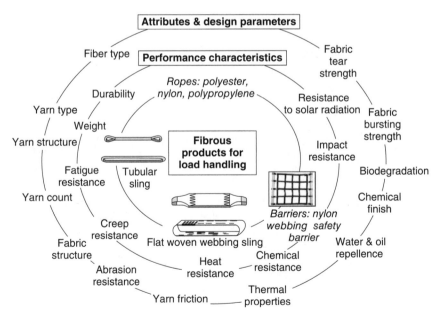

13.3 Performance characteristics and related attributes of fibrous products used for load handling.

13.11 Materials for agriculture applications

Obviously, millions of materials are used in various agricultural systems and applications. When the focus is on fibrous materials, the term agro-fiber is commonly used to distinguish these materials from others used in the field. Agro-fiber products can be made from many fiber types including polyolefin (high density polyethylene and polypropylene), polyester and polyamide. They can also be made in different structural forms including woven, mesh knit or nonwoven fabrics. The most common application is the protection of the plants by providing a better growing environment and protecting cultivation (e.g. preventing ground desiccation, facilitating recollection and treatments, and reducing the use of fertilizers, agrochemicals and water).

In general, three categories of protective meshes are used:[71] (1) weather resistant agro-fiber products (woven and nonwovens), (2) against insects and animals (woven and knits) and (3) against solar radiation. Figure 13.4 shows basic performance characteristics associated with these product categories and key design parameters and attributes corresponding to these characteristics.

Not only can fibrous material be used to protect plants, it can also be used to simulate plants. This is demonstrated by artificial turfs that are

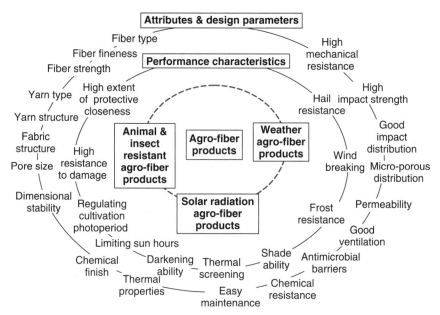

13.4 Performance characteristics and related attributes of fibrous products used for agriculture applications.

used in many sport fields to provide important features such as easy cleaning, low cost maintenance and convenient replacement. Artificial turfs are commonly made from tufted fabrics. In principle, tufted fabrics consist of yarns inserted into fabric backing to form loops or cut loops (piles). Yarns are fixed in place using latex or some other forms of adhesive substances that in essence form a perforated secondary backing element to ease drainage and water permeability.[72] The nature of turf requires that yarns should have special structures such as slit films or fibrillated filaments, monofilament yarns or textured yarns. These yarns are made from synthetic filaments with added chemicals during extrusion to enhance its resistance to climate change and UV effects. In addition, different cross-sectional shapes can be made giving different grass appearances.

From a cost viewpoint, slit films have the advantage in that they can be extruded at higher output rates than monofilaments. Slit films or fibrillated polyethylene yarns used for sports applications may reach a height of up to 60 mm. When turf is used for decorative purposes, lower yarn heights are used. Backing is mainly made from raffia woven fabrics of polypropylene or polypropylene combined with weft polyester multifilament. A nonwoven component, commonly spun-bonded, can also be added to the primary support to offer more dimensional stability. Furthermore, the grass structure can be filled with marble sand, rubber particles or other materials, depending on the use, in order to give yarns more stability. Different performance characteristics of artificial turfs and attributes or design parameters that are critical for meeting these characteristics are shown in Fig. 13.5.

13.12 Smart materials

Two categories of smart fibrous materials can be identified: (a) materials utilizing information, computers and communication applications and (b) materials utilizing nanoparticles and special finishes. Both types share one common capability, which is to detect and react in a spontaneous fashion to external stimulus by making a self-adjustment that is useful for that particular application. The first category is typically a result of incorporating non-fibrous components such as electronics and light reactors into a fabric structure in order to perform specific functions. This incorporation opens the door to numerous advanced applications in different areas such as telecommunication biology, medicine, transportation and military.[73-75]

According to Aubouy et al.,[75] smart fabrics can be designed in a wide range of categories from simple fabrics that operate by external stimuli to fabrics that are capable of detecting stimuli by themselves, analyzing the signals and performing appropriately in accordance with the stimuli and

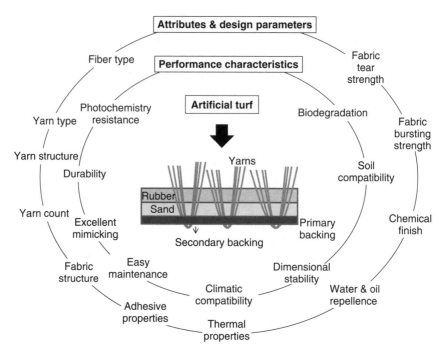

13.5 Performance characteristics and related attributes of fibrous products used for artificial turf.

to the benefit of the wearer. This capability requires a combination of sensors and performance systems using physical, chemical and telecommunication phenomena. For example, some E-fabrics use the physics of storing and producing electricity without disturbing the wearer by the achievement of conductivity and capacity concepts. Conductivity features are fulfilled using yarns and/or finishing techniques in which conductive materials such as silver, copper, conductive polymers and carbon are incorporated. Common conductive polymers include polyanilines, polypyrroles and polythiophenes. Electricity is usually provided by ultrafine or flexible batteries. Some smart fabrics generate their own electricity by means of flexible photovoltaic plates and kinetic or piezoelectric generators.

The second category relies on the special characteristics of particular materials (e.g. nano materials or special polymers) in providing a spontaneous response to external stimuli. In Chapter 11, examples of these materials were discussed. Given below is a list of examples of smart fabrics that have been used in the market.

- Smart T-shirts originally developed for use by US Navy for the purpose of identifying the exact location of a body wound and diagnose the physical problem and its level of seriousness, then transfer all the information to the receiver for appropriate action. This is basically a woven structure with inserts of electrical fibers that can detect and record heat and respiration rates, body temperatures and calories burned. Information can then be wirelessly dispatched to doctors and other medical responders. Potential applications of smart shirts include law enforcement officers, firemen, astronauts, military personnel, chronically ill patients, elderly persons living alone, infants and their parents, and professional and amateur athletes.

- Smart bras designed for the detection of breast tumors. A smart bra operates using an electric current that goes through the breast, monitoring the electromagnetic conduction differences between the tissue of the healthy breast and the tumorous one. By determining the difference in density between healthy and carcinogenic tissues, detection of the presence of carcinogenic cells is possible. One type of smart bra relies upon thermograph principles by detecting minor temperature changes that occur in the breast tissue. As tumors begin to grow, they demand a higher blood flow to develop. This increased blood flow will result in an elevated temperature, signifying the initial stages of cancerous growth.[75] In this regard, smart bras using this principle should be made from materials that are sensitive to passive microwaves, similar to those used in remote sensing technologies, to detect the thermal alterations. The cancer-diagnosing bra can be worn instead of a normal undergarment and can incorporate a discrete alarm feature to alert the wearer to seek medical attention if necessary. One obstacle facing thermograph principles is that blood flow rates could be increased for a number of reasons leading to false positive detections.
- Smart fabrics that allow thermoregulation using thermal sensors and thermoelectric principles to provide warmth or cold feeling to the wearers.
- Breathable and water repellent fabrics produced by nanotechnology self-cleaning fabrics based on the Lotus effect principle.
- High protective fabrics against UV obtained by applying inorganic nanoparticles.
- Antibacterial fabrics based on the application of silver TiO_2 and ZnO nanoparticles.
- Anti-odor fabrics obtained by means of the nanoencapsulation of different active compounds.
- Wrinkle resistant fabrics based on nanoparticles of TiO_2 or SiO_2.

13.13 Conclusions

In this chapter, general thoughts on the performance characteristics of function-focus fibrous products have been discussed and a review of various categories of function-focus materials and products has been presented with some attention to fibrous materials. In the context of fiber-to-fabric engineering, the objective of this chapter was to serve the information gathering phase of product development and design conceptualization by introducing various speciality applications in which fibrous materials are currently used or can potentially be used. More details of these applications can be found in numerous literature references and highly specialized publications. In Chapters 14 and 15, we will return to the discussion of product development and design conceptualization using specific examples of function-focus products.

13.14 References

1. CHANG W and KILDUFF P, *The US Market for Technical Textiles*, report submitted to Small Business & Technology Development Center, Raleigh, NC, http://www.sbtdc.org/pdf/textiles.pdf, May 2004.
2. BYRNE C, 'What are the technical textiles?', *Industrial Fabric Products Review*, 1997, 57–60.
3. SHISHOO R L, 'Technical textiles – Technological and market developments and trends', *Indian Journal of Fiber & Textile Research*, 1997, **22**, 213–21.
4. DAVID RIGBY ASSOCIATES, 'Hometech: An overview of developments and trends in the world market for technical textiles in home furnishings applications', *Nonwoven Industry*, 1999, 42–50.
5. SMITH W F, *Materials Science and Engineering*, in AccessScience@McGraw-Hill, http://www.accessscience.com, DOI 10.1036/1097–8542.409550, last modified: May 4, 2001.
6. MURRAY G T, *Introduction to Engineering Materials: Behavior, Properties, and Selection*, Marcel Dekker, New York, 1993.
7. CALLISTER W D JR, *Materials Science and Engineering: An Introduction*, 2nd edition, Wiley Interscience, John Wiley & Sons, New York, 1991.
8. FLINN R A and TROJAN P K, Engineering Materials and Their Applications, 4th edition, Academic Press, New York, 1990.
9. KROEMER H and KITTEL C, *Thermal Physics*, 2nd edition, W. H. Freeman Company, Los Angeles, CA, 1980.
10. CALLEN H B, *Thermodynamics and an Introduction to Thermostatistics*, 2nd edition, John Wiley & Sons, New York, 1985.
11. CUFF D J and YOUNG W J, *US Energy Atlas*, Free Press/Macmillan Publishing, NY, 1980.
12. KNIGHT J A, 'Pyrolysis of wood residues with a vertical bed reactor', *Progress in Biomass Conversion, Vol. 1*, Sarkanen KV and Tillman DA (editors), Academic Press, NY, 1979.
13. HONGU T and PHILLIPS G O, *New Fibers*, 2nd eition, Woodhead Publishing Limited, Cambridge, UK, 1997.
14. *Metals Handbook, Vol. 1*, American Society for Metals, Metals Park, Ohio, 1961, 185–7.

15. COMPTON W D and GJOSTEIN N A, 'Materials for ground transportation', *Scientific American*, 1986, **255** (4), 92–100.

16. WRIGHT K, 'The shape of things to go', *Scientific American*, 1990, **262** (5), 92–101.

17. FUNG W and HARDCASTLE M, *Textiles in Automotive Engineering*, Woodhead Publishing Limited, Cambridge, UK 2001.

18. MUKHOPADHAY S K and PARTRIDGE J F, 'Automotive textiles', *Textile Progress*, 1999, **29** (1/2) 68–87 (Manchester, The Textile Institute).

19. CHEEK M, 'Automotive carpets and fibers; an international perspectives', *Textiles in Automotive Conference*, Greenville, SC, October 1991.

20. CARRAHER C E JR, *Polymer*, in AccessScience@McGraw-Hill, http://www.accessscience.com, DOI 10.1036/1097-8542.535100, last modified: March 12, 2004.

21. BLOCK I, *Manufactured Fiber*, in AccessScience@McGraw-Hill, http://www.accessscience.com, DOI 10.1036/1097-8542.404050, last modified: May 6, 2002.

22. STEINBERG M A, 'Materials for aerospace', *Scientific American*, 1986 **255** (4), 66–72.

23. CHOU T-W, MCCULLOUGH R L and BYRON PIPES R, 'Composites', *Scientific American*, 1986, **255** (4), 192–203.

24. MAYO J S, 'Materials for Information and Communication', *Scientific American*, 1986, **255** (4), 58–66.

25. PATEL C K N, 'Materials and processing: core competencies and strategic resource', *AT&T Technical Journal*, 1990, **69** (6), 2–8.

26. BENSON K E, KIMERLING L C and PANOUSIS P T, 'Reaching the limits in silicon processing', *AT&T Technical Journal*, 1990, **69** (6), 16–31.

27. REICHMANIS E and THOMPSON L F, 'Challenges in lithographic materials and processes', *AT&T Technical Journal*, 1990, **69** (6), 32–45.

28. MITCHELL J W, VALDES J L and CADET G, 'Benign precursors for semiconductor processing', *AT&T Technical Journal*, 1989, **68** (1), 101–112.

29. NAKAD Z S, *Architectures for e-Textiles*, PhD Dissertation, Virginia Polytechnic Institute and State University, Blacksburg, VA , 2003.

30. PARKER R, RILEY R, JONES M, LEO D, BEEX L and MILSON T, 'STRETCH – An Etextile for large-scale sensor systems', *International Interactive Textiles for the Warrior Conference*, Cambridge, MA, Abstract for poster presentation, July 2002, 59.

31. JONES M, MARTIN T and NAKAD Z, 'A service backplane for E-textile applications', *Workshop on Modeling, Analysis, and Middleware Support for Electronic Textiles (MAMSET)*, Atlanta, GA, October 2002.

32. POST E R, ORTH M, RUSSO P R and GERSHENFELD N, 'E-ebroidery design and fabrication of textile-based computing', *IBM Systems Journal*, 2000, **39** (3&4), 2000.

33. MARCULESCU D, MARCULESCU R and KHOSLA P K, 'Challenges and opportunities in electronic textiles modeling and optimization', *Proceedings of the 39th Design Automation Conference*, Greenville, SC, 2002, 175–180.

34. FIROOZBAKHSH B, JAYANT N, PARK S and JAYARAMAN S, 'Wireless communication of vital signs using the georgia tech wearable motherboard', *2000 IEEE Con-*

ference on Multimedia and Expo (ICME 200) Raleigh, NC, 2000, Vol. 3, 30 July–2 Aug 2000, 1253–6.

35. CHANDRA M, JONES M T and MARTIN T L, 'E-textiles for autonomous location awareness', *Eighth IEEE International Symposium on Wearable Computers (ISWC'04)*, Raleigh, NC, 2004, 48–55.

36. LOCHER I, KIRSTEIN T and GERHARD T, 'Routing methods adapted to e-textiles', *Proceedings 37th International Symposium on Microelectronics (IMAPS 2004)*, Los Angeles, CA, November, 2004.

37. JUNG S, LAUTERBACH C H, STRASSER M and WEBER W, 'Enabling technologies for disappearing electronics in smart textiles', In *IEEE International Solid-State Circuits Conference*, San Francisco, CA, February 9–13, 2003.

38. KALLMAYER C, LINZ T, ASCHENBRENNER R and REICHL H, 'System integration technologies for smart textiles', *mst news*, 2005, **2**, 42–3.

39. LORUSSI F, ROCCHIA W, SCILINGO E P, TOGNETTI A and D E, ROSSI D, 'Wearable, redundant fabric-based sensor arrays for reconstruction of body segment posture', *IEEE Sensors Journal*, 2004, **4** (6), 807–18.

40. WESTERTERP-PLANTENGA M S, WOUTERS L and TEN HOOR F, 'Deceleration in cumulative food intake curves, changes in body temperature and diet-induced thermogenesis', *Physiology & Behavior*, 1990, **48**, 831–6.

41. DE VRIES J, STRUBBE J H, WILDERING W C, GORTER J A and PRINS A J A, 'Patterns of body temperature during feeding in rats under varying ambient temperatures', *Physiology & Behavior*, 1993, **53**, 229–35.

42. HEARLE J W S (editor), *High-performance Fibers*, Woodhead Publishing Limited, Cambridge, UK, 2001.

43. 'Materials science', *Encyclopædia Britannica*, 2006. Encyclopædia Britannica Online. 16 Oct. 2006 <http://search.eb.com/eb/article-32302>.

44. FULLER R A and ROSEN J J, 'Materials for medicine', *Scientific American*, 1986, **255** (4), 118–25.

45. BARENBERG S A, 'Abridged report of the committee to survey the needs and opportunities for the biomaterials industry', *Journal of Biomedical Materials Research*, 1988, **22** (12), 1267–92.

46. THOMSEN J L, CHRISTENSEN L, NIELSEN M, BRANDT B, BREITLING V B, FELBY S and NIELSEN E, 'Histologic changes and silicone concentrations in human tissue surrounding silicone breast prostheses', *Plastic and Reconstructive Surgery*, 1990, **85**, 38–41.

47. JACK C, FISHER 'The silicone controversy – when will science prevail?' *New England Journal of Medicine*, 1992, **328**, 1697.

48. CHU C C, VON FRAUNHOFER ANTHONY J and GREISLER H P (editors), *Wound Closure Biomaterials and Devices*, CRC Press, 1996.

49. PLACK H, DAUNER M and RENARDY M (editors), *Medical Textiles for Implantation, Proceedings of the 3rd International ITV Conference On Biomaterials*, Stuttgart, June 14–16, 1989.

50. SMITH A J, 'Textile biomaterials', Block 6.1 and 6.2, *Proceedings TECHTEXTIL Symposium 1999*, Frankfurt, Germany, 1999.

51. JONES K L, 'New organs grow in the laboratory', *Illustrerad Vetenskap*, 1999, No. 5, 10–19, (in Swedish).

52. ARNOLD A, DAUNER M, DOSER M and PLANCK H, 'Application of 3-dimensional structures for tissue engineering', *Proceedings TECHTEXTIL Symposium 1999*, Lecture No. 618, Block 6.1, Frankfurt, Germany, 1999.

53. DAUNER M and PLANCK H, 'Composite biomedical materials for surgical implants', *Textile Asia*, 1999, 33–7.
54. OEHR C, MILLER M and VOHRER U, 'Surface treatment of polymers improving the biocompability', *Proceedings TECHTEXTIL Symposium 1999*, Lecture No 615, Block 6.1, Frankfurt, Germany, 1999.
55. CHRISTEL P, CLAES L and BROWN S A, *Carbon-reinforced Composites in Orthopaedic Surgery in High Performance Biomaterials*, Szycher M (editor), Technomic Publishing, Lancaster, 1991, 499–518.
56. LINTI C, DAUNER M, MILWICH M and PLANCK H, 'Fiber reinforced polymers for implantation', *Proceedings TECHTEXTIL Symposium 1999*, Lecture No. 617, Block 6.1, Frankfurt, Germany, 1999.
57. Soil Behavior and Soft Ground Construction, (Geotechnical Special Publication No. 119) 2003, *American Society of Civil Engineers*, Germaine J T, Sheahan T C and Whitman R V (Editors), 2003.
58. HORROCKS A R and ANAND S C (editors), *Handbook of Technical Textiles*, Woodhead Publishing Limited, Cambridge, UK, 2000.
59. KOERNER R M and WELSH J P, *Construction and Geotechnical Engineering using Synthetic Fabrics*, Wiley-Interscience, John Wiley & Sons, NY, 1980.
60. ELMOGAHZY Y, GOWAYED Y and MAYO L, 'The frictional behavior of nonwoven geotextiles in granular soils', *International Nonwoven Journal*, 1995, **6** (4), 66–71.
61. EL MOGAHZY Y, GOWAYED Y and ELTON D, 'Theory of soil/geotextile interaction', *Textile Research Journal*, 1995, **64** (12), 744–54.
62. BROWN R, SHUKLA A and NATARAJAN K R, *Fiber Reinforcement of Concrete Structures*, Uritc Project No. 536101, University of Rhode Island Transportation Center, September 2002.
63. DANIEL J I, ROLLER J J and ANDERSON E D, *Fiber Reinforced Concrete*, Portland Cement Association, 1998, Chapter 5, 22–6.
64. DETRELL J, 'Air cleaning', *Journal of Textile Innovation-Technical Textile Guide*, http://www.technicaltextilesguide.com/pdf/aircleaning.pdf, 2007.
65. ORTEGA B and DONG L, 'Characteristics of mismatched twin-core fiber spectral filters', *Photonics Technology Letters, IEEE*, 1998, **10** (7), 991–993.
66. LAMB G E R, 'Cleaning of fabric filters', *Textile Research Journal*, 1987, **57** (8), 472–8.
67. LEUBNER H and RIEBEL U, 'Pulse jet cleaning of textile and rigid filter media – characteristic parameters', *Chemical Engineering and Technology*, 2004, **27** (6), 652–61.
68. OVERINGTON M S and LEECH C M, 'Modelling heat buildup in large polyester ropes', *International Journal of Offshore and Polar Engineering*, 1997, **7** (I), March.
69. MANDELL J F, 'Modeling of marine rope fatigue behavior', *Textile Research Journal*, 1987, **57** (6), 318–30.
70. DETRELL J, 'Load elevation', *Journal of Textile Innovation–Technical Textile Guide*, http://www.technicaltextilesguide.com/pdf/loadelevation.pdf, 2007.
71. FERNÁNDEZ O, 'To protect or to ensure cultivations?' *Journal of Textile Innovation–Technical Textile Guide*, http://www.technicaltextilesguide.com/pdf/protectensure.pdf, 2007.

72. DETRELL A, 'Artificial grass: more ecological than natural grass', *Journal of Textile Innovation–Technical Textile Guide*, http://www.technicaltextilesguide.com/pdf/artificialgrass.pdf , 2007.

73. JUNG S, LAUTERBACH C H, STRASSER M and WEBER W, 'Enabling technologies for disappearing electronics in smart textiles', In *IEEE International Solid-State Circuits Conference*, San Francisco, CA, February 9–13 2003.

74. KALLMAYER C, LINZ T, ASCHENBRENNER R and REICHL H, 'System integration technologies for smart textiles', *MST News*, 2005, **2**, 42–43.

75. AUBOUY L, JIMÉNEZ J and SÁEZ J, 'The third generation of smart fabrics', *Journal of Textile Innovation–Technical Textile Guide*, http://www.technicaltextilesguide.com/pdf/thirdgeneration.pdf, 2007.

14

Development of textile fiber products for transportation applications

Abstract: There are numerous transportation applications for many types of fibers and yarns and styles of fabric. In this chapter, product-development aspects associated with transportation applications are discussed. The advantages of using fibrous materials in these applications are emphasized. These include abrasion resistance, UV resistance, physical and thermal comfort, flame resistance and compatibility between fibrous materials and other types of materials including metals and foams. The chapter emphasizes three critical transportation products: safety airbags, safety seatbelts and transportation seats.

Key words: air bags; air cushion restraint system; inflator; propellant; crash sensor; diagnostic units; blister-inflation technique; foam-in-place technique; direct joining method.

14.1   Introduction

For many years, fibrous and polymeric products have been used for transportation applications such as for all vehicle interiors, exteriors or body parts, seat belts, airbags, tires, engine filters, belts, hoses, and aircraft seats and carpets. Indeed, it is difficult to imagine any form of transportation that does not involve one form or another of fibrous assemblies. This makes transportation one of the largest beneficiaries of function-focus fibrous products by volume and fiber consumption.[1] Figures 14.1 and 14.2 illustrate the main fibrous components that are used in automobiles and aircraft, respectively.

Transportation applications also use numerous types of fibers including natural, synthetic and high performance fibers.[2] Yarns used in these applications can be spun yarns, continuous filament yarns, textured yarns and specialty core/sheath yarns. Fabrics can also be of all styles including narrow fabrics, wide fabrics, woven, circular knitted, and flat knitted fabrics. In addition, many specialty fabrics such as pile and raised constructions are used. Transportation applications also open the door for more use of fiber composites as discussed in Chapter 13. This is primarily due to the need for light weight and fuel economy with acceptable durability.

Utilization of fibrous materials in transportation applications generates a number of common design and development challenges that are imposed

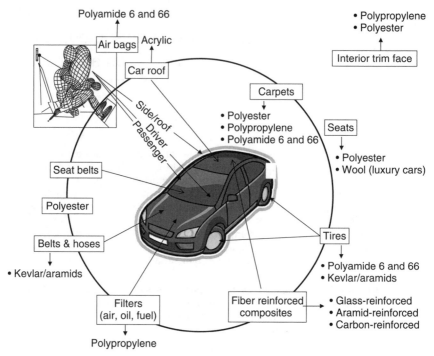

14.1 Examples of fibrous components used in transportation vehicles.

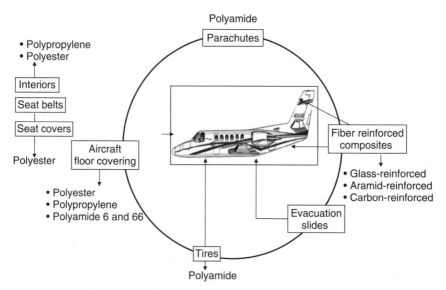

14.2 Examples of fibrous components used in aircraft.

by a number of critical performance characteristics of transportation products. These are as follows:

- the need for high abrasion resistance as a result of the continuous interaction between the users and transportation units (e.g. seats, carpets and interiors);
- the need for high UV resistance as a result of the continuous exposure of transportation units to sunlight;
- the need for physical and thermal comfort as a result of the long duration of interaction between the users and the transportation units, and the restricted movement of transportation units' occupants;
- the need for high flame resistance as a result of the potential exposure to heat and flame under normal traveling conditions or during a crash;
- the need for high compatibility between fibrous materials and other types of materials including metals and foams.

A good review of fibrous products used in transportation was made by Walter Fung and Mike Hardcastle in their book titled *Textiles in Automotive Engineering*.[2] Engineers in this area are encouraged to read this review. In this chapter, the focus will be on three important examples of products: safety airbags, seatbelts and transportation seats. Discussions of these products will follow closely the concepts of product development and design conceptualization discussed earlier in Chapters 3 through 5, and the tools of design analysis discussed in Chapter 6. In addition, the discussion will not be limited to the role of fibrous products but rather extended to the integrated concept of the end product assembly. This will include discussion of the deployment mechanism of airbags, seatbelt mechanisms and seating concepts. The purpose of this extension is to create mutual understanding by fiber and polymer engineers as well as engineers from other fields of the benefits of knowledge integration in the process of product development.

14.2   Development of fabrics for airbags used in automotive safety systems

An airbag is an inflatable cushion designed to protect vehicle occupants from serious injuries in crashes or serious collisions. It is only one component of an inflatable restraint system, also known as an air cushion restraint system (ACRS) or an airbag supplemental restraint system (SRS). Following the basic steps of product development discussed in Chapter 3 (Fig. 3.1), an idea for a new airbag fabric system or a modified version of an existing system should primarily stem from sound realization of the relationships between the desired performance characteristics of an airbag and the attributes of the various components forming the airbag system. These relationships should be based on good knowledge of the airbag system. A design engineer should understand three basic aspects associated with airbags:[3,4] (1) the basic components of an airbag system, (2) the deployment mechanism and (3) safety-related issues. These aspects are discussed below.

14.2.1 Basic components of an airbag system

An airbag module has three main parts (see Fig. 14.3): the airbag, the inflator and the propellant. The airbag fabric is basically a woven construction that can be made in different shapes and sizes depending on specific vehicle requirements. The inflator canister or body is commonly made from either stamped stainless steel or cast aluminum. Inside the inflator canister is a filter assembly commonly consisting of a stainless steel wire mesh with ceramic material sandwiched in between. When the inflator is assembled, the filter assembly is surrounded by metal foil to maintain a seal that prevents propellant contamination. The propellant, in the form of black pellets, is conventionally sodium azide combined with an oxidizer and is typically located inside the inflator canister between the

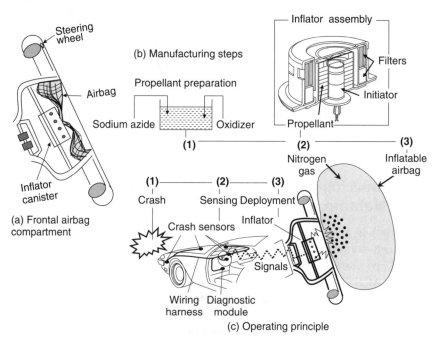

14.3 Basic components of airbag.[2-6]

filter assembly and the initiator. Other propellant materials will be discussed later.

The conventional manufacturing process used for making the airbag module consists of three different separate assemblies (Fig. 14.3b): (1) propellant manufacturing, (2) inflator components assembly and (3) airbag cutting and sewing. These operations can be performed at three different manufacturing sites producing components that can be assembled later into a complete airbag module. As indicated above, the conventional propellant consists of sodium azide mixed together with an oxidizer, a substance that helps the sodium azide to burn when ignited. The sodium azide is supplied by outside vendors and inspected to make sure it conforms to requirements. After inspection, it is placed in a safe storage place until needed. At the same time, the oxidizer is also supplied by outside vendors, inspected and stored. Different manufacturers use different oxidizers. From storage, the sodium azide and the oxidizer are then carefully blended under a sophisticated computerized process control. Because of the possibility of explosions, powder processing takes place in isolated bunkers. Production occurs in several redundant smaller facilities so that if an accident occurs, production will not be shut down, only decreased. After blending, the propellant mixture is sent to storage. Presses are then used to compress the propellant mixture into disk or pellet form.

Inflator components are the metal canister, the filter assembly and initiator (or igniter). These components are also manufactured separately and inspected prior to the final assembly using an automated production line. The final assembly consists of the inflator components combined with the propellant.[3] Laser welding (using CO_2 gas) is used to join stainless steel inflator sub-assemblies, while friction inertial welding is used to join aluminum inflator sub-assemblies. Laser welding involves using laser beams to weld the assemblies together, while friction inertial welding involves rubbing two metals together until the surfaces become hot enough to join together. The inflator assembly is then tested and sent to storage until needed.

The airbag woven fabric is manufactured in a weaving facility. It is then inspected carefully, die cut to the appropriate shape and sewn both internally and externally to join its sides. The normal design of the driver's airbag is two circular pieces of fabric sewn together. The passenger bag is tear-drop shaped, made from two vertical sections and a main horizontal panel. The sewing thread should be chosen carefully. Nylon 66, polyester, and aramid fibers can be used for sewing. When sewn the airbag is folded inside its cover like a parachute with extreme care to ensure smooth deployment. These folds are of various types including[4,5] accordion folds, reversed accordion folds, pleated accordion folds and overlapped folds. After sewing, tests are performed to check for leak and seam imperfections. The airbag is then mounted on the tested inflator assembly, folded and secured with a breakaway plastic horn pad cover.

In addition to the basic airbag module components discussed above, other components such as crash sensors, diagnostic monitoring units, steering wheel connecting coils and indicator lamps are combined with the airbag module during vehicle assembly. All the components are connected via a wiring harness, as shown in Fig. 14.3c.

Airbags are classified by their location in a car, which determines the size of the airbag.[5] The driver and the passenger airbags are designed to protect against frontal collisions. Typically, the driver airbag module is located in the steering wheel hub and the passenger airbag module is located in the instrument panel. When fully inflated, the driver airbag is approximately the diameter of a large beach ball. The passenger airbag can be two to three times larger since the distance between the right-front passenger and the instrument panel is much greater than the distance between the driver and the steering wheel. New cars are also equipped with side airbags to protect against moderate to severe side impact crashes. These airbags are generally located in the outboard edge of the seat back, in the door or in the roof rail above the door. Seat and door-mounted airbags all provide upper body protection. Some also extend upwards to provide head protection. Two types of side airbags, known as inflatable tubular structures and inflatable curtains, are specifically designed to reduce the risk of head injury and/or help keep the head and upper body inside the vehicle.[6] These airbags are claimed to reduce injuries and ejection from the vehicle in rollover crashes. Side airbags are typically smaller in size than front airbags and they deploy more rapidly.

14.2.2 Airbag deployment mechanism

In a vehicle crash, the frontal crash sensor is activated sending an impulse to the diagnostic unit. This unit evaluates the strength of the input signal and triggers an electric impulse of similar strength, which causes the central igniter inside the airbag to fire. Igniter fire penetrates the propellant chamber; this ignites the propellant, and produces and expels hot gas (e.g. nitrogen gas, comprising of 78% ambient air). This gas passes through a filter and enters the system's bag through inflator ports to inflate the airbag. This entire process occurs in a fraction of a second (i.e. within 0.06 s).

In principle, the impact of a collision process can be described using Newton's law which says that 'a body in motion will stay in motion until it is acted upon by an outside force'. Travelers driving vehicles may be driving within city limits at low speeds or on highways at very high speeds. In either situation, the driver does not typically feel that his/her body is moving at the same speed as that of the vehicle. Only when a sudden brake is performed, or in the event of a crash, does the occupant of a vehicle feel the speed almost at an equal rate as the original speed of the vehicle. The problem here is that this sense is not just a mental one, but more seriously a physical one, as the body reacts to the sudden stop or crash by a fast movement that in many situations can result in an occupant flying out of the seat, if not restrained by a seat belt. This is a body movement with a momentum that is equivalent to the initial speed of the vehicle.

The above illustration clearly indicates two critical aspects of safety for vehicle occupants: body movement during a crash and serious injuries resulting from this movement. In response to these two aspects, an occupant will need a restraint from movement and a restraint from impact (or a cushion against sharp objects). The former is achieved by the car safety seatbelt and the latter is achieved by the safety airbag, which not only prevents the impact but also distributes the impact force over a larger body area.

In light of the above discussion, there are three consecutive steps involved in airbag deployment:[5,6] crash, sensing and deployment. Under a severe frontal crash, crash sensors send a signal to the inflator unit within the airbag module. An igniter starts a reaction, which produces a gas to fill the airbag, making the airbag deploy through the module cover. Some airbag technologies use nitrogen gas to fill the airbag while others may use argon gas. The gases used to fill airbags are supposedly harmless. The time duration from the onset of the crash to the entire deployment and inflation process is only about 1/20th of a second. This is an essential requirement since a vehicle changes speed very quickly in a crash. As the occupant contacts the bag, deflation begins immediately and the gas escapes through pores in the fabric. The bag is fully inflated for only one-tenth (1/10) of a second and is nearly deflated by three-tenths (3/10) of a second after impact. Talcum powder or corn starch is used to line the inside of the airbag and is released from the airbag as it is opened. Once deployed, the airbag cannot be reused and should be replaced by an authorized service department.

Crash sensors located in the front of the vehicle and/or in the passenger compartment are typically activated by forces generated at the moment of crash impact. They measure deceleration, which is the rate at which the vehicle slows down.[6] Accordingly, the vehicle speed at which the sensors activate the airbag varies with the nature of the crash. It is important to point out that airbags are not designed to activate during a sudden brake or on a rough or bumpy road. Indeed, the maximum deceleration generated in the severest braking is only a small fraction of that necessary to activate the airbag system. The diagnostic unit detects the severity of the impact and transmits the signal to the airbag activation system. This unit is always activated when the vehicle's ignition is turned on. It also detects any defect in the airbag system and gives a warning light to alert the driver of the need for system examination and repair. Most diagnostic units contain a device which stores enough electrical energy to deploy the airbag if the vehicle's battery is destroyed very early in a crash sequence.

The variable nature of a car crash adds to the complexity of the design of crash sensors. Some estimate a near frontal and a frontal collision to be comparable to hitting a solid barrier at approximately 8 to 14 miles per hour (mph) (12–23 kmph). These estimates are obtained from laboratory testing or individual road incidents and they may not reflect the true impact of some crashes.[7] The variability of a car crash impact will depend on many factors including:[5,6] the size of the car involved in the crash, the relative speed between a striking and struck vehicle, the collision angle and the distribution of crash forces across the front of the vehicle. These factors are likely to create actual collisions with much higher impacts than equivalent barrier crashes. The speeds and pressures developed in airbag deployment are difficult to measure; some popularly quoted values for airbag deployment speed are in the order of 400 mph (over 640 kmph) and internal cushion pressures regularly rise above 100 kPa. In addition to these substantial mechanical stresses, the airbag cushion may also be exposed to high temperatures from inflation gases and explosive pressure. Temperatures of up to 5000°F (or over 2700°C) have been cited in the technical literature.

14.2.3 Airbag safety-related concerns

As indicated in Chapters 3 and 4, a product model is typically a result of design analysis that aims to provide the best compromise solution and not necessarily the best optimum solution. In the case of an airbag model, the focus of design analysis was on the deployment mechanism and all factors that can lead to accurate and reliable sensing, timely deployment and timely deflation of the airbag. These are largely deterministic aspects that can easily be handled by design engineers and manufacturers. However, some deficiencies in the airbag model were discovered after some time in use, some of which were direct results of the inherent mechanism used and others which were due to probabilistic factors that were not accounted for in the initial design analysis. The latter factors were mainly human–airbag interaction related, as will be clarified shortly. Unfortunately, flaws and deficiencies in products often create negative feelings and attitudes toward the products by consumers which overshadow their true benefits, and airbags were no exception. As indicated in Chapter 4, the difficulty of accounting for probabilistic outcomes in design analysis stems from the fact that many of these outcomes are seldom realized in the design conceptualization phase, or even in preliminary tests, and only real-world situations can reveal them. When these outcomes are the result of highly variable passenger and driver behavior, matters become more complex.

Safety-related concerns recognized in the initial testing phase

Some of the safety-related issues associated with airbags that were realized in the initial testing phase include the release of dust-like particles into the vehicle's interior during deployment. Most of this dust consisted of corn-starch or talcum powder, which was used to lubricate the airbag during deployment. Small amounts of sodium hydroxide may initially be present. This chemical can cause minor irritation to the eyes and/or open wounds; however, with exposure to air, it quickly turns into sodium bicarbonate (common baking soda). Depending on the type of airbag system, potas-sium chloride (a table salt substitute) may also be present. For most people, the only effect the dust may produce is some minor irritation of the throat and eyes. Generally, minor irritations only occur when the occupant remains in the vehicle for many minutes with the windows closed and with no ventilation. However, some people with asthma may develop an asth-matic attack from inhaling the dust.

Other problematic issues that were raised in the initial design were common aspects that are typically handled in the design conceptualization phase of a product through reliable statements of broad and specific problem definitions, for example, the considerable inflation force of the airbag. This high impact force was found to cause injuries to car occupants. However, the initial realization was that these contact injuries, when they occur, are typically very minor abrasions or light burns and they should be considered to be insignificant in view of the obvious merits of an airbag. Another issue was the choice of raw materials, particularly the fiber type, and the seaming characteristics. These are all safety-related issues, as an inappropriate material may not deploy properly particularly after long storage and in different environments. Seam quality is also critical in con-nection with the overall integrity of an airbag during deployment and in the storage state.

Safety-related issues not recognized in the initial testing phase

As indicated earlier, unrecognized design issues are mainly a result of probabilistic or random behavior by vehicles' occupants and variation in human ability to accommodate impact forces. Although airbags are designed to save lives, they can also be the cause of severe or even fatal injuries. As indicated earlier, an airbag acts as a cushion against body impact. However, in order for an airbag to assume this function, it must inflate rapidly and at high impact. In an unstrained (no seatbelt) situation, the faster the body momentum at the time of crash the greater the impact of the airbag against the body. The combination of body momentum and airbag strike can indeed cause injuries to the occupant. The complexity of this combination stems from the multiplicity of variables that can result in a wide range of injuries from a mild one to a fatal one.

Statistics associated with these injuries and deaths were reported in Chapter 2. Most airbag injuries result from occupants being too close to, or in direct contact, with an airbag module when the airbag deploys.[7] Some injuries may be sustained by unconscious drivers who are slumped over the steering wheel, unrestrained, or improperly restrained who slide forward on the seat during pre-crash braking. Objects attached to, or near, an airbag module can also be propelled with great force against the occupant's body during airbag deployment.

Prior to airbag development for automobiles, the only safety component was the familiar safety seatbelt mounted in all vehicles. Safety belts were originally designed for aircraft in the 1920s and later used in all transportation vehicles and enforced by law. This is essentially a safety harness designed to restrain the occupant of a vehicle from free movement during a sudden brake or a collision that can result in body impact inside the vehicle against sharp objects or other passengers. It can also prevent an occupant from being thrown from the vehicle in a severe impact.

The coordination between seatbelts and airbags in safety aspects was largely subjective and was not accounted for in the overall design of airbags. As a result, safety organizations had to make several recommendations regarding the use of safety belts. In general, airbags are supposed to supplement the safety belt by reducing the chance that the occupant's head and upper body will strike some part of the vehicle interior. They also help reduce the risk of serious injury by distributing crash forces more evenly across the occupant's body. However, an unrestrained or improperly restrained occupant can be seriously injured or killed by a deploying airbag. As a result, the National Highway Traffic Safety Administration (NHTSA) recommended that drivers sit with at least 10 inches (25 cm) between the center of their breastbone and the center of the steering wheel, and children 12 and under should always ride properly restrained in a rear seat. In addition, safety belts should always be worn with the lap belt low and snug across the hips and the shoulder belt across the chest. Shoulder belts should never be placed under the arm or behind the back. Front seat drivers and passengers should sit upright against the back of the seat. Finally, drivers should adjust the seat such that they position themselves away from the airbag module, while maintaining the ability to safely operate all vehicle controls. These are not design issues but they represent critical supporting elements for the safety performance of airbags. From an engineering viewpoint, this may be called 'regulation-supported design', a concept that should be stated clearly in the design conceptualization phase and implemented in all safety-related products.

In recent years, the option of deactivating an airbag (by switching it on or off) was given to drivers by NHTSA. This was an interesting regulatory rule that assumed that some drivers can make appropriate decisions about whether an airbag is suitable for their safety. This rule was made based on three conditions: (a) occupants with medical conditions that place them at specific risk, (b) occupants who cannot adjust their driver's position to keep at least 10 inches (25 cm) from the steering wheel and (c) occupants who cannot avoid situations that require a child 12 or under to ride in the front seat.

The safety concerns associated with airbag deployment have resulted in considerable research effort to develop more safety-accommodating airbags. This effort is mostly made by the auto industry and in confidential environments. Examples of this effort include:

- airbags that exhibit gentler inflation and less abrasive fabric material;
- a smart airbag that has the ability to sense the size and weight of the seat occupant, or even if the seat is unoccupied, and deploy accordingly;
- outward-deploying airbags (from the seat belt);
- a hybrid inflator that uses a combination of pressurized inert gas (argon) and heat from a propellant to expand the gas's volume significantly. This system would have a cost advantage, since less propellant could be used;
- elimination of sodium azide propellant (toxic in its undeployed form). Pyrotechnic inflation technology has changed over the past few years from reliance on sodium azide to the use of organic propellants in order to minimize the environmental impact of the propellant and increase efficiency;
- appropriate airbag fabric coating or uncoated airbags;
- the use of an energy absorbing material such as polyurethane foam and polypropylene foam with optimum properties for absorption of impact energy in surface vehicle interior.

14.2.4 Design aspects of safety airbags

The above information shed a great deal of light on the basic performance characteristics of an automotive airbag. From a design viewpoint, this product provides a unique opportunity for engineers of different disciplines to work together and exchange ideas and thoughts on the best way possible to develop an airbag that can operate reliably in the event of crashes, through understanding the principle of the deployment mechanism and the different safety considerations discussed above. The interdisciplinary aspect of this development is a result of the fact that this product consists of different types of materials including fibers, polymers, stainless steel, cast aluminum and ceramics. In addition, it requires electrical and computer engineers to design the sensors and diagnostic units, mechanical engineers to design the deployment mechanism, polymer and fiber engineers to design the airbag fabric, and chemical engineers to design the gas and chemical flow systems. Our discussion of the design aspects will be limited to fabric engineering for the airbag module.

Performance characteristics of fabrics used in airbag systems

The design of a suitable fabric for airbag systems represents the most critical phase in the development of the airbag module. This is due to the fact that the fabric is the primary component that comes into direct contact with the vehicle occupant at the time of crash; all other components work to support this immediate contact. The fabric is also the component that must withstand the force of the hot propellant chemicals used in the inflation process, allow gas release, disallow any penetration of chemicals that could burn the skin of the car occupant and finally, fold and unfold easily particularly after a long storage period. The great flexibility associated with fabric design and raw material selection has made it common among many airbag design engineers to believe that it is far easier and less costly to adjust the bag and fabric to meet the needs of a particular inflator than to develop mechanisms and inflation processes that are fabric friendly.

In light of the above discussion, the primary performance characteristics of the fabric component of an airbag module should be as follows (see Fig. 14.4):

- the ability to withstand the force of the hot propellant chemicals used in the inflation;
- the ability to allow gas release and disallow any penetration of chemicals through the fabric that could burn the skin of the car occupant;
- the ability to fold and unfold easily, particularly after a long storage period;
- the ability to provide a brief cushioning effect on the occupant in the event of a crash.

The above performance characteristics can be translated into the following material attributes:

- The first performance characteristic implies the need for fabric with high tear strength, high tenacity and high toughness. In addition, the resistance to sudden energy during airbag deployment requires a material that can withstand the high temperatures of inflation.

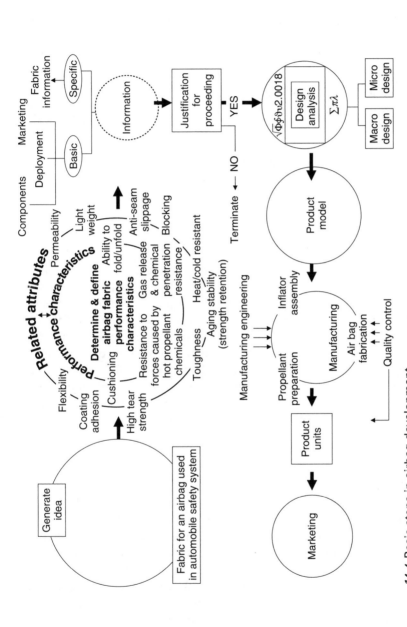

14.4 Basic steps in airbag development.

- The second performance characteristic implies a closed fabric construction, optimum permeability, optimum blocking and high anti-seam slippage (in case of fabrics with seams).
- The third performance characteristic implies light weight and high flexibility as well as high aging stability.
- The fourth performance characteristic implies a smooth and soft surface.

The discussion above clearly reveals the magnitude of effort that must be made in developing an airbag fabric that can meet the different performance characteristics of an airbag system.

Optimum levels of these attributes can be met using a design cycle that begins with the appropriate fibers and end with the appropriate fabric. Figure 14.4 lists some of the basic features of fibers and fibrous assemblies used in airbag systems. These are discussed in the following sections.

Fiber selection for safety airbag fabrics

Before fibers can be considered with respect to their attributes in relation to the deployment characteristics of an airbag fabric, it is important to consider a key aspect that is directly related to the nature of use of safety airbags, namely the aging aspect. An airbag may be stored folded in a vehicle airbag compartment for many years without deployment; yet, it is expected to deploy reliably and immediately at the time of crash and overcome the mechanical stresses imposed by the rapid inflation process. Prior to the time of use, an airbag could be exposed to extremes of temperature and humidity over a significant period of time. These factors can significantly influence the strength retention of fabrics. Obviously, a substantial reduction in strength retention over time can result in a failure in airbag deployment. It is important, therefore, to select a fiber with high strength retention over time.

The two primary fiber candidates that have been examined for airbags are polyester and polyamide fibers. Within the polyamide fiber category, nylon 66 is the most commonly used fiber in airbag fabric, but nylon 6 has been examined for special uses. Experimental comparative analysis of fibers suitable for airbags[8,9] revealed that nylon 66 has superior strength retention over polyester fiber over time at a temperature of 85°C and relative humidity of 95%. These results are shown in Fig. 14.5. According to DuPont, these results were also supported by extensive testing under simulated vehicle lifetime aging cycles in laboratory conditions as well as real-life observations over many years.

Nylon 66 was also found to be superior to polyester fibers in many of the attributes related to the performance characteristics of safety airbags.

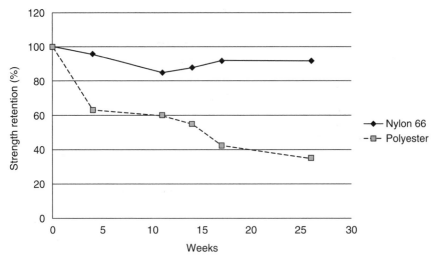

14.5 Comparison of uncoated Nylon 66, and polyester fabric tensile strength retention after an aging test (modified from DuPont data[8–11]).

Tables 14.1 and 14.2 show a comparison between the levels and values of these attributes reported by DuPont.[9] Note that although nylon 66 and polyester have similar melting points, the large difference in specific heat capacity causes the amount of energy required to melt polyester to be about 30% less than that required to melt nylon 6,6. According to DuPont, in any inflation event that uses a pyrotechnic or pyrotechnic-containing inflator, cushions made from polyester yarn are far more susceptible to burn or melt through in the body of the cushion or at the seam.[10,11]

The low bulk density of nylon 66 (1.14 g cm^{-3}) in comparison with that of polyester fiber (1.39 g cm^{-3}) represents another advantage in favor of nylon 66. It implies that fabrics with identical yarn and weave structures made from the two fibers will have a different weight, with the polyester fabric being about 20% heavier than the nylon 66 fabric. As indicated earlier, a lighter airbag provides many advantages including higher flexibility in folding and unfolding, lower vehicle weight and lower kinetic energy of impact on the vehicle's occupants. In addition, the difference in density between the two polymers leads to polyester yarns usually being of higher denier or decitex (weight per unit length) than nylon 66, generating the same filament diameter. This will result in reduced fabric coverage, higher gas permeability and weaker seam strength of fabrics made from polyester filaments. According to DuPont, using polyester yarn, the cushion fabric is more open to gas permeation. This reduces thermal protection for the vehicle occupants and makes it more difficult for the cushion designer to

Table 14.1 General comparison of nylon 66 and polyester fibers (DuPont data[9])

Fiber properties	Nylon 66	Polyester
Air permeability	Low (is better)	High
Abrasion resistance	High	Low
Thermal resistance	High	Low
Energy absorption	High	Low
Stiffness in cushion	Low	High

Table 14.2 Comparison of nylon 66 and polyester fibers (DuPont data[9])

Fiber properties	Nylon 66	Polyester
Density (g cm^{-3})	1.14	1.39
Melting point (°C)	260	258
Softening point (°C)	220	220
Specific heat (J g^{-1} °C^{-1})	1.67	1.30
Specific heat of fusion (J g^{-1})	1.88	117
Total heat to melt (J)	587	427
Total heat to soften (J)	523	377

control the bag deployment dynamics. In addition, since seam strength is strongly dependent on cover factor, there is a negative impact on seam performance. This is particularly important since seam leakage of hot gas is one of the principal concerns in engineering airbags and the potential for an increase in leakage combined with a reduction in thermal resistance is critical.

Development efforts continue to examine alternative fibers that can provide better performance than that witnessed for nylon 66. The main obstacle to these efforts is cost. Examples of alternative fibers that are being evaluated include aramid fiber, nylon 4,6 and some modified polyester films. It should also be pointed out that although polyester fiber is not widely used in airbag systems, it occupies over 75% of the market share of fibers used in automotive applications as a result of its dominant use in automotive interiors.

Yarns used for safety airbags

In determining the yarn structure suitable for use in airbag fabrics, yarn count represents the most critical design factor. This is because of the well-known effects of yarn count (in denier or dtex) on critical fabric attributes

Table 14.3 Typical values of some properties of nylon 66 filaments used in airbag yarns (DuPont Technical Report[9])

Parameter	235-68-T749	350-105-T749	350-140-T749	470-140-T749	585-140-T749	700-210-T769
Dtex/ filament	3.5	3.3	2.5	3.4	4.2	3.3
Tenacity (cN tex^{-1})	84.8	81.2	81.2	81.2	81.2	75
Elongation at break (%)	20	22	22	20	20	23
Hot air shrinkage (%)	7	6.2	6.2	6.5	6.5	6

such as fabric strength, elongation, flexibility, weight and covering power. Yarns used in airbag fabrics are high tenacity continuous multi-filaments with counts commonly ranging from 200 to 1000 dtex. The ability to use a certain yarn denier will primarily depend on the denier per filament used. In other words, the key parameter determining yarn count is the filament fineness or dtex (denier). In general, it is desirable to use filaments with low dtex, which can result in finer yarn counts (lighter, flexible and high covering power fabric) and more filaments per yarn cross-section (better control of gas permeability). Table 14.3 lists filament fineness (dtex) and other properties of commercial nylon 66 fibers used in airbag yarns.[9] It should be pointed out that the production of fine count filaments, or low denier-per-filament yarns will be likely to cost more than heavy count filaments as a result of the use of smaller spinnerets and better control on filament flow and orientation during spinning.

Fabrics used for safety airbags

Airbag fabrics are normally tight plain woven structures with weights ranging from 170 to 220 g m^{-2} and thickness from 0.33 to 0.4 mm. Following the expression of fabric specific volume given by Equation (12.1) (Chapter 12), these values correspond to a fabric specific volume ranging from 1.94 to 1.82 cm^3 g^{-1} (or a fabric density from 0.52 to 0.55 g cm^{-3}). When nylon 66 fiber is used, the range of fiber fraction expressed by Equation (12.2) (Chapter 12) of the airbag fabrics will be from 0.45 to 0.48. A corresponding range using polyester fibers in the fabric will be from 0.37 to 0.40. One can see from this comparison that fabrics made from nylon 66 will exhibit a higher fiber fraction leading to lower air and gas permeability than fabrics of the same construction made from polyester fibers.

The basic construction of airbag fabric will largely depend on the yarn denier used. Common weave constructions used in airbag fabrics include 840×840 denier filament yarn, 98×98 plain weave, 60 inch width (152.4 cm), and 420 × 420 denier filament yarn, 193 × 193 plain weave, 60 inch width. Some passenger airbag fabric with 41 × 41 plain-weave may be made from 630 denier nylon 66 yarn while a 49 × 49 plain-weave is typically made from 420 denier nylon 66 yarn.

The airbag fabric is not dyed, but it has to be scoured to remove impurities. In addition, coating of airbag fabric is a critical design issue, particularly in relation to fabric protection and permeability aspects. In this regard, developments have been made using two main approaches: the development of high performance coating material and the development of uncoated fabrics. In the first approach, new coating substances such as silicone coating have been used to replace the traditional Neoprene coating.[12] Silicone fabric coatings have been used for many years in many industrial fabrics such as conveyor belts, electrical and protective sleeving and welding blankets. Silicone's high heat resistance and long-term aging stability makes it appropriate for coating airbag fabrics.[13] It also results in lighter, thinner and softer fabrics because of the smaller amounts needed in comparison with Neoprene coating (almost twice the amount at the same level of heat protection). Another benefit of silicone coating is its compatibility with nylon fabric. Some experts suggest that Neoprene coating generates hydrochloric acid during aging, which actively damages the fiber. The silicone coating provides a protective layer against hydrolysis and also remains chemically inert. As is well-known, uncoated nylon can be attacked by moisture (hydrolysis). Key performance characteristics of coating material for airbags include[12,13] good adhesion with the fabric surface, anti-blocking, long term flexibility, resistance to cyclic temperature changes (−40 to 250°F), ozone resistance, long term stability and low cost.

Developments toward uncoated airbag fabric were driven by the need for lighter and smaller thickness airbags that can easily fold and unfold. In addition, uncoated airbags can be more easily recycled than coated airbags. In order to develop an uncoated airbag that can meet the desired performance discussed above, a great deal of effort in both yarn and fabric design must be made. In this regard, the key parameter is pore size and its distribution over the fabric area. Commercially, driver airbags are normally coated and are made from low-denier yarns. Passenger airbags may be uncoated and they are larger in size, as indicated earlier. This is justified on the basis that they impose lower gas pressure and have longer inflation times. These airbags can also be made from heavier denier yarns.

Design analysis of airbags

In light of the discussion above, one can imagine the magnitude of effort that must be made in developing an airbag fabric that can meet the different performance characteristics of an airbag system. Figure 14.6 illustrates the different design parameters that can be controlled in an airbag fabric and the different performance characteristics related to these parameters. In a step-wise procedure, appropriate fibers must be selected, suitable yarn deniers must be used and fabrics with optimum construction and finish must be designed.

An optimum design for an airbag fabric will certainly require modeling analysis of its performance characteristics in relation to the various potential factors that may influence them. This analysis should yield analytical and simulative models that can ultimately be verified by extensive experimental analysis. Such models will be highly computation intensive as they should accommodate most of the parameters listed in Fig. 14.6 and different levels of these parameters. In addition, objective measures of performance outcomes or response variables should be established. These include the exact range of forces of the hot propellant chemicals used in the inflation, the threshold range of pores that will allow gas release and disallow penetration of chemicals and the time required to unfold an airbag during deployment.

The issue of modeling airbag fabric performance was addressed in a study by Keshavaraj *et al.*[14] In this study, the authors indicated that the material properties of engineering fabrics that are used to manufacture airbags cannot be modeled easily by the available nonlinear elastic–plastic shell elements used by the auto industry, on the basis that a nonlinear membrane element that incorporates an elaborate tissue material model is highly computation intensive and does not differentiate between the various physical properties of the fabrics, like fiber denier, polymer fiber and weave pattern. Instead, the authors introduced a new modeling technique that uses artificial neural networks. Experimental permeability data for fabrics under biaxial strain conditions were obtained through a blister–inflation technique and were used to train the proposed network architecture. In this training environment, various properties of the fabric were incorporated and the network was trained to generalize relative to the environment. Typically, once a network is trained, a cause-and-effect pattern can be assimilated by the network with appropriate weights to produce a desired output. Fabrics tested in this study included nylon 66 fabrics with three different fabric deniers: 420, 630 and 840 and two types of weave, and two 650-denier polyester fabrics having different calendering effects. The predictions obtained from this neural network model agreed very well with the experimental data. This indicates that neural nets can be

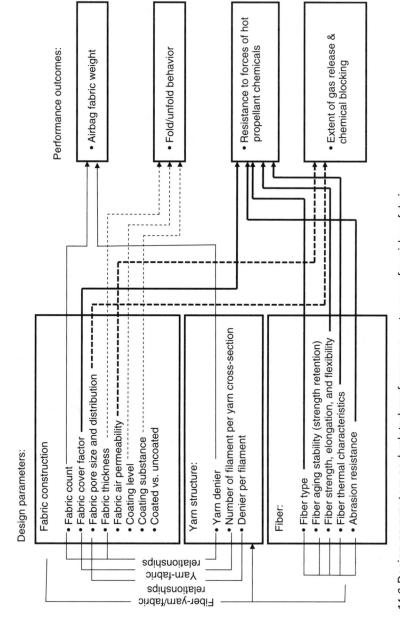

14.6 Design parameters and related performance outcomes for airbag fabric.

considered as a serious design tool in determining permeability and biaxial stress–strain relationships for textile fabrics used in airbags.

As indicated above, the intensity of modeling computational analysis to simulate and explore airbag fabric performance can be overcome using advanced computation tools such as neural network analysis. The challenge, however, will stem from the high variability associated with the airbag module performance in real-life road crash situations. This challenge cannot be met fully in laboratory simulative testing and a database generated from actual crash incidents will be necessary.

14.3   Development of fabrics for safety seatbelts

As indicated earlier, the safety seatbelt is another important safety component that is used in all vehicles and aircraft. It is essentially an energy absorbing device that is designed to restrain the occupant's momentum and movement during a crash, maintaining a safe distance between the occupant and harmful interior objects and reducing the load imposed on occupants during a crash down to survivable limits. Unlike the safety airbag, the seatbelt is locked in place at the occupant's choice, which requires good awareness by the occupant of the safety merits of the seatbelt. In most countries, using seatbelts is enforced by law. Unfortunately, the use of seatbelts by all vehicles occupants is still a long way from becoming a reality. It appears that a combination of the restraining nature of seatbelts, lack of awareness of its safety merits by some and the law enforcement associated with its use represents significant obstacles against its widespread use. From an engineering design viewpoint, these obstacles can be overcome by providing a combination of comfort and a friendly restraining system.

Some historians trace back the origin of the seatbelt invention to the 1800s and its introduction in aircraft to 1913. But recent history clearly indicates that seatbelts were first noticeably in American automobiles in the early 1900s. Today, new ideas are still being introduced to improve further the performance of seatbelts. The common seatbelt used today is the so-called three-point belt, which is secured by two fittings on the floor and a third on the sidewall or pillar of the vehicle.

14.3.1 Performance characteristics and component attributes of safety seatbelts

The key performance characteristics of a seatbelt primarily stem from its basic functions, as shown in Fig. 14.7. Along the length of the seatbelt, high strength is a key requirement. An optimum level of stretchability and

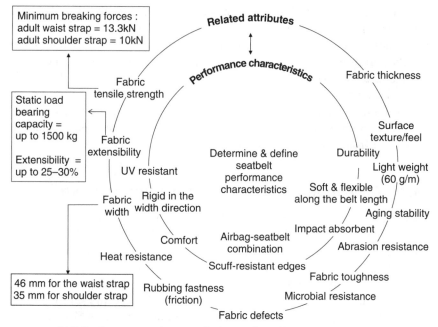

Minimum breaking forces :
adult waist strap = 13.3kN
adult shoulder strap = 10kN

Static load
bearing
capacity =
up to 1500 kg

Extensibility =
up to 25–30%

46 mm for the waist strap
35 mm for shoulder strap

Related attributes

Performance characteristics

Fabric
tensile strength

Fabric
extensibility

UV resistant

Fabric   Rigid in the
width   width direction

Heat resistance

Comfort

Rubbing fastness
(friction)

Fabric thickness

Surface
texture/feel

Determine & define   Durability
seatbelt
performance   Light weight
characteristics   Soft & flexible   (60 g/m)
along the belt length
Aging stability

Impact absorbent
Airbag-seatbelt
combination   Abrasion resistance

Scuff-resistant edges
Fabric toughness

Microbial resistance

Fabric defects

14.7 Performance characteristics and attributes of safety seatbelts.

flexibility is also important in this direction, but they must be limited by the tightness effect that the seatbelt must impose in case of a crash. In the width direction, rigidity is required to allow easy sliding through buckles. This is particularly important at the edges and the thickness direction of the seatbelt. UV degradation resistance and aging stability also represent key performance characteristics of seatbelts.

Comfort is a key performance attribute of seatbelts which deserves more attention in the design process. This is due to the fact that a seatbelt not only imposes restraint on the occupant's body in the event of a sudden brake or a crash, but also imposes pressure on the waist, shoulders and neck during driving, which can last for many hours and in various environmental conditions, particularly hot and humid climates. In a survey made by the present author of the reasons why many people do not use a seatbelt in these climates, discomfort was one of the main causes.[15] People with extreme body sizes (too small or too large) were among those who complained most about seatbelt discomfort, but a significant number of people express an unpleasant feeling about seatbelts including dissatisfaction with the idea of being restrained and the inconvenience associated with fastening and unfastening the seatbelt. Although this is not considered to be a classic case of comfort–safety trade off, it is an aspect that deserves some design attention since a seatbelt that provides comfortable touch and feel against the human body will certainly attract many passengers to wear it.

From a design viewpoint, comfort–safety relationships should be viewed in terms of the use-duration period. For example, an airbag is only felt for few moments but other products may be associated with a use-duration period of a few minutes to a few hours. These include very heavy bullet-proof vests used in high-risk operations, face masks used for contamination leak or medical operations and fire-resistant uniforms worn during fire fighting. The urgency associated with these products makes safety an overwhelming aspect in comparison with comfort. In case of seatbelts, this urgency is not normally felt. Yet, seatbelts are required, as they should be, to be worn for different use-duration periods that can last for many hours. The pressure against the vehicle occupant's body imposes restraint against light body movement during traveling and is often associated with irritating rubbing against the body and around the neck–shoulder area.

Obviously, a great deal of this discomfort can be the result of poor adjustment of the seatbelt mount and, in many cases, it was found to be dependent on the extent of the geometric design of the seatbelt in matching the occupant's anatomical characteristics (e.g. shoulder belt fit, seat belt upper anchorage location and seatback angle). However, the constant contact between the seatbelt and the human body represents a significant source of irritation. This aspect provides an opportunity for fabric design engineers to develop narrow fabrics that can generate a pleasant (cushioning) feel against human body with minimum contact pressure. Generating ideas in this direction requires a fundamental design trade-off, which is the need for wider (more contact area) seatbelts to distribute the pressure against the human body, or ultimately the impact force during a crash over a larger area of the body, versus smaller seatbelts that exhibit minimum contact with the body. This is an area that requires more research investigation as this may lead to the best design compromise.

14.3.2 Fiber selection for safety seatbelts

In order to meet the performance characteristics of a safety seatbelt, the fabric must exhibit optimum levels of the attributes shown in Fig. 14.7. The most critical attributes are fabric strength, extensibility and stiffness. These are the attributes that are directly translated into seatbelt performance during traveling and in the event of a crash or sudden brake. The key design parameter to meet optimum levels for these attributes is fiber selection. In this regard, many fiber types can be used to make seatbelts but the most commonly used fibers are polyester and nylon fibers. Polyester fibers are commonly preferred to nylon fibers on the basis of their relatively lower extensibility.

In recent years, Honeywell Inc® have developed a new polyester composite fiber, called 'Securus™ fiber' which was designed to meet many of the fiber attributes required for high-performance seatbelt fabrics.[16,17] This fiber was manufactured by a proprietary process to provide a unique combination of strength and elongation of seatbelt fabric that is claimed to yield optimum energy absorption. Securus fiber introduces a new category of synthetic fibers, PELCO(TM), based on a patented polyester-caprolactone block copolymer. According to the developer, 'Securus™ fibers deliver a three-step reaction in an accident: first, the high-strength fibers hold passengers tightly in place at impact; then, Securus™ fibers relax or stretch as needed to limit the force on the occupant and cushion the body's movement into the airbag; finally, the fibers hold again as the car halts and prevents the passenger from hitting the steering wheel, dashboard or windshield as the airbag deflates'. In response to different seatbelts performance requirements, different levels of the different physical attributes of Securus™ fiber can be tailored as shown in Table 14.4.

14.3.3 Seatbelt yarns and fabrics

Fabrics used for seatbelts are typically twill-woven narrow fabrics (46 mm for the waist strap and 35 mm for shoulder strap for adults) made from continuous polyester filament yarns.[17] Common fabric constructions include[17–19] 320 ends of 1100 dtex yarns and 260 ends of 1670 dtex yarns. In the filling direction, typical yarns are 550 dtex. Seatbelts made from nylon filaments are typically woven from 180 dtex yarns in the warp direction and 470 or 940 dtex yarns in the filling direction. These constructions are chosen because they allow maximum yarn packing within a given area for maximum strength and good abrasion resistance.

Commercial needle looms used to weave seatbelt fabrics can accommodate six weaving stations simultaneously side by side. The weft is inserted

Table 14.4 Different value levels of different physical attributes of Securus™ seatbelt fiber[16]

Parameter				
Denier	500	1000	1350	1500
Decitex	550	1100	1500	1670
Filament count	35	70	100	100
Breaking strength (kg N^{-1})	2.8/27	7.0/68.6	9.5/92.7	10.5/102.8
Tenacity (g denier^{-1})/(cN dtex^{-1})	5.5/4.9	7.0/6.2	7.0/6.2	7.0/6.2
Elongation at break (%)	27	27	27	27
Toughness (g denier^{-1})/(cN tex^{-1})	0.9/0.8	0.9/0.8	0.9/0.8	0.9/0.8
Thermal shrinkage at 177°C (%)	16	20	16	16

at right angles to the warp direction from one side of the loom and a selvedge is formed. The other side of webbing is held by an auxiliary needle, which manipulates a binder and a lock thread. Once these are combined with the weft yarn, a run-proof selvedge is created. Special care is taken when constructing the selvedge to ensure high integrity. Woven seatbelts are typically transferred under tension to a dyeing and finishing range where the grey webbing is dyed and heat set.[17,18] The purpose of heat setting is to control the extent of extensibility in the seatbelt. Linear fabric weight in the grey state is about 50 g m^{-1}, and upon finishing it is increased to about 60 g m^{-1} as a result of the shrinkage in the length direction induced by heat setting to improve the energy absorption properties. Seatbelts may also be lightly coated to improve cleanability, durability, ease of passage in and out of housing and to impart some antistatic properties.[17]

14.3.4 Development of integrated seatbelt–airbag (inflatable seatbelt) systems

As indicated above, airbags and seatbelts represent two essential safety components in a vehicle which, if activated simultaneously at the time of crash, can certainly save lives and reduce serious injuries associated with car crashes. In recent years, much effort has been made to develop integrated seatbelt–airbag systems.[17,18] The common term used to describe an integrated seatbelt–airbag system is inflatable seatbelt. Basically, an airbag–seatbelt system is a tubular seatbelt that can inflate at the time of the crash to provide a protective cushioning effect for the occupant. In some developments, the seatbelt is held by weak stitches that burst open when the seatbelt is inflated. Under impact the surface seatbelt increases by over 400% in surface area more than the normal flat seatbelt. These seatbelts are designed to be fitted in the rear seats of the vehicles and they can replace conventional airbags in this area.

One example of inflatable seatbelts is the model introduced by the Ford Motor Company, which consists of a tubular inflatable bag inside the shoulder belt which inflates upon a vehicle crash, protecting the passenger's head and neck by limiting forward motion. Another example is the AAIR® aviation inflatable restraint developed by AM-SAFE® for use in aircraft. This system comprises a standard lap belt modified with an inflatable bag that is connected to a cold gas generator and an electronic sensor activation box. The entire unit is self-contained for each seat. The sensor determines when to activate the airbag in order to provide the required protection.

14.4   Development of fabrics for transportation seats

Seats are essential parts of human life. In transportation, millions of passengers use seatbelts every day on the ground or through the air. During this experience, they are intimately in contact with vehicle or aircraft seats for many hours. In order to make this inevitable intimacy a pleasant experience, transportation seats should be designed in such a way so as to allow maximum comfort in steady-state travel conditions, maximum durability during rough drive or extreme air turbulence and maximum safety in the event of sudden braking, air pumps and crashes. In the following sections, key performance characteristics of transportation seats are discussed and the requirements to meet these characteristics (attributes, material and constructions) are reviewed.

14.4.1 Basic seat components

Since the apparent function of transportation seats is to keep passengers in place during the duration of a trip, the seat construction consists of the following basic components:[2,20,21] structural frame members and non-structural frame members. These are discussed below.

Structural frame member

The structural frame member is usually made of steel that has been formed into a tabular configuration or of stamped or rolled sheet metal. This provides the weight support for both the seat and the occupant and forms the shape and the cushioning aspects of the seat. From a design viewpoint, the obvious trade-off here is safety and strength versus energy saving. In this regard, the anchoring frame strength typically requires a heavy weight and stabilized anchoring geometry while energy saving typically requires a low frame weight. The frame member commonly consists of a seat base and a backrest component. Commonly, seat bases are made of a shallow steel trough, roughly 1 mm thick. Some seat frames being developed are designed using an aluminum seat base and a backrest structure made of rolled high-tensile steel sections. Obviously, a steel construction is stronger than aluminum or magnesium, which is important for providing better energy absorption in the event of a crash. However, as weight is a consideration for energy saving, it makes a combination of steel and aluminum an appropriate way to achieve a considerably lower weight for the backrest (roughly 25% less), since light-alloy structures need a much greater wall thickness for the same strength. These developments aim at a target weight for automotive front seats ranging from 11 to 18 kg. Composite structures have also been highly considered for the design of seats.

Non-structural seat components

These are cushions, springs and upholstery that provide optimum contact, appropriate load distribution between the occupant and the seat frame and comfort features via optimum contour and seating geometry. Our focus in this section is on the seat fabric component. In this regard, three basic components associated with any seat should be considered: (1) seat cover laminate (face fabric, foam and scrim), (2) seat back (squab) and (3) seat bottom (cushion).

Under normal steady-state rides, the seat supports or anchorages transmit compression, tension and shear forces from the seat to the floor or side structure. In the event of collision or sudden braking, the resulting forces are transmitted in a reverse direction from the floor to the seat. As a result, the seat anchorage structures and attachments require a design with adequate strength to accommodate seat and occupant inertial forces.

A seat cover can be considered to be a triple-laminate system.[22-24] The fabric is normally laminated to polyurethane foam (typically 2–10 mm thickness) to keep it in an uncreased state. This lamination process allows easy cleaning of seat covers and imparts a soft touch to the fabric. In addition, it allows deep attractive sew lines in the seat cover. It helps the seat cover slide along the sewing machine surface during sewing and assists sliding when the made-up cover is pulled over the seat structure. A scrim fabric is also laminated to the other side of the polyurethane foam. This helps control the stretch properties of the seat cover especially when knitted fabrics are used. The polyurethane foam to which the cover fabric is laminated can be of two general types, polyester polyurethane foam and polyether polyurethane foam. Both types can be made into different flame-retardant (FR) grades. Polyether polyurethane foam needs to be modified slightly with certain additives to make it flame retardant.

Different methods that were utilized to fabricate automotive seat covers were reported by Fung and Hardcastle.[2] The traditional method of fabricating a seat cover system involves cutting and sewing panels of the seat cover laminate (face fabric/foam/scrim) into a cover, which is then pulled over the squab (seat back) and cushion (seat bottom) and fixed in place using a variety of clips and fastenings. In addition to the traditional method, many other methods are implemented by different companies. These are summarized here:

- Foam in place technique:[20-22] This method combines the processes of foam cushioning and squab molding with fixing the seat cover in place over the premolded foam. Panels of the seat cover laminate are cut and sewn into a 'bag' and the liquid foam components are poured in to form the solid foam by internal reaction. A polyurethane barrier is also included to prevent the liquids seeping through the fabric cover laminate before the reaction is complete.
- Direct joining method:[23-25] This technique is based on joining the cover fabric laminate directly to the squab and cushion. It results in a smaller

thickness of the laminate foam, which assists in fabricating seats with curvaceous and rounded contours. Direct joining is achieved using either hot-melt adhesive films or solvent spray adhesives. The basic flaw with this method is that it can adversely affect the pile in velvet fabrics.

- Hook-and-loop fastenings: This method is based on using hook-and-loop fastening components (e.g. Velcro products) made from raised, knitted nylon 66 (or polyester) fabrics. The hook part of the fastener is attached to the seat foam cushion and the loop part sewn to the cover. When brought together, a very strong join is produced.
- 3-D knitting of car seat covers:[26-28] This method is based on knitting fabric in one piece to avoid panel cutting and sewing and reduce material waste. The entire operation is computer-controlled to form a predesigned single 3-D shaped piece.

14.4.2 Performance characteristics of transportation seats

As indicated earlier, the key performance characteristics of transportation seats are durability, comfort and safety (see Fig. 14.8). These characteristics are discussed in some detail in the following sections. An optimum balance between these characteristics represents a true design challenge,

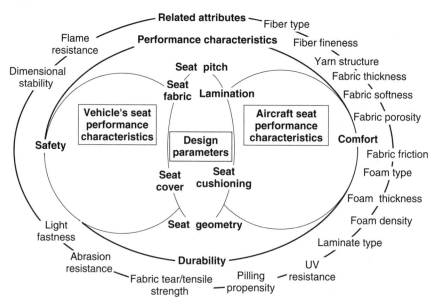

14.8 Performance characteristics and attributes of transportation seats.

particularly in view of the critical need for sufficiently light weight to achieve fuel economy and minimize inertial stresses encountered during a crash.

Durability aspects of transportation seats

Although transportation seats are expected to survive over the entire service life of the vehicle or the aircraft, they may fail or deteriorate under many circumstances, due to the effects of many factors including excessive external stresses, harsh environmental factors, flame or high heat, and stain or dirt. The most noticeable potential cause of seat failure is the loading of the seat by the occupant's body mass, particularly in the rearward direction. Seat failure is also expected in the event of a crash owing to the relative acceleration of the occupant's body mass during a collision. More serious than the failure of the transportation seat is the impact of this failure on the occupants. For example, it is commonly known that when an automobile front seat fails in a rear-end collision causing the seat back to move suddenly rearward, serious hazardous conditions can be encountered including[29,30] loss of control (exposing occupants to multiple crashes), seat ejection and interior impact. These aspects can be handled in the design phase of transportation seats using stress–strain analysis of all the potential forces applied to the seat under both static and dynamic conditions and the failure analysis discussed in Chapter 7. One of the key factors in this analysis, which is often overlooked, is the variability aspect resulting from variation of occupant size and weight, and the variable scenarios of a vehicle or aircraft crash. These aspects should be handled through probabilistic design in which these variables are simulated by some form of probability distributions.[31]

In the context of seat cover performance, key attributes related to mechanical durability include fabric tear strength, fabric tensile strength, abrasion resistance and pilling propensity. Different levels of these attributes can be selected and controlled through appropriate selection of fiber type, yarn structure, fabric construction and fabric finish. These attributes have been a part of the basic specifications of upholstery fabrics used for transportation interiors for many years.

Most transportation units are likely to be subjected to harsh environmental conditions during their service life. The need for lighter vehicles has resulted in new styles and body shapes that has led to more exposure of vehicles interiors to the surrounding environment.

Examples of these styles include slanting glass and larger windows, which in turn have led to more exposure to sunlight. Through glass, which is transparent to visible light, vehicles interiors can be heated to temperatures well above 100°C. Vehicle interiors can also be exposed to levels of relative humidity ranging from 0 to 100%. Severe heat and humidity conditions can result in substantial degradation of seat covers, reflected in color fading and surface deterioration that can in turn influence the mechanical resistance of vehicle seats. In the design analysis, the key factors in dealing with the environmental effects are fiber type, yarn structure, fabric construction and finish treatment.

Flame resistance represents a durability and safety factor not just for transportation seats but for the entire interior from carpets to sidewalls and roof interiors. As a result, most design analysis involves a great deal of emphasis on this aspect. There are many ways of minimizing the impact of flames by containing their propagation that can be implemented in the design of transportation units. Similarly, stain effects are handled through the design of fibrous structures that are stain repellent as well as by the use of stain-resistant finish.

Comfort aspects of transportation seats

In recent years, consumer demands for comfort have increased as a result of the increasing tendency to reduce the size of transportation units, or the space allocated per occupant (e.g. in aircraft), and the exponential increase in aircraft passenger population. Under normal traveling conditions, restraining an occupant in a single location in the vehicle is a mild source of discomfort, although many children would consider it a severe restraint. As travel duration increases, the extent of discomfort will be likely to increase, resulting in driver fatigue which could contribute to road safety hazards. For some professions (e.g. truck drivers, salesmen and postage or package carriers), 50 or more hours of driving per week is common. This amounts to over 2300 hours driving per year.

In aircraft travel, long hours of flying associated with seating restrictions, sometimes under severe conditions of air turbulence is an experience that most of us are familiar with. Some airlines would like to reduce seat pitch and seat width to fit more seats per aircraft for obvious economical reasons. On the other hand, people are getting bigger in size and obesity is becoming a global epidemic. Various sources of statistics on overweight and obesity suggest a significant climb in the world overweight population. Some estimates indicate that 30% of the world's people today is overweight. In the USA alone, the 2007 statistics revealed that about 58 million people are overweight, 40 million obese, and 3 million morbidly obese. These are issues that must be considered in the design of airplane and car seats, otherwise they will be left to inconvenient regulations.

Automotive and aircraft seating comfort has been discussed in many investigations in which two main human-related aspects are commonly addressed:[32-38] anthropometry and ergonomics. Anthropometry is the study of human body measurements in designing ergonomic furniture in view of body variability. Ergonomics deal with the design for maximum comfort, efficiency, safety, and ease of use, both in the workplace or in areas in which a long stay is expected. As a result, both vehicles and aircraft seats should be developed in an iterative manner in which subjective feedback from occupants is used to support the analysis. Obviously, the time and cost associated with the iteration analysis could be justified if the process was guaranteed to produce a comfortable seat.[34,35] Unfortunately, most design analyses of transportation seats tend to overlook this matter, relying totally on objective and measurable laboratory standards.

In order to incorporate comfort-related features in the design of automotive and aircraft seats, it will be important to understand the specific potential causes of discomfort associated with them. These can be divided into two main categories: physical causes and thermal causes. The first category deals with human body interaction with the seat and the second deals with heat and moisture flow between the body and the seat cover.

Physical comfort of transportation seats

A transportation seat should provide support for the occupant's body as well as a pleasant physical interaction with sensitive parts of the body such as the back, neck and waist. Accordingly, the key physical aspects of seating comfort are seat/body interface pressure distribution, pressure change rate and body vibration. These are anthropometric-related aspects that have been studied extensively by many researchers. Design engineers should refer to the numerous findings in this area as part of their information gathering phase. For example, the characteristics of pressure distribution on a rigid seat under whole-body vehicular vibrations have been thoroughly studied by many researchers.[36-38] Some of these studies[38] developed ways of measuring and predicting the seat pressure in the context of the driver's discomfort in order to provide engineers and manufacturers with key design information. Techniques for measuring seat pressure distribution and area pressure change were also proposed for truck seating in the context of evaluating the cushioning effect of transportation seats on the driver's comfort.[39]

One of the benefits of evaluating the physical comfort of transportation seats is the introduction of alternative cushioning systems that can apply gentler and more acceptable pressure to the human body. For example, air-inflated (or pneumatic) seats were found to provide a significant improvement in pressure distribution at the seat cushion–occupant interface compared with the traditional foam cushion.[39-41] Most studies revealed substantial differences between the dynamics of foam cushions and those of air-inflated cushions, particularly in terms of their interface with the human body. Other advantages of using pneumatic seats over foam seats included reduced weight (resulting from the use of pneumatic actuators instead of heavy drive motors), the avoidance of using flammable substances and the possibility of more variable adjustment options for seat height, inclination, shape and surface hardness. Pneumatic cushions also open up the possibility of additional options such as body massaging and temperature control in the seating system.

Body vibration represents another cause of physical discomfort, particularly vertical vibration caused by road irregularities or severe air turbulence in the case of airplanes. A significant portion of this vibration is transmitted to the body (the buttocks and back of the occupant) through the seating system. Typically, the natural frequency of the human trunk falls in the range of 4–8 Hz. This is commonly the frequency range within which the whole body is expected to vibrate.[41,42] From a design viewpoint, seat geometry is the key parameter in dealing with the mechanical energy absorption characteristics of the seated occupants under vertical vibration. Some studies suggested that the absorbed power quantity increases in a quadratic manner with the exposure level of the person and that the absorbed power is strongly dependent upon individual anthropometry variables such as body mass, fat and mass index.[43]

When the focus is on the physical comfort provided by the seat cover system, key design attributes will include fiber type, yarn structure, fabric thickness, fabric surface texture, fabric frictional characteristics, foam type, foam thickness and foam density. These parameters should be incorporated in the design analysis, particularly in the simulation and modeling phase. In this regard, design engineers of seat cover systems should consult some of the models developed to simulate the dynamics of seating systems.[44–51]

In addition to the above factors, it is also important to consider the fitting aspects of seats. In this regard, seat design parameters can be divided into three categories:[43–47]

1. fit parameter levels, determined by the anthropometry of the occupant population and including such measures as the length of the seat cushion;
2. feel parameters relating to the physical contact between the sitter and the seat and include the pressure distribution and upholstery properties;
3. support parameters that can affect the posture of the occupant and include seat contours and adjustments.

Thermal comfort of transportation seats

Analysis of the thermal comfort provided by a transportation seat is a very complex one, primarily owing to the multiplicity of interface media in a vehicle or an aircraft. These media include body–garment interface, garment–seat interface and seat–surrounding interface. Analysis of these interfaces is further complicated by the dynamic media change during traveling. This means that an occupant may suffer from multiple heat or cold sources simultaneously in a vehicle or an aircraft.

In order to deal with the problem of the thermal comfort of a transportation seat in design analysis, it is important that engineers understand the basic concepts of thermo-physiological comfort. These concepts were discussed in Chapter 12 and they largely hold for the situation of transportation seating. However, interpretation of these concepts should be modified in view of the dynamic and transient nature of the thermal environment in a vehicle or an aircraft. A comprehensive modeling analysis of seating thermal comfort should account for many factors including: (1) a double fabric layer system consisting of the garment worn by the occupant and that of the seat cover, (2) the squab and cushion underneath the seat cover, or the role of foam material and foam thickness as additional barriers to the escape of perspiration and (3) the role of foam material as a heat insulator. From a design viewpoint, critical factors that should be considered in relation to thermal comfort include fiber type, fiber fineness[52] (e.g. normal or micro-denier), fabric structure[52] (e.g. flat, velvet and Jacquard) and laminate type.[53]

It should be pointed out that despite significant developments made in recent years, thermal comfort still represents a design challenge in both vehicle and aircraft seating systems. The seats of modern vehicles are still hot and sticky in a hot and humid environment and they provide low insulation in a cold environment. This challenge has led to the consideration of alternative components such as 'bead' seats to provide a cooler sensation in a hot environment as a result of the air gaps they create between the skin and the seat thus allowing some air circulation and sweat evaporation.[2] However, these alternatives do not provide optimum tactile comfort in comparison with fabric material.

Safety aspects of transportation seats

A seating system in a vehicle can provide safety features that are as important as those provided by safety air bags or seat belts. In frontal and rear-end crashes, the seat frame and seat back contain the occupant and keeps him/her in an upright position and the seat cushioning absorbs some of the energy generated by the impact. The key issue here is for the seat not to collapse in the event of a crash.

From a design viewpoint, the safety criteria of a seat should be addressed in conjunction with both the air bag and the seat belt mechanisms discussed earlier. It should also be addressed in relation to the other seat criteria, namely durability and comfort. For example, the trend of replacing traditional foam material with pneumatic seat systems has been mainly based on meeting comfort and light weight specifications, particularly under normal traveling conditions. When safety is considered, particularly in the event of a crash, the question becomes whether pneumatic seats are crashworthy components or not.

14.4.3 Fibers used in transportation seats

The selection of fiber type for transportation seat fabrics has evolved over the years with the progress in fiber development and discoveries. Initially, seats were made from natural fibers such as wool and cotton. These fibers suffer low abrasion resistance and they were highly moisture absorbent. Later, synthetic fibers, from rayon to polyester, dominated this market because of their superior durability and moisture absorption features. Today, the most commonly used fiber for seat fabric is polyester. This is primarily because of its high strength, high abrasion resistance, relatively high UV degradation resistance, good mildew resistance, and good resilience and crease resistance. The main problem associated with polyester fabric is its low moisture absorption. This problem is overcome using a hydrophilic surface finish, or by blending polyester with other fibers such as wool fibers. Polyester fabric is typically treated with a special finish to improve its UV resistance. In addition to common rounded polyester fibers, multichannel polyester fibers are also evaluated in the design of transportation seats because of their good moisture transfer, easy dyeability and stain protection.[2,53]

Acrylic fibers are also used in smaller quantities in some vehicles. This fiber has superior UV degradation resistance but suffers poor abrasion resistance in comparison with polyester fiber. Polypropylene fibers have been also tried because of their light weight and easy recyclability. However, problems such as dyeability and low abrasion resistance have hindered this fiber type from being widely used in transportation seat material.

In recent years, some developments have been made in using polyester and wool fibers including recycled fibers in a nonwoven assembly as a replacement for polyurethane laminate foam.[2,54,55] Another development towards replacing foam material by fibrous material was the use of spacer knit fabrics, which is a fabric construction in which yarns are raised perpendicular to the fabric plane and sandwiched by two knit layers. Design challenges associated with using fibrous assemblies instead of laminate foam material include significant loss in thickness under compression and the difficulty of duplicating the pure isotropic structure of foam, as fibrous structures often exhibit partial anisotropy.

In addition to the above design problems associated with the use of fibrous materials in replacement of other parts of the seat, resilience to high temperatures and geometrical integrity in comparison with poly-urethane foam also represent difficulties.[2] One attempt to overcome these problems was by using polyester fiber clusters with a coiled and fluff con-figuration, introduced by DuPont. The clusters are put into a mold made of perforated metal and hot air is applied which bonds the clusters together. Claims by DuPont regarding this development include:[2,56]

- weight savings of up to 30 to 40% resulting from using polyester clusters instead of foam
- equivalent seating support to that of foam associated with easier disas-sembly and recycling
- Better comfort through increased breatheability.

Another development made by Toyobo is 'BREATH AIR™', which consisted of random continuous loops of a thermoplastic elastomer. This development was also associated with similar claims of comfort and recy-clability.[57] In addition, some natural fibers such as jute, sisal and kapok, are also being considered for use in seats as alternatives to polyurethane foam.[2]

Developments to replace urethane foam in seat cushions with fibrous materials were also extended to aircraft seats. In this regard, the objectives were to achieve superior flame resistance, light weight and durability. One of these efforts was based on using pitch based carbon fiber manufac-tured by Osaka Gas as a substitute for urethane foam. This fibrous material was highly incombustible, produced no harmful gas and exhibited light weight.

14.4.4 Yarns and fabrics used in transportation seats

In order to meet the performance characteristics of transportation seats, both the yarn structure and the fabric construction should be optimized with respect to key attributes such as tensile strength, abrasion resistance, pilling propensity, porosity, moisture absorption and resilience. From a design viewpoint, optimum levels of these attributes should be selected in such a way as to create a good balance between the key performance char-acteristics of seat fabrics, namely durability, comfort and safety. In this regard, yarns can be made with different structures from spun to continu-ous filaments and with a wide range of counts and twist levels. However, spun or twisted yarns typically suffer poor abrasion resistance.

In practice, air-textured yarns dominate the market in transportation seats. The main reason for this domination is the common agreement among most seat designers that air-textured polyester yarns meet the balance of durability and comfort. Air-textured polyester yarns are made from a wide range of deniers depending on prespecifications associated with lamination and the foam materials used. Many yarns can be doubled during air texturing to create core/effect structures. Yarns are typically dyed in the yarn form. This means that precision winding of yarns and minimum yarn defects are key requirements.

Seat cover fabrics can be made from different fabric constructions including[58] flat woven fabric at 200 to 400 g m^{-2}, flat-woven velvet at 350 to 450 g m^{-2}, warp-knit tricot with a pile surface of 280 to 380 g m^{-2} and circular knit fabric with a pile surface of 160 to 230 g m^{-2}. For flat woven fabrics, partially oriented yarns (POY) are typically used. Typical basic yarns used for woven fabrics are 167 dtex, 48 filaments. These yarns are typically quadrupled to form heavier yarns of 668 dtex, 192 filaments and 835 dtex, 240 filaments made from five ends of basic yarns. For knit fabrics, lighter yarns of about 300 dtex are used.

As indicated in Chapter 10, woven fabrics typically have higher strength and higher abrasion resistance than knit fabrics. However, they exhibit lower stretchability, which makes them less flexible in seat cover laminate fabrication. From a design viewpoint, consideration of some elastomeric materials may be an option for enhancing the stretchability of woven fabrics. Another option is to use another fiber type with higher stretchability than the traditional polyester fibers which share the same positive features such as polybutylterephthalate, PBT. This type of fiber is significantly more expensive than regular polyester.[58] The most expensive fabrics used in seat covers are flat woven velvet fabrics. In addition to stretchability, knitted fabrics with raised surfaces provide more of the desirable softness to the touch than flat woven fabrics.

14.4.5 Finish applications in transportation seats

As indicated earlier, yarns used in seat cover fabrics are typically package dyed. This omits the need for dyeing in the fabric form. However, different forms of finishing must be applied to the fabric form to meet the desired performance characteristics.[2,58] In this regard, woven fabrics may be scoured, stentered, finished and laminated before cutting and sewing. Warp knit fabrics are typically brushed, stentered, scoured, finished and laminated. Weft knit fabrics on the other hand are typically sheared, scoured, stentered, finished and laminated. Three-dimensional knits are typically heat set before fitting to provide structural stabilization. Note that stentering is an essential finishing phase for all seat cover fabrics. This is because stentering provides a stable, flat, tension free substrate for lamination and eventual seat fabrication.[2] It is also important that any finish used should be applied in such a way that will not hinder the adhesion capability during lamination. This is one of the reasons why silicon-based finishes are often avoided in seat cover fabrics.

Seat cover fabric can also be coated, either to improve some performance characteristics or to maintain the integrity of fabric structure. For example, some woven fabrics may be coated with acrylic or polyurethane resin to improve flame resistance and abrasion resistance. On the other hand, woven velvets are coated to improve pile pull-out properties.[58] Finally, the continuous need for styles and appearance changes in modern cars has led to more attention being given to the printing aspect of seat cover fabrics.

14.5   References

1. BYRNE C, 'Technical textiles market – an overview', *Handbook of Technical Textiles*, Horrocks A R and Anand S C (editors), Woodhead Publishing and The Textile Institute, Cambridge, UK, 2000.
2. FUNG W and HARDCASTLE M, *Textiles in Automotive Engineering*, Woodhead Publishing Limited, Cambridge, UK, 2001.
3. ROSS H R (AlliedSignal), *A Technical Discussion on Airbag Fabrics*, Stay-Gard™ nylon 6, Technical Information brochure, AlliedSignal, 1993.
4. MUKHOPADHYAY SK and PARTRIDGE JF, 'Automotive textiles', *Textile Progress*, 1999, **29**(112), 68–87 (Manchester, The Textile Institute).
5. CHAIKIN D, 'How it works – airbags,' *Popular Mechanics*, 1991, **June**, 81–3.
6. GOTTSCHALK M A, 'Micromachined airbag sensor tests itself', *Design News*, 1992, October, 26–8.
7. NATIONAL HIGHWAY TRAFFIC SAFETY ADMINISTRATION. *Effectiveness of Occupant Protection Systems and Their Use*: Fifth/sixth report to Congress. US Department of Transportation, Washington, DC, 2001.
8. *DuPont Automotive TI leaflets* H-48030 and H 48032 (USA).
9. *DuPont Technical Documents*, http://www2.dupont.com/Automotive, 2007.
10. BARNES J A and RAWSON N, 'Melt-through behavior of nylon 6.6 airbag fabrics', *Proceedings 'Airbag 2000'*, Fraunhofer Institut für Chemische Technologie, Karlsruhe, Fraunhofer Press, November, 1996, 26–7.
11. BARNES J A, 'Experimental determination of the heat resistive properties of airbag fabrics', *Proceedings 8th World Textile Congress, Industrial, Technical and High Performance Textiles*, University of Huddersfield, July 15–16, 1998, 329–38.
12. ROSS H R (AlliedSignal), *New Future Trends in Airbag Fabrics*, IMMFC, Dornbirn, 17–19 September 1997.
13. BOHIN F and LADREYT M, 'Silicone elastomers for airbag coatings', *Automotive Interiors International*, 1996, **5**(4), 66–71.
14. KESHAVARAJ R, TOCK R W and HAYCOOK D, 'Airbag fabric material modeling of nylon and polyester fabrics using a very simple neural network architecture', *Journal of Applied Polymer Science*, 1998, **60**(13), 2329–38.
15. EL MOGAHZY Y, *Developing a Design-Oriented Fabric Comfort Model*, Project Final Report, National Textile Center, Project S01-AE32, 2003.
16. http://www.honeywell.com/sm/index.jsp
17. MORRIS WJ, 'Seat belts', *Textiles*, 1988, **17**(1), 15–21.

18. ROCHE C, 'The seat belt remains essential', *Technical Usage Textiles*, 1992, **3**, 63–4.

19. RUNE A and BÄCKSTRÖM C-G, *The Seat Belt*, Swedish Research and Development for Global Automotive Safety, Kulturvårdskommittén Vattenfall AB, Stockholm, 2000, 12–16.

20. GRANT P, 'Textile laminates for the foam in fabric process', *Urethanes '90, Plastics and Rubber Institute Conference*, Blackpool, 16–17 October 1990.

21. OERTEL G, *Polyurethane Handbook*, Macmillan Publishing, New York, 1985.

22. KOEPPEL R C, 'Developments in lamination and laminates in foam in place seating', *Textiles in Automotives Conference*, Greenville, NC, 29 October 1991.

23. *HCTM Process*, Astechnologies Technical Information Leaflet, Roswell, Georgia 1997.

24. BRAUNSTEIN J, 'Whole lot goin' on underneath', *Automotive & Transportation Interiors*, 1997, 22–32.

25. ANON, 'Seat systems', *Automotive Engineering*, 1994, 25–35.

26. ANON, 'Knitted car upholstery', *Knitting International*, 1993, **100**(1194), 34.

27. ROBINSON F and ASHTON S, 'Knitting in the third dimension', *Textile Horizons*, 1994, 22–4.

28. GARDNER C, 'CAD and CAE, balancing new technology with traditional design', *Inside Automotives International*, 1994, **June**, 17–20.

29. Mercedes-Benz, comments to *NHTSA docket 89–20, Notice 1*, December 7, 1989.

30. General Motors submission to *NHTSA, docket no. 89–20, Notice 1*, December 4, 1989.

31. EL MOGAHZY Y, *Statistics and Quality Control for Engineers and Manufacturers: from Basic to Advanced Topics*, 2nd edition, Quality Press, Atlanta, GA, 2002.

32. KOLICH M, 'Predicting automobile seat comfort using a neural network', *International Journal of Industrial Ergonomics*, 2004, **33**(4), 285–93.

33. SMITH DR, ANDREWS DM and WAWROW PT, 'Development and evaluation of the automotive seating discomfort questionnaire (ASDQ)', *International Journal of Industrial Ergonomics*, 2006, **36**(2), 141–9.

34. KOLICH M, WAN D, PIELEMEIER W J, MEIER R C, JR and SZOTT M L, 'A comparison of occupied seat vibration transmissibility from two independent facilities', *Journal of Vibration and Control*, 2006, **12**(2), 189–96.

35. KOLICH M and TABOUN S, 'Ergonomics modeling and evaluation of automobile seat comfort', *Ergonomics*, 2004, **47**(8), 841–63.

36. FAIL T C, DELBRESSINE F and RAUTERBERG M, 'Vehicle seat design: state of the art and recent development', *Proceedings World Engineering Congress*, Federation of Engineering Institutions of Islamic Countries, Penang Malaysia, 2007, 51–61.

37. BOILEAU P E and RAKHEJA S, 'Vibration attenuation performance of suspension seats for off-road forestry vehicles'. *International Journal of Industrial Ergonomics*, 1990, **5**, 275–91.

38. GYI D E, PORTER J M and ROBERTSON K B, 'Seat pressure measurement technologies: considerations for their evaluation'. *Applied Ergonomics*, 1998, **27**(2), 85–91.

39. SEIGLER M and AHMADIAN M, 'Evaluation of an alternative seating technology for truck seats', *Heavy Vehicle Systems*, 2003, **10**(3), 188–208.

40. NA S, LIM S, CHOI H and CHUNG M K, 'Evaluation of driver's discomfort and postural change using dynamic body pressure distribution', *International Journal of Industrial Ergonomics*, 2005, **35**, 1085–96.

41. HINZ B, RUTZEL S, BLUTHNER R, MENZEL G, WOLFEL H P and SEIDEL H, 'Apparent mass of seated man – First determination with a soft seat and dynamic seat pressure', *Journal of Sound and Vibration*, 2006, **298**, 704–24.

42. VAN NIEKERK J L, PIELEMEIER W J and GREENBERG J A, 'The use of seat effective amplitude transmissibility (SEAT) values to predict dynamic seat comfort', *Journal of Sound and Vibration*, 2003, **260**, 867–88.

43. WANG W, RAKHEJA S and BOILEAU P E, 'The role of seat geometry and posture on the mechanical energy absorption characteristics of seated occupants under vertical vibration', *International Journal of Industrial Ergonomics*, 2006, **36**, 171–84.

44. BOUAZARA M and RICHARD M J, 'An optimization method designed to improve 3-D vehicle comfort and road holding capability through the use of active and semi-active suspensions', *European Journal of Mechanics A, Solids*, 2001, **20**, 509–20.

45. DE CUYPER J and VERHAEGEN M, 'State space modeling and stable dynamic invension for trajectory tracking on an industrial seat test rig', *Journal of Vibration and Control*, 2002, **8**, 1033–50.

46. GILLBERG M, KECKLUND G and AKERSTEDT T, 'Sleepiness and performance of professional drivers in a truck simulator-comparison between day and night driving', *Journal of Sleep Research*, 1996, **5**, 12–15.

47. MAVRIKIOS D, KARABATSOU V, ALEXPOULOS K, PAPPAS M, GOGOS P and CHRYSSO-LOURIS G, 'An approach to human motion analysis and modeling', *International Journal of Industrial Ergonomics*, 2006, **36**, 979–89.

48. SONG X and AHMADIAN M, *Study of Semiactive Adaptive Control Algorithms with Magnetorheological Seat Suspension*; SAE technical paper no. 2004-01-1648. Society of Automotive Engineers, Warrendale, PA, USA, 2004.

49. VERVER M M, VAN HOOF J, OOMENS C W J, WISMANS J S H M and BAAIJENS F P T, 'A finite element model of the human buttocks for prediction of seat pressure distributions', *Computer Methods in Biomechanics and Biomedical Engineering*, 2004, **7**(4), 193–203.

50. VERVER M M, DE LANGE R, VAN HOOF J and WISMANS J S H M, 'Aspects of seat modeling for seating comfort analysis', *Journal of Applied Ergonomics*, 2005, **36**, 33–42.

51. REBELLE J, 'Development of a numerical model of seat suspension to optimize the end-stop buffers', *35th United Kingdom Group Meeting on Human Responses to Vibration*, ISVR, University of Southampton, Southampton, England, 13–15 September, 2000.

52. CENGIZ T G and BABALIK F C, 'An on-the-road experiment into the thermal comfort of car seats', *Applied Ergonomics*, 2007, **38**, 337–47.

53. FUNG W, 'How to improve thermal comfort of the car seat', *Journal of Coated Fabrics*, 1997, **27**, 126–45.

54. SCHMIDT G and BOTTCHER P, 'Laminating nonwoven fabrics made from or conntaining secondary or recycled fibres for use in automotive manufacture', *Index Conference, Session 3A*, Geneva, Brussels, EDANA, 1993.

55. FUCHS H and BOTTCHER P, 'Textile waste materials in motor cars – potential and limitations', *Textil Praxis International*, 1994, **April**(4), II–IV.
56. GARDNER C, 'Interiors industry's one stop shop – DuPont Automotive', *Inside Automotives International*, 1995, **March/April**, 40–5.
57. TANKA H (Toyobo), '*Highly Functional Cushion Material, BREATH AIR*', IMMFC, Dornbirn, 1997, **September**, 17–19.
58. FUNG W, 'Textiles in transportation', *Handbook of Technical Textiles*, Horrocks A R, and Anand S C (editors), Woodhead Publishing Limited, Cambridge, UK, 2000, 490–528.

15

Development of textile fiber products for medical and protection applications

Abstract: When a wound is open, there is no other material that can prevent many microorganisms from contacting and infecting the wound better than a fibrous material in various applications assembled in products such as gauzes, bandages, sutures, hydro gels, hydro colloids, surgical drapes, gowns and clean air suits. Function-focus fibrous products also save the lives people who work in risky and extreme climate conditions. This chapter focuses on the superior features of health-related and protection fibrous products. The development of fibrous products for health-related applications faces a number of challenges. In this chapter, the basic performance characteristics and corresponding attributes of health-related products are reviewed. Also in this chapter, protective clothing and protective systems are discussed from a product development viewpoint.

Key words: collagen; alginates; chitin; gauze; sterilization wrap; incontinence products; bandages; non-implantable products; implantable products; extracorporeal devices; healthcare/hygiene; vapor-permeable adhesive films; hydro-gels; hydrocolloids, dressings; silicone meshes; sutures; micro porous membranes; uniforms; helmets; tents; protective clothing; antiflash hoods and gloves; survival suits; ropes and harnesses; awareness–fuzziness diagram.

15.1 Introduction

The need for fibrous materials in medical, hygiene and healthcare applications stems from a number of unique attributes including[1-3] light weight, flexibility, softness, manipulability, strength, air permeability, absorbency, wickability, biodegradability and non-toxicity. These attributes contribute significantly to the key functional performance of health-related products. In addition, fibrous components can be utilized in all forms and in different assemblies including spun, monofilament, and multifilament yarns, woven, knitted and nonwoven fabrics, and fiber-reinforced components.

The development of fibrous products for health-related applications faces a number of challenges. The first challenge to understand the functional characteristics of the product, which requires a great deal of multi-disciplinary knowledge and cooperation between product developers, polymer scientists and physicists. This knowledge should be based on a full realization of the liability and safety regulations associated with these products that must be satisfied at appropriate costs. This issue makes it important to consider design aspects such as the performance/cost ratio and safety/cost ratio.[4] The second challenge stems from the increasing awareness of the effects of disposable products on the environment and the cost associated with handling these products (e.g. landfill capacity and waste management cost). This issue occupies a great deal of the development efforts of fibrous products in this field as a result of substantial quantities of disposable health-care products. Options such as the use of biodegradable fibers and reusable disposables are being evaluated in many health-related products. More options and ideas in this direction will certainly deserve a great deal of attention as they can both preserve resources and minimize waste materials.

In this chapter, basic performance characteristics and corresponding attributes of health-related products are reviewed. Also in this chapter, protective clothing and protective systems are discussed from a product development viewpoint. As many of the readers are fully aware, both health-related and protective products have a great deal in common as they often serve the same purpose either for prevention or for remedy of human pain.

15.2   Fiber types used for health-related products

In health-related applications, fiber type and fiber characteristics represent a critical design aspect. This is primarily due to the fact that most performance characteristics of these products are initially driven by the fiber type and further enhanced by the structural features of the fiber assembly used. Figure 15.1 illustrates examples of fiber types used in the medical field categorized in two different ways: by source (natural, synthetic and speciality fibers) and by body reaction or absorption (biodegradable and non-biodegradable). In most applications, fibers should originate from polymeric structures that are linear, long and flexible, their side groups should be simple, small, or polar and their chains should be capable of being oriented and crystallized. Fibers must be non-toxic, non-allergenic, non-carcinogenic and have the ability to sterilize without imparting any change to their physical or chemical characteristics.

Natural fibers used in medical applications include cotton, long-vegetable fibers and silk. These fibers are mainly used in non-implantable products. Regenerated cellulosic fibers such as viscose rayon are also widely used in non-implantable products and healthcare/hygiene products. For some special applications, natural fibers can be treated by chemicals to enhance their performance. For example, microbiocidal compositions that inhibit the growth of microorganisms can be applied to natural fibers as coatings. Most synthetic fibers are used for implantable as well as other high-performance medical products. Commonly used synthetic materials include polyester, polyamide, polytetrafluoroethylene (PTFE), polypropyl-

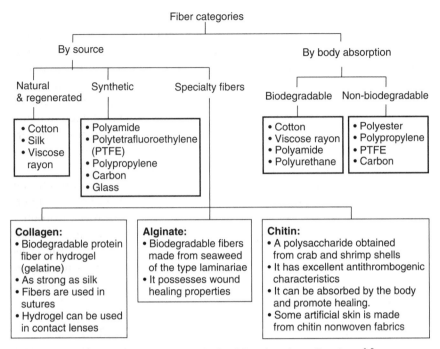

15.1 Fiber categories used in health-related applications.[1-3]

ene, carbon and glass. Again, microbiocidal compositions that inhibit the growth of microorganisms can be incorporated directly into these.

When body absorption is of primary concern, the extent of fiber biodegradability becomes a critical aspect in the development of health-related products. Biodegradable fibers are those which can be absorbed by the body within two to three months after implantation.[2] These include cotton, viscose rayon, polyamide, polyurethane, collagen and alginate. Fibers that are slowly absorbed within the body and take more than six months to degrade are considered to be non-biodegradable. These include polyester, polypropylene, PTFE and carbon.

Specialty fibers are those developed specifically to be absorbed in the human body shortly after medical applications. Examples of these fibers, which have been proved to contribute significantly to the healing process, include collagen, alginate and chitin fibers.[1,2,5,6] Collagen fibers are commonly used as sutures for surgery inside the human body. Collagen is a protein substance obtained from bovine skin and available either in fiber or hydro-gel (gelatin) form. The fibers can be used as sutures because of their strength and biodegradability. The transparent hydro-gel, formed when collagen is cross-linked in 5–10% aqueous solution, has high oxygen permeability and can be processed into soft contact lenses. Calcium alginate fibers are used for wound dressings owing to their non-toxicity and hemostatic properties as well as their biodegradability.[5] These fibers are produced from seaweed of the type Laminariae. Chitin is a polysaccharide obtained from crab and shrimp shells.[2] It has excellent anti-thrombogenic characteristics and can be absorbed by the body. Chitin nonwoven fabrics can be used as artificial skin owing to their ability to adhere to the body, stimulating new skin formation which accelerates the healing rate and reduces pain. When chitin is treated with alkali, it yields chitosan, which can be spun into filaments of similar strength to viscose rayon.

The development of fibers suitable for medical applications is likely to continue for many years to come as a result of the wide variety of medical applications and the continuing need for fibers that meet critical health-related criteria such as biodegradability, biocompatibility and biosafety. Waste disposal and environmental friendliness have also become essential criteria for such materials. The naturally derived fibers mentioned above (collagen, alginate and chitin fibers) are now used in many applications. More fibers from natural sources are being developed. For example, poly-lactic acid or lactide, generically known as PLA fiber, is derived from natural raw materials (starch and cellulose) and is completely biodegradable. Researchers in Japan have used poly L-lactic acid (PLLA) fibers to develop a drug delivery system for surgical implants. Medical textile research groups at the Institute of Textile Technology, RWTH Aachen were able to spin polyvinylidene fluoride (PVDF) and poly D-lactide (PDLA) multifilaments which can be converted into staple fibers to make needle-punched nonwovens for scaffolds for tissue regeneration of periodontal defects. In addition, nano fibers are likely to have a solid place in medical applications leading to new directions in the development of medical fibers.

15.3   Fibrous structures used for health-related products

As will be seen in the following sections, fibrous structures used for health-related products can be any of the four basic forms: woven, knitted, braided and nonwoven. The first three structures require yarn formation and the last one can be made directly from fibers, or even directly from polymers (e.g. Gore-Tex based products or electrostatically spun materials from polyurethane). Among all the fibrous structures used, nonwovens represent a significant proportion of health-related products. This is directly a result of the fact that a high percentage of these products are disposable, or short-lived. Nonwoven structures can be made into a wide variety of products including:[7,8] (1) operating room apparel such as surgical packs, pants and gowns, and accessory operating theater apparel such as caps, masks, shoe and covers, gauze replacements (sponges and bandages) and sterilization wrap (CSR wrap) and (2) disposable bedding, incontinence products, wound dressings and bandages.

In addition to the economical aspects associated with making nonwovens, particularly in comparison with making costly woven or knit structures, nonwoven structures allow unlimited design options to meet healthcare functions. These include[7,8] different raw material selections or blends, different ways of web formation, different bonding techniques and different possibilities for incorporating fibrous with non-fibrous materials to form nonwoven composite structures. These design options can result in a wide range of weight, strength, flexibility and comfort. For example, nonwovens can be made very light weight, 5–15 g m^{-2}, which is important for many healthcare products such as tissues, transmission, barriers or wicking layers in composites. Heavier nonwovens are used in absorbent dressings, incontinence use, insulation and filtration. They can also provide strength for durable products such as slings or moderate strength for products such as padding bandages. When filtration and protection is required, nonwovens can be made un-stretchable (low elongation) and when support and retention are required, they can be made stretchable with high elastic recovery.

Perhaps the most critical performance characteristic of nonwovens is controlled wetting and fluid absorbency. In this regard, nonwovens can perform in many different ways, depending on the raw material used, the structural density and the type of finish used. Nonwovens can also be made with a wide range of thermal properties, from highly insulative, which is good for wound dressing (to avoid heat loss), to moderate insulation associated with good breathability. As filters, nonwovens can also be made with different levels of pore size and pore distribution depending on whether large particles or small particles need to be isolated.

Another key advantage of using nonwovens is the option for developing nonwoven composite structures that combine the attributes of different materials including fibrous and nonfibrous substances. In the healthcare field, numerous forms of composite nonwovens have been developed over the years and for various applications. These include[7,8] (a) padding layers (for absorbency, super absorbency, and thermal/mechanical insulation and protection), (b) barrier layers for prevention of liquid bacterial strike through and (c) carrier layers for other materials such as super absorbent powders, activated carbon, antimicrobials, adhesives and microspheres.

15.4   Categories of health-related products

In general, health-related products utilizing fibrous components are divided into four main categories:[1,2,5,6] (a) non-implantable products, (b) implantable products, (c) extracorporeal devices and (d) healthcare/ hygiene products. These products are described in the following sections.

15.4.1 Non-implantable products

Non-implantable products are typically used to provide protection against infection, to absorb blood and exudates and to promote healing. The term non-implantable is used generally to indicate surface wound treatments of different parts of the human body. This means that items such as wipes and swabs are also included in this category. However, the main products of this category are wound dressings or gauzes and bandages.

Wound dressings are used in the medical field to provide the critical functions that collectively aim to promote wound healing. These functions are[9–12] protection, absorption, compression, immobilization and esthetics. Protection is the primary function of wound dressing since exposed wounds can be subjected to further trauma and additional tissue loss caused by external forces (i.e. severe environments, touching objects or direct interaction). Wound dressing acts as a barrier against these forces. The fact that most open wounds generate blood and exudates requires wound dressing that can absorb these substances to reduce the risk of bacterial proliferation and subsequent infection. A key aspect of the absorption function is quick release of these liquid substances so that they do not accumulate at the interface between the skin and the wound dressing. This is achieved through a combination of wicking and absorption mechanisms. Covering the wound entirely with a wound dressing often involves undesirable compression against the open wound. It is important, therefore, that the wound dressing provides a cushioning feel rather than a pressing feel to the wound. Immobilization is also a key function for some wounds as tissue movement can delay the healing process. Finally, wound appearance and esthetics have become important, as wounds naturally draw attention. In this regard, wound dressings should appear neat and tidy, particularly on the outer surface.

Most wound dressings consist of multiple layers of components that assist in promoting wound healing.[2,9] As shown in Fig. 15.2, the first layer is the direct-contact layer. This layer should not adhere to the wound and should allow easy removal of the wound dressing without disturbing new tissue growth. This layer may be made from a very light knitted, woven or nonwoven structure using fibers such as silk, polyamide, viscose or polyethylene. Non-adherence is typically achieved because of the nature of the fabric structure and by using non-adhesive coating material. The next layer should attract any drainage that might exude from the wound. This is achieved by the use of an absorbent pad that provides a combination of wicking and absorption. This pad is typically made from a cotton or viscose

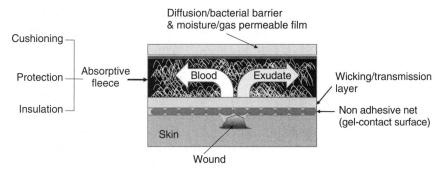

15.2 Basic structure of a wound dressing laminate (modified after Rigby and Anand[2] and Lin *et al.*[9]).

nonwoven structure. The final layer is a flexible diffusion layer attached to the absorbent pad by an adhesive.

In most situations, the direct-contact wound dressing is wrapped by other dressings such as soft gauze rolls to secure it in place. One common dressing is non-elastic roll gauze. Stability can be provided to this dressing by reinforcement with adhesive tape. Obviously, as more layers are added, compression may escalate and this should be avoided. When additional immobilization is required, a plaster splint or prefabricated splint might be required. This should be well padded so that additional adverse pressure is not applied.

Other wound dressing products include gauze, lint and wadding. Gauze is an open weave absorbent structure which, when coated with paraffin wax, is used for the treatment of burns and scalds. For surgical applications gauze serves as an absorbent material when used in pad form (swabs). Lint is a plain weave cotton fabric that is used as a protective dressing for first-aid and mild burn applications. Wadding is a highly absorbent material that is covered with a nonwoven fabric to prevent wound adhesion or fiber loss.

It should be pointed out that no single dressing is suitable for all types of wounds and a combination of different types of dressing may be used during the healing process of a single wound. In recent years and beginning in the mid-1990s, many new synthetic dressings were introduced. These include[9–12] vapor-permeable adhesive films, hydro-gels, hydrocolloids, alginates, synthetic foam dressings, silicone meshes, tissue adhesives, barrier films and silver- or collagen-containing dressings.

In light of the above discussion, the development of wound dressing products should be based on a clear understanding of the specific healing function of the product. In this regard, wound dressings can be classified into three broad categories[2,5,6]: passive dressing, interactive dressing and bio-interactive dressing. A passive dressing represents the simplest and least expensive structure as it primarily aims to provide a cover over a wound. An interactive dressing is typically a polymeric film, which is typically transparent, permeable to water vapor and oxygen, and non-permeable to bacteria. Examples of this type of dressing include hyaluronic acid, hydrogels and foam dressings. A bio-interactive dressing is a structure that can deliver the functions of the simple and interactive dressings but also deliver substances active in wound healing, for example hydrocolloids, alginates, collagens and chitosan. Figures 15.3 and 15.4 illustrate some commercial wound dressing examples, their basic functions and typical brand names.

Bandages are familiar products that are used in many medical applications, primarily to hold dressings in place over wounds. They can be made in many different forms such as woven, knitted or nonwoven fabrics using cotton or viscose fibers. The fabrics are typically cut into strips then scoured, bleached and sterilized. Some bandages should exhibit elastic and stretch characteristics so that when applied under sufficient tension, the recovery of stretch provides support for sprained limbs. Bandages used for simple applications can be non-elastic. Some bandages are made in a knitted tubular form with different diameters. These are inherently elastic and stretchy. Most woven bandages are used for light support in the management of sprains or strains. In this case, the elastic nature of the fabric is induced by weaving cotton crepe yarns with high twist levels or by weaving two warp yarns at different tension levels.

For applications such as the treatment and prevention of deep vein thrombosis, leg ulceration and varicose veins, compression bandages are used. These bandages are designed to exert certain levels of compression on the body when applied at a constant tension. For leg and ankle treatments, compression bandages are classified by the amount of compression they can exert at the ankle by extra-high, high, moderate and light compression. These types of bandages can be made from tubular or customized knitted fabrics or from woven fabric that contains cotton and elastomeric yarns.

Another type of bandage are the so-called orthopedic cushion bandages. These are used under plaster casts and compression bandages to provide padding and prevent discomfort. Some nonwoven orthopedic cushion bandages are produced from either polyurethane foams, polyester, polypropylene fibers or blends of natural and synthetic fibers. In this regard, loftiness and bulkiness are key characteristics that can be controlled using light needle-punching.

Performance characteristics of non-implantable products

The key performance characteristics of non-implantable products, particularly wound dressings, are shown in Fig. 15.5. Any wound dressing should

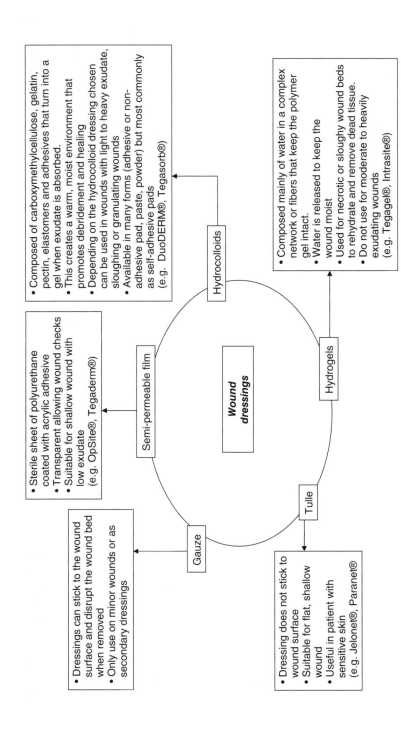

- Composed of carboxymethylcellulose, gelatin, pectin, elastomers and adhesives that turn into a gel when exudate is absorbed.
- This creates a warm, moist environment that promotes debridement and healing
- Depending on the hydrocolloid dressing chosen can be used in wounds with light to heavy exudate, sloughing or granulating wounds
- Available in many forms (adhesive or non-adhesive pad, paste, powder) but most commonly as self-adhesive pads
  (e.g. DuoDERM®, Tegasorb®)

Hydrocolloids

- Composed mainly of water in a complex network or fibers that keep the polymer gel intact.
- Water is released to keep the wound moist
- Used for necrotic or sloughy wound beds to rehydrate and remove dead tissue.
- Do not use for moderate to heavily exudating wounds
  (e.g. Tegagel®, Intrasite®)

Hydrogels

Semi-permeable film

Wound
dressings

- Sterile sheet of polyurethane coated with acrylic adhesive
- Transparent allowing wound checks
- Suitable for shallow wound with low exudate
  (e.g. OpSite®, Tegaderm®)

Tulle

Gauze

- Dressings can stick to the wound surface and disrupt the wound bed when removed
- Only use on minor wounds or as secondary dressings

- Dressing does not stick to wound surface
- Suitable for flat, shallow wound
- Useful in patient with sensitive skin
  (e.g. Jelonet®, Paranet®)

15.3 Typical types of commercial wound dressing.[1,2,5,6]

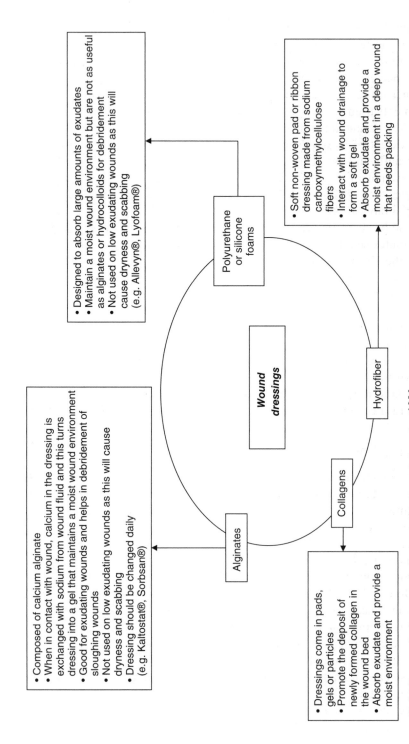

- Designed to absorb large amounts of exudates
- Maintain a moist wound environment but are not as useful as alginates or hydrocolloids for debridement
- Not used on low exudating wounds as this will cause dryness and scabbing (e.g. Allevyn®, Lyofoam®)

- Soft non-woven pad or ribbon dressing made from sodium carboxymethylcellulose fibers
- Interact with wound drainage to form a soft gel
- Absorb exudate and provide a moist environment in a deep wound that needs packing

Polyurethane or silicone foams

Wound dressings

Hydrofiber

Collagens

Alginates

- Composed of calcium alginate
- When in contact with wound, calcium in the dressing is exchanged with sodium from wound fluid and this turns dressing into a gel that maintains a moist wound environment
- Good for exudating wounds and helps in debridement of sloughing wounds
- Not used on low exudating wounds as this will cause dryness and scabbing
- Dressing should be changed daily (e.g. Kaltostat®, Sorbsan®)

- Dressings come in pads, gels or particles
- Promote the deposit of newly formed collagen in the wound bed
- Absorb exudate and provide a moist environment

15.4 More typical types of commercial wound dressing.[1,2,5,6]

maintain a moist environment at the wound/dressing interface and should absorb excess exudate without leakage to the surface of the dressing. In addition, wound dressings should provide thermal insulation, mechanical and bacterial protection. These characteristics can be optimized using numerous design parameters reflected in the various attributes of the fibrous elements used for making non-implantable products. As indicated earlier, the key parameter is fiber type. This is the parameter that will determine the extent of biodegradability, the mechanical properties of the product (strength and elasticity), the absorption or wicking characteristics and the weight of the product. As fibers are transformed into yarns and/or fabrics, key attributes that contribute largely to the performance of non-implantable are those that are structural related as illustrated in Fig. 15.5.

The performance characteristics listed in Fig. 15.5 and their associated attributes should be viewed on the basis of the wound-related criteria as specified by medical experts.[9–12] These criteria may include type of wound, size of wound, location of wound (e.g. poor vascular areas or areas under tension heal slower than areas that are highly vascular), age of wound (e.g. fresh surgical wounds versus chronic wounds), presence of wound contamination or infection (e.g. bacterial contamination slows down the healing process), age of the patient (e.g. the older the patient, the slower the wound heals), general condition of the patient (e.g. malnutrition slows down the healing process) and medication (e.g. anti-inflammatory drugs may slow down the healing process). Failure to account for these criteria can result in products that are of high liability and questionable safety.

15.4.2 Implantable products

Implantable products, often called biomaterials, are used to assist in wound closure (e.g. sutures) or replacement surgery (e.g. vascular grafts, artificial ligaments, etc.). Figure 15.6 shows examples of implantable products and associated fiber types and fibrous structures. In recent years, a dramatic increase has occurred in the number of implantable products in which numerous types of biomaterials are utilized. Many of these products have become common in most operating theaters.

Descriptions of different implantable products are presented in numerous instances in the literature.[2,5,6,9–12] Among all implantable products, sutures represent the most commonly used products. Sutures used for wound closure can be made from monofilament or multifilament yarns.[13–18] For internal wound closures, biodegradable sutures should be used. These can be made from monofilament braided collagen, polylactide and polyglycolide. To close exposed wounds, non-biodegradable and removable sutures can be used. These can be made from monofilament braided polyamide, polyester, PTFE and polypropylene.

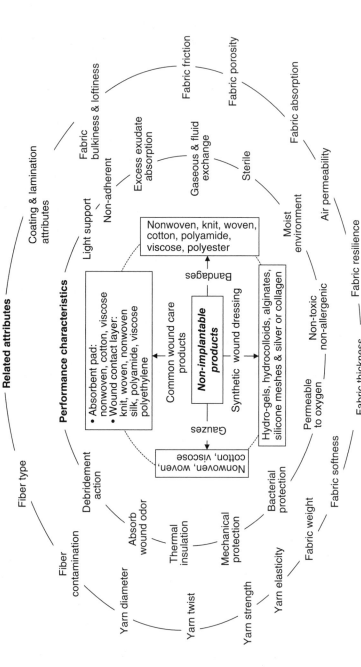

15.5 Performance characteristics and related attributes of non-implantable products.

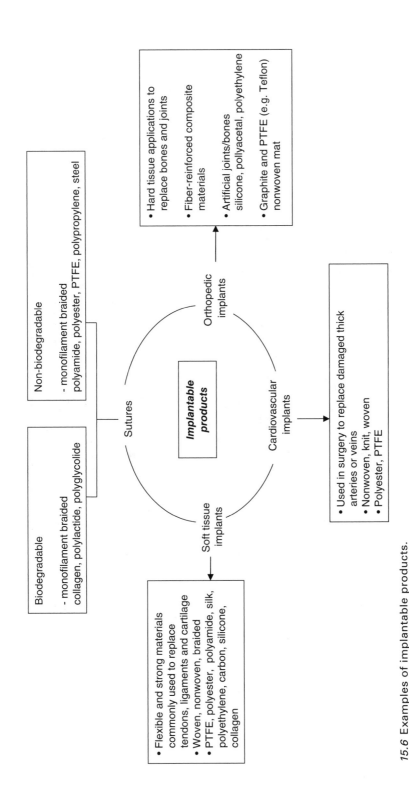

15.6 Examples of implantable products.

Another important category of implantable products is soft-tissue implants.[1,2] These are flexible strong materials commonly used to replace tendons, ligaments and cartilage in both reconstructive and corrective surgery. Artificial tendons are typically inelastic cords or bands of tough white fibrous connective tissue that attach a muscle to a bone or another part. They are typically made from woven or braided porous meshes or tapes surrounded by a silicone sheath. During implantation, the natural tendon can be looped through the artificial tendon and then sutured to itself in order to connect the muscle to the bone. Implantable materials used to replace damaged knee ligaments (anterior cruciate ligaments) include braided polyester artificial ligaments and braided composite materials containing carbon and polyester filaments. Cartilage is a tough elastic tissue that is found in the nose, throat and ear and in other parts of the body and forms most of the skeleton in infancy, changing to bone during growth. Some cartilages are hard and dense (e.g. hyaline cartilage). These typically contain no flexible fibers as rigidity is important for this material. Others are elastic and flexible and provide protective cushioning. Low density polyethylene is used to replace facial, nose, ear and throat cartilage. Carbon fiber-reinforced composite structures are used to resurface the defective areas of articular cartilage within synovial joints (knee, etc.) as a result of osteoarthritis.

Orthopedic implants are typically used for hard tissue applications to replace bones and joints.[2,5,6] Fixation plates can also be considered under this category; these are used to stabilize fractured bones. For these components, fiber-reinforced composite materials are the best candidates as they can be designed to meet biocompatibility and high strength requirements. Traditionally, metal implants have been used for these applications. One of the advantages of using fibrous materials in these applications is the promotion of tissue in-growth around the implant using a graphite and PTFE (e.g. Teflon) nonwoven mat. This acts as an interface between the implant and the adjacent hard and soft tissue.

Another category of implantable products is the so-called cardiovascular implants (or vascular grafts). These are used in surgery to replace damaged thick arteries or veins. Materials used in this category include[2] knitted or woven constructions made from polyester or PTFE structures. Knitted fabrics are useful because of their flexibility and porous structures that allow the graft to become encapsulated by new tissue. However, porosity can also allow blood leakage (hemorrhage) through the pores after implantation. Woven structures, on the other hand, can be made of lower porosity but this can hinder tissue ingrowth. Artificial blood vessels with an inner diameter of 1.5 mm have been developed using porous PTFE tubes, which consist of an inner layer of collagen and heparin to prevent blood clot formation and an outer biocompatible layer of collagen with the

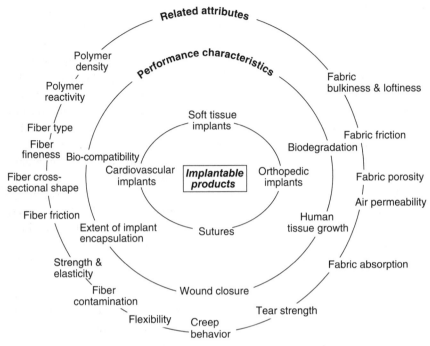

15.7 Performance characteristics and related attributes of implantable products.

tube itself providing strength. Artificial heart valves, which are caged ball valves with metal struts, are covered with polyester fabrics in order to provide a means for suturing the valve to the surrounding tissue.

Performance of implantable products

Figure 15.7 illustrates key performance characteristics of implantable products and their associated attributes. The most critical challenge associated with the design of implantable products is body compatibility or biocompatibility. From a medical viewpoint, an implantable product should be developed with regard to the tissue bed in which it is placed.[5,6] The simplest definition of body tissue is that it is a layer or group of similarly specialized cells that conjointly perform certain functions. There are many different types of tissue throughout the body and they differ markedly in composition, structure, strength and function. Body skin and many internal parts exhibit very soft tissues that have to be treated carefully when wound closure products such as sutures are used. In this regard, critical design aspects should include biocompatibility, biomaterial tensile strength, flexibility and tear resistance with reference to tissue strength. It should also be pointed out that even within the body soft tissues, some are stronger than others. For example, bladder tissues are relatively weaker than skin tissues.

Safety and side effects represent key aspects that should be associated with implantable products, particularly when biocompatibility is a major concern. The implantation of biomaterials may initiate both an inflammatory reaction in the injury as well as mechanisms to induce healing. For instance, implantation of non-resorbable biomaterials can cause a permanent alteration in the microenvironment surrounding the implant and in the tissues into which they are implanted. The extent to which the inflammatory reaction and healing mechanisms are activated is a measure of the host reaction to the biomaterial and ultimately may lead to impairment of functional capacity or the permanent biocompatibility of the implant. It is important, therefore, to develop an accurate understanding of the biological response to implantable products or biomaterials.

The reaction of the human body to implantable substances is determined by a number of key attributes. The most important factor is porosity, which determines the rate at which human tissue will grow and encapsulate the implant. Fiber fineness is another key factor with small and uniform diameters being better in terms of encapsulation with human tissue. Toxicity is also an important factor as it can dangerously influence body reaction to implantable products. Release of toxic substances, such as harmful chemicals, lubricants and finish by the fiber polymer can be fatal. Finally, biodegradability should be considered in any implantation application.

Sutures performance

Since sutures represent the largest volume of implantable products, it will be useful to dwell on their performance characteristics. A suture is defined as a thread that can perform one of two functions:[13,14] (a) joins adjacent cut surfaces of the wound through replicating and maintaining tissues until the natural healing process has provided a sufficient level of wound strength and (b) compresses blood vessels to stop bleeding. Commonly, suture sizes are given by a number representing the diameter ranging in descending order from 10 to 1 and then 1-0 to 12-0, 10 being the largest and 12-0 being the smallest with a diameter smaller than a human hair;[13] the more 0s that describe a suture size, the smaller the suture diameter and the weaker the suture. For example, a suture of 4-0 or 0000 will typically have about 0.2 mm diameter and a tensile strength of about 7.5 N. A 3-0 or 000 will typically have about 0.3 mm diameter and 12 N tensile strength and a '1' will typically have about 0.5 mm diameter and a typical tensile strength of about 37 N.

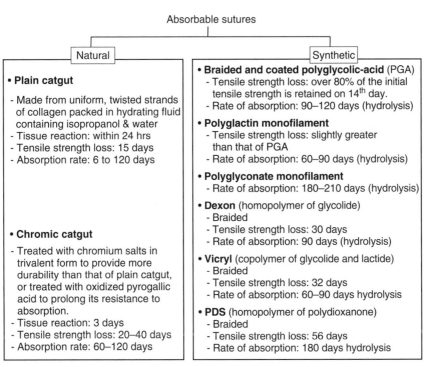

15.8 Classification of absorbable sutures.[13–17]

As indicated earlier, sutures used for wound closure can be made from monofilament or multifilament yarns. For internal wound closures, biodegradable sutures should be used. These are commonly called absorbable sutures as they do not have to be removed and they are absorbed or decomposed by body reactions such as hydrolysis. To close exposed wounds, non-biodegradable, also called non-absorbable sutures, can be used. Figures 15.8 and 15.9 illustrate examples of these two types of suture.

According to the nature of their basic functions, surgical sutures should exhibit a number of key performance characteristics (see Fig. 15.10): absorption rate, strength loss and tissue reaction for absorbable sutures, and strength, knot security and tissue reaction for non-absorbable sutures. Design efforts to meet these characteristics have been focused on critical aspects such as[13–19] (a) for braided yarns: improving the structure of the braids (e.g. spiral versus lattice braided materials), (b) for core/sheath yarns: reducing the difference in the elongation properties between the core and the sheath yarns, and using finer denier filaments in the sheath yarns, and (c) for knot performance: improving knot security and performance (e.g. exposing a two-throw square knot to laser beam energy).

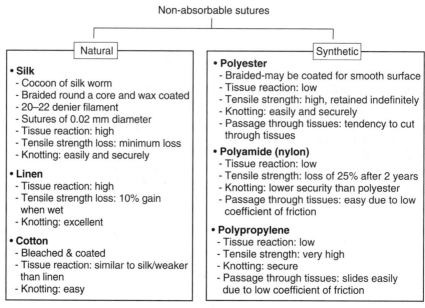

Non-absorbable sutures

Natural

• **Silk**
  - Cocoon of silk worm
  - Braided round a core and wax coated
  - 20–22 denier filament
  - Sutures of 0.02 mm diameter
  - Tissue reaction: high
  - Tensile strength loss: minimum loss
  - Knotting: easily and securely

• **Linen**
  - Tissue reaction: high
  - Tensile strength loss: 10% gain when wet
  - Knotting: excellent

• **Cotton**
  - Bleached & coated
  - Tissue reaction: similar to silk/weaker than linen
  - Knotting: easy

Synthetic

• **Polyester**
  - Braided-may be coated for smooth surface
  - Tissue reaction: low
  - Tensile strength: high, retained indefinitely
  - Knotting: easily and securely
  - Passage through tissues: tendency to cut through tissues

• **Polyamide (nylon)**
  - Tissue reaction: low
  - Tensile strength: loss of 25% after 2 years
  - Knotting: lower security than polyester
  - Passage through tissues: easy due to low coefficient of friction

• **Polypropylene**
  - Tissue reaction: low
  - Tensile strength: very high
  - Knotting: secure
  - Passage through tissues: slides easily due to low coefficient of friction

15.9 Classification of non-absorbable sutures.[13–17]

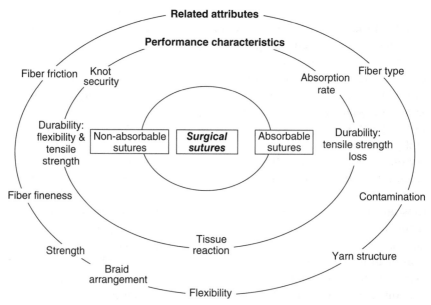

15.10 Performance characteristics and related attributes of surgical sutures.

15.4.3 Extracorporeal devices

Figure 15.11 illustrates the main examples of extracorporeal devices, their performance characteristics and associated attributes. Extracorporeal devices are mechanical organs that are primarily used for blood purification. They include[1,2] the artificial kidney (dialyzer), the artificial liver and the mechanical lung. The function of an artificial kidney is achieved by circulating the blood through a membrane that retains the unwanted waste materials. This membrane may be either a flat sheet or a bundle of hollow regenerated cellulose fibers in the form of cellophane. Waste material can also be removed using multilayer filters composed of numerous layers of needle-punched fabrics with varying densities. The artificial liver utilizes hollow fibers or membranes similar to those used for the artificial kidney to perform their functions. The microporous membranes used for mechanical lungs possess high permeability to gases but low permeability to liquids and function in the same manner as the natural lung allowing oxygen to come into contact with the patient's blood.

15.4.4 Healthcare and hygiene

Figure 15.12 illustrates the main examples of healthcare and hygiene products, examples of their performance characteristics and some associated attributes. Healthcare and hygiene products include[2] hospital gowns and uniforms, clothing and wipes, surgical covers, masks, caps and hospital bed products. These products should exhibit a number of key characteristics such as cleanness, contamination-free and infection control. In a typical hospital environment, infection represents a critical issue as it can result in further complications in the patient's condition.[2] For this reason, traditional muslin materials have been replaced by barrier materials in hospitals in many developed countries. Pollutant particles shed by hospital staff may carry bacteria that can result in an infection to the patient. Hospital gowns, particularly in surgery rooms, should be designed in such a way that they can prevent the release of pollutant particles into the air. As a result, the fiber material and the fabric structure of these gowns should be carefully selected so that they do not act as dust and contamination traps and they do not easily release these particles to the surroundings. In this regard, most surgical gowns are made from disposable nonwoven fabrics.

Surgical masks should be of light weight and they should have a high filter capacity and high level of air permeability. A typical surgical mask will consist of a three-layer structure:[5] outer and inner layers made from acrylic bonded parallel-laid or wet-laid nonwoven, and a very fine middle layer of extra fine glass fibers or synthetic microfibers. Disposable surgical

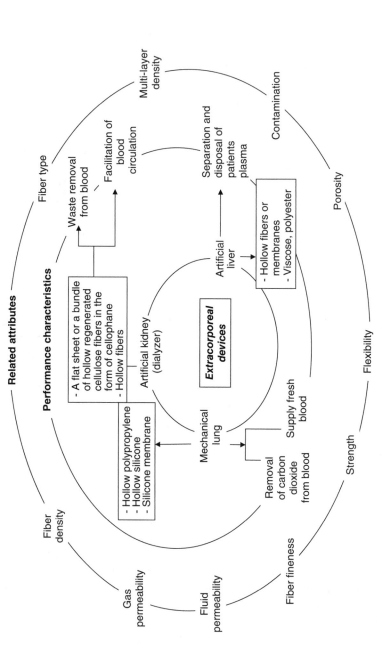

15.11 Performance characteristics and related attributes of extracorporeal devices.

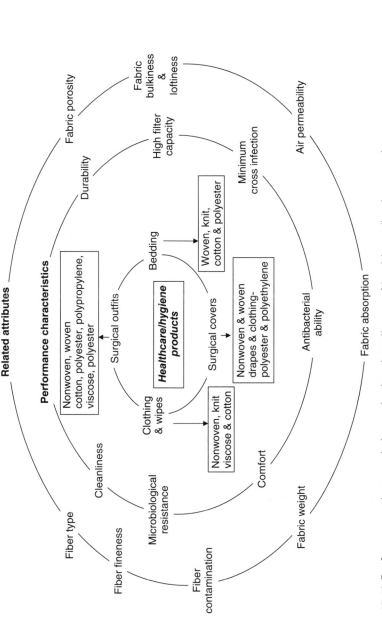

15.12 Performance characteristics and related attributes of healthcare/hygiene products.

caps are usually parallel-laid or spun-laid nonwoven materials based on cellulosic fibers.

Operating theater disposable products and clothing include[2] surgical drapes and cover cloths that are used in the operating theater either to cover the patient (drapes) or to cover working areas around the patient (cover cloths). Nonwoven structures are used extensively for drapes. Cover clothes are typically composed of films that are completely impermeable to bacteria, backed on either one or both sides with nonwoven fabrics that are highly absorbent to both body perspiration and secretions from the wound. Bacteria barrier characteristics are also achieved using hydrophobic finishes. Another development in surgical drapes is the use of loop-raised warp-knitted polyester fabrics that are laminated back to back and contain micro-porous PTFE films in the middle for permeability, comfort and resistance to microbiological contaminants. Bedding, clothing, mattress covers, incontinence products, cloths and wipes represent common products that are available in all hospitals and medical facilities. Most hospital blankets are made from cotton leno woven structures to reduce the risk of cross infection. The yarns used are typically soft-twisted, twofold yarns to provide desirable durability, good hand, appropriate thermal characteristics and which can easily be washed and sterilized.

Disposable diapers used in hospitals typically consist of a three-layer composite article:[7,8] (1) an inner covering layer, which is either a longitudinally orientated polyester web treated with a hydrophilic finish, or a spun-laid polypropylene nonwoven material, (2) an absorbent layer and (3) an outer impermeable layer.

15.5   Protective fibrous systems

The concept of protection of human being has evolved over the years from merely protection against the forces of nature and wild animals to a protection against numerous hazardous sources, most of which have been created by humans as a result of the many requirements of modern life. During of this evolution, fibrous products have contributed immensely to the protection of human beings. Today, numerous protection systems are available for use by humans both in regular living and in the workplace. These systems may be divided into two main categories:[3,20] protective clothing and protective gear. The term 'protective clothing', which refers to clothing items used specifically for protection, is often used interchangeably with the term 'protective gear', which refers to more general forms of protective systems including helmets, masks, guards and shields. Examples of protective clothing and protective gears are listed in Tables 15.1 through 15.5. These include protective garment systems, body armor, body protec-

Table 15.1 Examples of protective garment systems[21–24,30,31]

Protective system	Product example/purpose	Components/material
Protective garments Protects the wearer against various forms of hazardous substance including: biological, chemical, radiation, flash, bombs and other objects	**Hazmat suit** • A fully encapsulating garment worn as protection from hazardous materials or substances • Generally combined with breathing apparatus or protection and may be used by firefighters, emergency personnel responding to toxic spills, researchers, or specialists cleaning up contaminated facilities • Sometimes confused with or referred to as an NBC (nuclear, biological, chemical suit), which is a military version intended to be usable in combat	• Can be made from various materials and fibers that have inherent capabilities with respect to chemical, nuclear, or other hazardous substances • Fibers can also be treated to perform these protective functions • To protect against chemical agents, appropriate barrier materials like Teflon, heavy PVC, or rubber, are used
	Space suit • Complex system of garments, equipment and environmental systems designed to keep a person alive and comfortable in outer space • Basic criteria of a space suit are stable internal pressure, mobility, breathable oxygen, temperature regulation, shielding against ultraviolet radiation, protection against small micrometeoroids, a communication system, means to recharge and discharge gases and liquids, means to maneuver, dock, release, and/or tether onto spacecraft, and means of collecting and containing solid and liquid waste	The different materials which are used to make spacesuits are: • nylon tricot • spandex • urethane-coated nylon • Dacron • neoprene-coated nylon • Mylar • Gortex • Kevlar (material in bullet-proof vests) • Nomex

tion systems (arm/shoulder protection and gloves), protective masks and protective helmets.

The extent of protection provided by fibrous products will primarily depend on three main factors:[20] (a) the hazardous source, (b) the extent of danger (applications) and (c) the time of exposure. At the extreme levels of these factors, a single fibrous system may not be sufficient and a multiple-layer system or alternating specialized systems may be required. In addition,

Table 15.2 Examples of body armor

Protective system	Product example/purpose	Components/material
Body armor Also called a ballistic vest or bullet-proof vest. It is an item of armor that absorbs the impact from gun-fired projectiles and explosive fragments fired at the torso	**Soft vests** • Soft vests are commonly worn by police forces, private citizens and private security guards • Protects wearers from projectiles fired from handguns, shotguns and shrapnel from explosives such as hand grenades • When metal or ceramic plates are used with a soft vest, it can also protect wearers from shots fired from rifles **Hard body armor** • Hard-plate reinforced vests are mainly worn by combat soldiers as well as armed response police forces • Used to protect against many objects including stabs and punctures. Impacts from bats, clubs, and sticks are dispersed over a wide area, mitigating effects of blunt force trauma	• Soft vests made from layers of tightly-woven fibers • Kevlar armor • Use ceramic or metal plates inserts
	Interceptor body armor system Interceptor multi-threat body armor system is made up of two modular components: the outer tactical vest and small-arms protective inserts, or plates	• The outer tactical vest consists of a Kevlar weave aiming to stop 9 mm pistol rounds • Webbing on the front and back of the vest permits equipment like grenades, walkie-talkies and pistols to be attached • The small arms protective insert (SAPI) is made of a boron carbide ceramic with a spectra shield backing (extremely hard material) • It stops, shatters and catches any fragments up to a 7.62 mm (0.00762 m) round with a muzzle velocity of 2750 feet per second (838 m s^{-1}) • Also harder than Kevlar
	Dragon skin Dragon skin is distinguished by its silver dollar-sized circular disks that overlap like scale armor, creating a flexible vest that allows a greater range of motion and can absorb more hits than standard military body armor.	Composed of a silicon carbide ceramic matrices and laminates, much like the larger ceramic plates in other types of bulletproof vests

Table 15.3 Examples of body protection systems

Protective system	Product example/purpose	Components/material
Arm/shoulder protection Various components that aim to protect body parts particularly arms and shoulders	**Shoulder pad** Piece of protective equipment used in American football	• Most modern shoulder pads consist of a shock absorbing foam material with a hard plastic outer shell. The pieces are usually secured by rivets or strings that the user can tie to adjust the size • A related piece of protective equipment is the rib protector. It is attached to the shoulder pads and wrapped around the player's midsection. It is designed to protect the ribs, stomach and back areas
	Hand guards (grips) • Gymnastics equipment that aids both male and female athletes in artistic gymnastics (e.g. high bar, still rings and parallel bars) • Can also be used in sports exercises	• Consist of a strip of leather and a wrist strap of either Velcro or a buckle. The strip of leather is about 5 cm across and has finger holes at the top for the third and fourth fingers • Gymnasts normally choose to wear something soft under the wrist strap
	Gloves • A type of garment which covers the hand of a human • They can serve to protect and comfort the hands of the wearer against cold or heat, physical damage by friction, abrasion or chemicals, and disease; or in turn provide a guard for something a bare hand should not touch • Gloves have separate sheaths or openings for each finger and the thumb • If there is an opening but no covering sheath for each finger they are called 'fingerless gloves' • Fingerless gloves with one large opening rather than individual openings for each finger are sometimes called gauntlets • Gloves which cover the entire hand but do not have separate finger openings or sheaths are called mittens • Mittens are warmer than gloves made of the same material because of the extra air inside	Gloves may be made from: • latex- or natural rubber latex • vinyl- or polyvinyl chloride PVC nitrile- or acrylonitrile and butadiene • polyurethane Example: latex, nitrile rubber, or vinyl disposable gloves, often worn by health care professionals for hygiene and contamination protection

Table 15.4 Examples of protective headgear (masks)

Protective system	Product example/purpose	Components/material
Masks A form of protective gear used mainly to protect human face	**Filter mask** • Protection of the wearer from harmful airborne substances • Usually covers only the mouth and nose. It limits the course of air so that it must flow through a filter which removes harmful dusts or toxic gases	• A dense, fine natural or synthetic fiber mesh • To aid particulate filtration, the mesh is sometimes coated with substances that enhance the tendency of particulates to adhere to the fibers • For gas filtration, mask cartridges are filled with activated carbon or certain resins to absorb substances such as volatile organic compounds (VOCs), eliminating them from the air breathed
	Gas mask • Worn over the face to protect the wearer from inhaling airborne pollutants and toxic materials • It forms a sealed cover over the nose and mouth, but may also cover the eyes and other vulnerable soft tissues of the face	This type of mask performs three basic functions: filtration, absorption or adsorption, and reaction and exchange
	Surgical mask Worn by health professionals during medical operations to catch the bacteria shed in liquid droplets and aerosols from the wearer's mouth and nose	Modern surgical masks are made from paper or other non-woven material and are discarded after each use
	Diving mask Allows scuba divers, free-divers, snorkelers, to see clearly under water while protected from water penetration	Durable, tempered glass plate in front of the eyes and a 'skirt' of rubber or silicone to create a watertight seal with the diver's face. A strap is used to keep the mask in position

Table 15.4 Continued

Protective system	Product example/purpose	Components/material
	SCBA Self contained breathing apparatus. Sometimes referred to as a compressed air breaching apparatus (CABA) or simply breathing apparatus (BA). It is worn by rescue workers such as enforcement personnel and firefighters to provide breathable air in a hostile environment	• An SCBA typically has three main components: a high-pressure tank (e.g. 2200 psi to 4500 psi), a pressure regulator and an inhalation connection (mouthpiece, mouth mask or face mask), connected together and mounted on a carrying frame • Air cylinders used in these masks are made of aluminum, steel, or of a composite construction, usually carbon-fiber wrapped, which provides light weight
	Cold weather ski mask It has several features for comfort: • The mouth holes are designed to allow maximum airflow while still providing complete protection from the wind and cold. They allow water vapor to escape while warming the air you inhale • The eye openings are designed for unobstructed vision and to allow glasses or goggles to be worn • The ear holes allow clear hearing while still maintaining total protection from the wind and cold	It can be made from a single layer or multiple-layer fabric system, depending on the severity of application.

fibrous materials may require special treatments to modify their characteristics, or they may be combined with other materials (e.g. composites or multiple-layer systems) to serve particular applications for protection.

With regard to hazardous sources that require protection for humans, numerous sources can be listed including[20] fire, smoke and toxic fumes, weapons of various types (e.g. ballistic projectiles, nuclear, chemical,

Table 15.5 Examples of protective headgear (helmets)

Protective system	Product example/purpose	Components/material
Helmets A form of protective gear worn on the head, which uses a mechanical energy absorption principle for head protection There are two main types of helmet: hard shell and soft/micro shell In both types, impact energy is absorbed as a stiff foam liner is crushed, up to the point where the liner is crushed to its minimum thickness, or the helmet shatters, after which no further energy is absorbed. Collision energy varies with the square of impact speed	Head protection used in military, mining, industrial, construction and sports applications **Batting helmet** • Protective headgear worn by batters in games such as baseball or softball • It protects the batter from stray pitches thrown by the pitcher **Bicycle helmet** • Worn while riding a bicycle • They are designed to attenuate impact to the head of a cyclist in falls while minimizing side effects such as interference with peripheral vision **Football helmet** Protective device used primarily in American football	• Old helmets were primarily made from metals • Modern helmets can be made from a combination of materials including plastics, resins and fibers (e.g. Kevlar and Twaron) • The key component of most modern bicycle helmets is a layer of expanded polystyrene (EPS), essentially the plastic foam material used to make inexpensive picnic coolers • This material is sacrificed in an accident, being crushed as it absorbs a major impact • Bicycle helmets should always be discarded after any accident It consists of a hard plastic top with thick padding on the inside, a facemask made of one or more metal bars and a chinstrap used to secure the helmet.

biological), drowning, hypothermia, molten metal, chemical reagents, toxic vapors, foul weather, extreme cold, rain, wind, chemical reagents, nuclear reagents, high temperatures, molten metal splashes, microbes and dust. The extent of danger will obviously depend on the application in question. Applications or products in which fibrous structures are used for protection are numerous. These include military uniforms, mine-worker's clothing, firefighters' uniforms, police officers' uniforms, sportswear, helmets, tents, sleeping bags, survival bags and suits, heat-resistant garments, turnout coats, ballistic-resistant vests, biological and chemical protective clothing, blast-proof vests, antiflash hoods and gloves, molten metal protective clothing, flotation vests, submarine survival suits, immersion suits and dive skins, life rafts, diapers, anti-exposure overalls, arctic survival suits, ropes and harnesses.

15.5.1 Protection against extreme climate conditions

As discussed earlier, fibrous products can be designed in such a way as to provide protection against very hot or very cold weather. Figure 15.13 illustrates examples of performance characteristics and related attributes

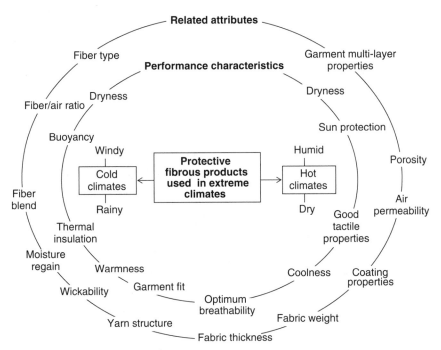

15.13 Performance characteristics and related attributes of protective fibrous products used for protection in extreme climates.

for fibrous products used for protection in extreme climates. These are only examples and they should be made more specific when a protective system is being designed for a specific application.

With regard to hot weather, people in many areas of the world have to endure many days at of 90°F (32°C) plus temperatures every year. When high temperatures are associated with high humidity and high physical activity levels (e.g. military operations, sports and rescue missions), discomfort becomes a serious issue and the need for protection becomes a necessity. Basic requirements in a clothing system in this type of environment are dryness (sweat absorption, wicking and evaporation), coolness (body insulation from surrounding temperature), sun protection and good tactile properties that allow humans to perform high-level physical activities with minimum resistance.

The other extreme of climate conditions is cold or very cold weather. This can be observed at low to extremely low temperatures and it may be multiplied by rainy or windy conditions. As indicated earlier, this situation may result in hypothermia, which is a condition occurring when the heat lost from the body exceeds that gained from food, exercise and external sources. Many approaches have been taken to minimize the effects of hypothermia. These include the use of appropriate insulative clothing and garment components, the use of flotation (in cold water), thermal protection devices and work wear coveralls which provide buoyancy and thermal insulation in case of accidental and emergency immersion in cold water.

Design for protection against extreme climates should be based on understanding the heat and vapor transfer mechanism of a clothing system. In these environments, clothing should act as a barrier to heat and as a means of transporting vapor between the skin and the environment. This barrier is formed both by the clothing materials themselves and by the air they enclose and the still air that is bound to its outer surfaces. In Chapter 12, basic concepts of heat transfer through clothing systems were introduced. The governing equations showing the effect of clothing on heat and vapor transfer are:[21]

$$\text{dry heat loss} = \frac{(t_{sk} - t_a)}{l_T} \tag{15.1}$$

where t_{sk} is the skin temperature, t_a is the air temperature and l_T is the clothing insulation, including air layers.

$$\text{evaporative heat loss} = \frac{(P_{sk} - P_a)}{R_T} \tag{15.2}$$

where P_{sk} is the skin vapor pressure, P_a is the air vapor pressure and R_T is the clothing vapor resistance, including air layers.

As discussed in Chapter 12, the primary modes of heat transfer through clothing systems are conduction and radiation. In the light of the above equations, since the volume of air enclosed is typically far greater than the volume of the fibers, factors such as fabric thickness and the amount of entrapped air will represent the most critical design factors associated with thermal insulation. Fiber type, on the other hand, can influence the amount of radiative heat transfer, as certain fibers can reflect, absorb or re-emit radiation.

When the hot environment is also associated with high levels of humidity and high levels of physical activity, the importance of fiber type becomes significant. For example, a clothing system made from 100% cotton is likely to absorb moisture (sweat) but it will also keep it, slowing the evaporation of sweat (a good cooling mechanism). However, wet clothing will also be inappropriate when the temperature goes down as it can result in undesirable cooling. Accordingly, a cotton/synthetic blend may provide a better alternative, with the natural cotton providing quick absorption of body sweat and the synthetic material providing quick release of moisture via a wicking effect so that quick drying and cooling evaporation can occur. However, the role of the fiber will largely depend on fabric thickness since thickness also determines the major part of clothing vapor resistance. Since the volume of fibers is usually low compared to the enclosed air volume, the resistance to the diffusion of water vapor through the garments is mainly determined by the thickness of the enclosed still air layer. Only with thin fabrics does the fiber type become important as it can affect the diffusion properties more than the thickness of fabric. When coatings, membranes or other treatments are added to the fabrics, this will have a major effect on vapor resistance since diffusion of vapor molecules becomes an important factor.

When fabrics are made into garments, fabric thickness and fiber type are manifested in a multiple-layer model of the skin–garment interface or fabric/fabric/skin interfaces[22] (in the case of multiple-layer garments). In this model, other aspects such as layer thickness, thickness of air attached to the fabric surface and degree of fit (loose or tight fit) become important. In a multiple-layer system, the air entrapping capacity between layers becomes a critical design aspect for thermal insulation.

In general, each fabric layer will have a still air layer attached to its outer surface. Outside this layer, the air is insufficiently bound and will move owing to temperature gradients. The nature and the thickness of this air layer are likely to be influenced by a number of factors including fabric surface construction (porosity, surface texture, fiber type and fabric finish), fabric thickness and the distance between fabric layers. Therefore, it is expected that for multi-layer garments or clothing ensembles, the total insulation will be much higher than could be expected from the insulation of a single fabric layer. In addition, clothing design, body shape and fit may alter the way fabric layers are separated and, consequently, the amount of still air entrapped in the system. At the shoulders, for example, the layers will be directly touching and thus the total insulation will only be the sum of the material layers plus one air layer on the outer surface. When the clothing fits tightly, the ability to entrap air will largely depend on garment design, but in most situations less air may be included than when it fits loosely. Also, in the presence of wind, smaller amount of still air will be expected as a result of air movement, garment movement or wearer body movement.

Protection against extreme climates is most important in military operations. This is particularly true in cold/wet regions which tend to cause the most severe problems. In these regions, it is critical to achieve and maintain dry thermal insulation. Products that can be affected by these climates include clothing, sleeping bags and other personal equipment. Again, the key design factors in these systems are the fiber-to-air ratio (or fiber fraction), the fiber arrangement in the overall fibrous structure and fiber fineness. These factors determine the efficiency of the fibrous insulator. A good insulative fibrous system will typically have a fiber to air ratio of 0.1 to 0.2 (10–20% of fiber and 80–90% of air). With regard to moisture, some analysis suggests that the presence of 10–20% by weight of moisture is sufficient to cause up to 50% loss in the dry insulation value.[23,24] Fiber arrangement should largely assist in entrapping air in the fibrous structure. Fine fibers can result in accommodation of a larger number of fibers per unit weight of fabric. However, this number should be carefully selected to allow still air to be trapped in the pores of the fabric structure, leading to a high specific surface area and not to a high level of compactness, which can hinder flexibility.

In recent years, numerous fibrous products have been developed for protection against extreme climates. For example, DuPont developed the spun bonded polyolefin-fiber fabric, Tyvek®, aluminized and made into survival suits, survival bags and many other weather-resistant products. Overalls made from Tyvek® primarily aim to provide thermal insulation. These are useful in cold water situations and can be used by Navy ships and by airlines flying the polar route in case the aircraft is forced down onto the Arctic ice.

Another example of weather-resistant products is Gentex's Dual Mirror® aluminized fabric, which offers multiple industrial, military and commercial applications. Specific applications of this fibrous product include industrial heat shielding, molten metal splash protective clothing, radiant heat protective clothing, proximity fire fighting and non-fire resistant (FR) infrared heat and sunlight shielding. These applications resulted in the development of a wide range of styles of Gentex's Dual Mirror® aluminized

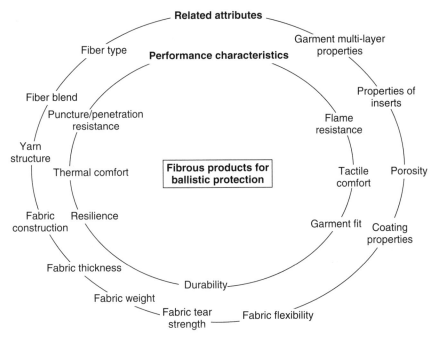

15.14 Performance characteristics and related attributes of fibrous products used for ballistic protection.

fabrics. This protective fabric consists of five layers: (a) an outer skin of aluminum, (b) a protective film, (c) a second layer of aluminum, (d) heat-stable adhesive and (e) a base fabric. These individual layers are then combined to form a single fabric.

15.5.2 Ballistic protection

Fibrous materials have been widely used for developing products that primarily aim to provide ballistic protection. Figure 15.14 illustrates examples of performance characteristics and related attributes for fibrous products used for ballistic protection. The idea is that these products should be able to absorb large amounts of energy owing to their high tenacity, high modulus of elasticity and low density.[23-26] The most common fibrous product used for ballistic protection is bullet-proof vests. This product is also one of the oldest components of protection used by humans over the years. Indeed, throughout recorded history, humans have used various types of material as body armor to protect themselves from external objects and injuries in combat situations. The first protective clothing and shields were made from animal skins. With the invention of firearms around 1500, other materials including wood and metal shields were also used for protection.

Although, these materials were effective in their protective aspects, they were too heavy and impractical for great physical actions, fast movement and battle maneuvering. This has resulted in the development of softer body armors.

One of the first recorded instances of the use of soft body armor was by the medieval Japanese, who used armor manufactured from silk. It was not until the late 19th century that the first use of soft body armor in the United States was recorded. At that time, the military explored the possibility of using soft body armor manufactured from silk. The project even attracted congressional attention after the assassination of President William McKinley in 1901. While the garments were shown to be effective against low-velocity bullets (i.e. traveling at 400 feet per second or 122 m s^{-1} or less), they did not offer protection against the new generation of handgun ammunition being introduced at that time (ammunition that traveled at velocities of more than 600 feet per second or 183 m s^{-1}). This deficiency associated with the prohibitive cost of silk made the concept unacceptable.

World War II was a turning point in the development of body armor with the introduction of the 'flak jacket' made from ballistic nylon. The flak jacket was very cumbersome and bulky. It provided protection primarily from ammunitions fragments, but was ineffective against most pistol and rifle threats. By the late 1960s, new fibers were discovered that have made their ways into today's modern generation of body armors. The invention of Kevlar by DuPont in the 1970s was another significant turning point in the development of body armor (e.g. Kevlar 29). Ironically, the fabric was originally intended to replace steel belting in vehicle tires. In 1988, DuPont introduced the second generation of Kevlar fiber, known as Kevlar 129, which offered increased ballistic protection capabilities against high energy rounds such as the 9 mm FMJ. In 1995, Kevlar Correctional was introduced, which provided puncture resistant technology for both law enforcement and correctional officers against puncture type threats.

The basic idea of body armor is a simple one. It is based on catching a bullet that strikes the body armor in a 'web' with a very strong fibrous assembly.[25] This assembly should absorb and disperse the impact energy that is transmitted to the vest from the bullet, causing the bullet to deform or 'mushroom'. Additional energy is absorbed by each successive layer of material in the vest, until such time as the bullet has been stopped. This principle requires a large area of the garment to be involved in preventing the bullet from penetrating to the body. Unfortunately, and despite the great progress in development, no structure exists which will prevent penetration of all ballistic objects and at the same time be wearable under all situations.

Typical bullet-proof vests are made from multiple layers of woven fabric, with the degree of protection being increased as the number of fabric layers increase. These layers are assembled into a 'ballistic panel', which is then inserted into the 'carrier', which is constructed from conventional garment fabrics such as nylon or cotton. The ballistic panel may be permanently sewn into the carrier or may be removable.[26] Although the overall finished product looks relatively simple in construction, the ballistic panel can be very complex. Even the manner in which the ballistic panels are assembled into a single unit can differ from one product to another. In some cases, the multiple layers are bias stitched around the entire edge of the panel; in others, the layers are tack stitched together at several locations. Some manufacturers assemble the fabrics with a number of rows of vertical or horizontal stitching; some may even quilt the entire ballistic panel. No evidence exists that stitching impairs the ballistic resistant properties of a panel. Instead, stitching tends to improve the overall performance, especially in cases of blunt trauma, depending upon the type of fabric used.

Plain woven fabric is more suitable for body armor. Neoprene coating or resination is also commonly used.[23] Needle punched nonwoven fabrics are also used for ballistic protection. These are typically made from high performance polyolefin fibers such as Dyneema polyethylene. The benefits of using nonwoven structures for these applications stems from their ability to provide protection against sharp fragments by absorbing projectile energy by deformation, rather than fiber breakage as is the case with woven fabrics. When needle punched nonwovens are used for ballistic protection, the felt structure should have a very low mass per unit area. However, as the mass increases, woven structures become more superior to nonwoven felts.[23] Nonwoven felts should also be designed in such a way that a high degree of entanglement of long staple fibers is achieved at a minimum degree of needling, since excessive needling can produce too much fiber alignment through the structure, which aids the projectile penetration.

In situations where high levels of protection (e.g. rifle fire) are required, body armor with either a semi-rigid or rigid construction should be used. These are typically multi-layer fibrous systems incorporating hard materials such as ceramics and metals. The heavy weight and high bulk of these body armors prevent their use in routine applications (e.g. by uniformed patrol officers or normal military operations), and restrict their use to tactical situations where it is worn externally for short periods of time when confronted with higher levels of threat.

The development of more effective body armor is unlikely to cease as a result of the continuing development of weapons with increasing power. As indicated earlier, the key aspect of development is the fibrous component from which body armor is made. The newest addition to the Kevlar line is Kevlar Protera, which DuPont made available in 1996. This is believed to be a high-performance fabric that allows lighter weight, more flexibility and greater ballistic protection in a vest design owing to the molecular structure of the fiber. Another development is the spectra fiber, manufactured by the former AlliedSignal (now Honeywell International), which is an ultra-high-strength polyethylene fiber used to make Spectra Shield composite. This basically consists of two unidirectional layers of spectra fiber, arranged to cross each other at 0° and 90° angles held in place by a flexible resin. Both the fiber and resin layers are sealed between two thin sheets of polyethylene film, which is similar in appearance to plastic food wrap. According to AlliedSignal, the resulting nonwoven fabric is incredibly strong, lightweight and has excellent ballistic protection capabilities. Spectra Shield is made in a variety of styles for use in both concealable and hard armor applications. Another product, also developed by the former AlliedSignal, uses the Shield Technology process to manufacture a shield composite called Gold Shield. This is made from aramid fibers instead of the Spectra fiber. Gold Shield is typically made in three types: Gold Shield LCR and GoldFlex, which are used in concealable body armor, and Gold Shield PCR, which is used in the manufacture of hard armor, such as plates and helmets.

Akzo Nobel has also developed various forms of its aramid fiber TWARON for body armor. This fiber uses more than 1000 fine spun single filaments that act as an energy sponge, absorbing a bullet's impact and quickly dissipating its energy through engaged and adjacent fibers. The use of many filaments is believed to disperse an impact more quickly and allow maximum energy absorption at minimum weights while enhancing comfort and flexibility.

15.5.3 Protection against flame

Fire-related hazards include[27–29] flames (convective heat), contact heat, radiant heat, sparks and drops of molten metal, and hot gases and vapors. A self-sustaining flame requires a fuel source and a means of gasifying the fuel, after which it must be mixed with oxygen and heat.[23] Flame-resistant fibrous products should be developed to minimize the effects of these hazards in a very wide range of applications including firefighting, military operations, offshore oil and gas rig operations, law enforcement rescue operations, aircraft and car crashes, and in many other situations where there are potential fuel spills and fire generation.

In general, flame resistance is the characteristic of a fabric that causes it not to burn in air. This can be achieved using fibers that are inherently or made to be flame resistant, or using fibers treated with flame-retardant chemical substances. Typical flame-resistant attributes measured on fabrics (ASTM D6413) are char length (or the length of fabric destroyed by the flame so that it will readily tear by application of a standard weight), duration (seconds) of visible flame remaining on the fabric after the ignition source has been removed and duration (seconds) of visible glow remaining on the fabric after all flaming has ceased. Passing the vertical flammability requirements is an essential criterion for protective clothing fabrics.

The key to developing efficient and effective flame-resistant fibrous products is to understand the elements constituting flammability. These include:[25,26]

- ease of ignition
- rate of burning and heat release rate
- explosion contents (e.g. coal, dust and methane)
- heat flux intensity levels ($kW\,m^{-2}$) and the way they vary during exposure
- duration of exposure, including the time it takes for the temperature of the garment to fall below that which causes injury after the source is removed.

It is also important to understand the factors associated with fibrous products that can assist in minimizing the effects of the above elements. These include:[23,25–27]

- the thermal and burning behavior of fibers (e.g. melting and shrinkage characteristics)
- the influence of fabric structure and garment shape on the burning behavior
- properties of flame-retardant treatments (i.e. non-toxicity, the extent of smoke associated with flame-retardant additives or finishes)
- properties of the protective garment (i.e. nature of usage and comfort aspects)
- extent of insulation between source and skin, including outerwear, underwear and the air gaps between them and the skin
- extent of degradation of the garment materials during exposure and the subsequent rearrangement of the clothing/air insulation
- condensation on the skin of any vapor or pyrolysis products released as the temperature of the fabric rises.

As the reader can imagine, the subject of flame resistance has been covered in numerous literature sources. In this section, we can only touch upon some of the main design-related aspects of fibrous products that can be useful for engineers involved in developing flame-resistant fibrous products. These are summarized below.

Basic performance characteristics of flame-resistant fibrous products

The process of fiber combustion is a complex one and is the result of the multiplicity of factors involved and their superimposed effects. This process

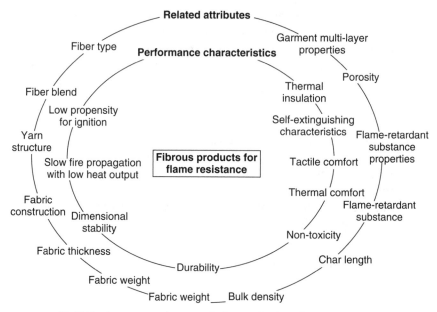

15.15 Performance characteristics and related attributes of fibrous products used for flame resistance.

typically involves heating, decomposition leading to gasification (fuel generation), ignition and flame propagation.[23] The key performance characteristics expected from flame-resistant fibrous products are low propensity for ignition from a flaming source, slowing fire propagation and generating low heat output (see Fig. 15.15). These criteria stem primarily from the fiber type used or the flame-retardant treatment applied. The role of the fabric or the garment assembly is also critical as they should act as a barrier against body skin burning by providing additional features including high thermal insulation and high dimensional stability (i.e. they should neither shrink nor melt). However, without an appropriate fiber type or a good flame-retardant treatment, these features will be impossible to meet. Accordingly, any development of a flame-resistant fibrous product should primarily focus on fiber type, fiber characteristics and flame-retardant treatments.

Unlike natural fiber, which tends to burn upon exposure to flame, synthetic fibers undergo noticeable physical and chemical changes under heat. In principle, thermoplastic fibers subjected to heat undergo two phases of changes: physical changes at their glass-transition temperature and melting temperature, and chemical changes at their pyrolysis temperature, where thermal degradation occurs. Upon pyrolysis, volatile liquids and gases, which are combustible, act as the fuels for further combustion. After pyrol-

Table 15.6 Typical values of melting, pyrolysis and combustion temperatures of some fibers[23,29]

Fiber	Melting temperature (T_m, °C)	Pyrolysis temperature (T_p, °C)	Combustion temperature (T_c, °C)
Nylon 6,6	265	420–477	530
Polyester	255		480
Polypropylene	165	469	550
Kevlar	560	590	>550
Nomex	375	310	500
Oxidized acrylic		>640	
PBI		>500	>500
Cotton		350	350
Wool		245	600
Viscose		350	420

ysis, if the temperature is equal to or greater than combustion temperature, flammable volatile liquids burn in the presence of oxygen to give products such as carbon dioxide and water.[23,29] Table 15.6 lists some typical values of melting, pyrolysis and combustion temperatures of some fibers.

Most thermoplastic fibers (e.g. nylon, polyester and polypropylene) tend to shrink away from flame, creating a self-contained spot. This feature is critical in minimizing the propensity for ignition and in slowing flame propagation. However, when these fibers are worn in the form of garments, concerns should be focused on skin burning when these fibers melt. Accordingly, conventional thermoplastic fibers fail to meet flame-resistance criteria. Instead, high-performance fibers such as aramid fibre (e.g. Nomex, DuPont), flame-retardant treated fibers (e.g. cotton or wool) and partially oxidized acrylic (Panox) fibers, and polybenzimidazole (PBI) fibers are used for flame-resistance applications. Despite their high performance, these fibers should be tested in the garment form in view of the flame-related factors listed earlier and the particular application in hand. For example, it was found that the aramid fibers, in spite of their high oxygen index and high thermal stability, are not suitable for preventing skin burns in molten metal splashes because of their high thermal conductivity.

In light of the above discussion, fiber type, fiber characteristics and flame-retardant finish are the most critical factors in determining the performance of flame-resistant fabric. In addition, some fabric-related attributes can also play significant roles. These include fabric thickness, bulk density and fabric integrity (construction). As indicated earlier, fabric thickness has a direct impact on thermal insulation. For flame-resistant applications, the greater the fabric thickness, the better the thermal insulation of fabrics made from flame-retardant treated fibers (e.g. cotton and wool). This is not the case with thermoplastic fibers whereas the thicker thermoplastic fiber fabrics produce more severe burns.[23,29] At a constant fabric thickness, low fabric density (lower fiber/air ratio) will result in higher thermal resistance. These effects are important, particularly when flame-retardant treated fibers such as cotton and wool are used.

With regard to fabric integrity, shrinkage or expansion in the plane of the fabric does not substantially change the thermal insulation of the fabric itself. However, the spacing between the fabric and the skin or between garment layers may alter, with a consequent change in overall insulation. For example, if the outer layer shrinks and pulls the garment on to the body, the total insulation is reduced and the heat flow increases. In addition, the most serious failure of flame-resistant garment is hole formation. When the fabric remains intact, its heat flow properties do not change greatly even when the component fibers are degraded, because heat transfer is by conduction and radiation through air in the structure and by conduction through the fibers (which is relatively small). Only when fibers melt or coalesce as a result of displacing the air, or when they bubble and form an insulating char, are heat flow properties substantially altered.[27]

15.5.4 Design aspects to meet an optimum trade-off between protection and comfort

Most protective fibrous systems distributed in the market are associated with some 'comfort' claim. This may seem to be a liability-free claim since protection in its totality often implies some form of relative comfort and comfort in its totality also implies some form of relative protection. The aspect of comfort has also been an integral part of most protective clothing analyses. Indeed, the literature is immensely rich with experimental evaluation of comfort parameters of protective clothing.[30–41] The challenge, however, is that when protective clothing is situated to use, which is typically under strenuous conditions, the perception of comfort is often confused and intermingled with the task of protection. This confusion raises the question of whether some level of comfort was accomplished during the protection process, or more precisely, to what extent of discomfort, was protection achieved?

The many years of research and development in protective clothing were put to the ultimate test during the September 11, 2001 terrorist attack on America and the results were significantly disappointing. Table 15.7 gives few examples of responses from people that had worked on the scene.[30] Typically, the risk involved and the strenuous situations often reflects nega-

Table 15.7 Some views on protective clothing expressed to the media after September 11, 2001[30]

Response comments	Source
Everything worked well when it was used for what it was intended	Most responses
My fire-resistant uniform suffered many cuts after only 10 minutes on the scene	Firefighter
My suit and my boots were too heavy, and too sticky, I could hardly move	Special-operations personnel
I was more comfortable inhaling dust than having to put on suffocating heavy respiratory equipment . . . fogging was too much and breathing was harder with the equipment . . . other people couldn't hear me	Person attempting to escape the scene
My protective clothing was OK for heat protection, but it often prevented me from moving or running freely or removing heavy objects or gravels	Firefighter
My boots became soaked with hot water leaking through, the seams failed	Firefighter
Clothing was extremely heavy and did not allow sufficient freedom of movement	FBI agent working at an anthrax site
Work shoes with steel reinforcements in the soles and toes protected feet against punctures by sharp objects but often could not be worn because they conducted and retained the heat, causing blistered or scorched feet	Law enforcement officer

tive views of almost all the surroundings including protective clothing and equipment; if you are facing the unpredictable, an unpleasant feeling will prevail. But the overall experience clearly reinforced the need for reliable, comfortable, protective clothing.

From a design conceptualization viewpoint, the protection requirement is typically straightforward as each form of protection is associated with design criteria that are largely deterministic or vary within pre-specified constraints or design boundaries. This point was clearly demonstrated in the examples of protective systems discussed above. Design for comfort represents a true challenge particularly when protection can only be achieved at the expense of comfort or at some level of inevitable discomfort. For example, to achieve high levels of protection, most protective clothing should be strong, stiff, heavy, thick, multiple-layered, largely pore-closed, often laminated, often coated, and often supported by inserts and additional components. These attributes promote the feeling of discomfort or an undesirable awareness of the protective system by the wearer.

Table 15.8 Examples of optimum comfort status[30]

Clothing type	Application	Anticipated optimum comfort criteria
Sportswear	Physical action: running walking jumping boxing wrestling swimming	• Minimum to moderate awareness • Minimum resistant intimacy of fabric to body movement • Reasonable to maximum physical protection • Reasonable to maximum environmental protection • Great breathability • Coolness/lightness at high level of physical activity • Warmth/heaviness (ice skating) • Minimum friction with outside media (swimming) • Optimum traction (ice skating, not too slippery and not too rough)
Fire-resistant	Specialty product	• Minimum awareness at low physical activity level • Moderate awareness at moderate to high physical activity level • Moderate breathability • Durability
Military clothing	Transitional physical activity and environmental conditions	• Minimum to moderate awareness at both resting mode and at high physical mode • Maximum physical protection • Maximum environmental protection • Great breathability • Coolness/lightness at high level of physical activity and in warm/hot environment • Warmth/heaviness in cold environment • Dust-free

One of the key psychological aspects of comfort is awareness. The research of the present author[30,41] revealed that awareness is a key reflection of discomfort by most wearers of both traditional and function-focus fibrous products. Table 15.8 gives examples of products with associated general applications and general anticipated comfort criteria derived from the author's research. As demonstrated in this table, the anticipated comfort criteria may vary substantially with different product types and applications. The key points revealed in this table are awareness and protection. When the wearer of protective clothing becomes fully aware of its burden, his/her feel of discomfort increases. As the level of awareness decreases, the wearer begins to act normally reaching a point where he/she feels relatively comfortable.

In dealing exclusively with the comfort phenomenon, it is well known that humans can easily recognize discomfort status. Indeed, a high percentage of people can easily realize discomfort when wearing very heavy, closed-structure, abrasive and stiff clothing. The problem lies in recognizing the comfort status. As indicated above, the low awareness of the clothing system is a good indication of neutral comfort. However, a maximum status of comfort is difficult to recognize because of the high variability in expressing this status from one person to another, and the complex factors determining maximum comfort, particularly the psychological factors. As a result, beyond a neutral stage where awareness of clothing is minimal, fuzziness in describing comfort increases. It is important therefore to consider two key aspects in determining the optimum comfort status: the degree of fuzziness and the level of awareness. These aspects are discussed below.

Awareness–fuzziness diagram of comfort

The meaning of optimum comfort may be partially described through examining possible relationships between the level of human awareness of objects that come into contact with human body (in this case clothing) and human realization of the comfort/discomfort status. These relationships are conceptually illustrated in Fig. 15.16 in which both the awareness level and the comfort status are characterized by a subjective rate-scale from 0 to 100%, with the comfort status represented on the horizontal

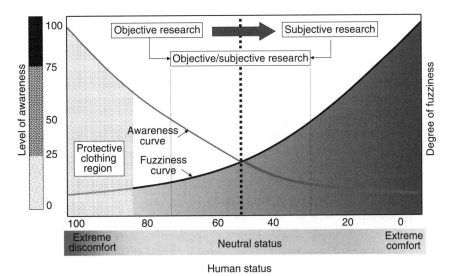

15.16 The awareness–fuzziness diagram of comfort.[20]

axis (0% being extremely comfortable and 100% being totally unbearable or extremely uncomfortable) and the extent of awareness represented on the vertical axis (0% being no awareness and 100% being full awareness).

The relationship between comfort and awareness is described by two curves: the awareness curve, which follows closely the awareness scale, and the fuzziness curve, which follows closely the comfort scale. At values of awareness level approaching 100%, discomfort is completely realized (close to 100%). At this point, the awareness curve is at its maximum point (full awareness) and the fuzziness curve is at minimum point (no fuzziness, or high clarity). As the awareness level decreases, fuzziness initially increases at a very low rate. Both the awareness curve and the fuzziness curve then meet at a point that we call 'neutral comfort status'. It is at this point that we feel that good protective clothing reaches optimum levels of both comfort and protection. Now, the fuzziness–awareness diagram is divided into two parts around the neutral comfort status. The left-hand part represents progressive increase in awareness associated with a progressive decrease in fuzziness. The right-hand part represents progressive decrease in awareness associated with a progressive increase in fuzziness as a result of the complexity associated with absolute comfort.

In connection with the awareness–fuzziness diagram, a number of key points should be made:

1. The awareness–fuzziness diagram focuses on the tactile aspect of comfort; however it may hold for the thermo-physiological aspect particularly in the left-hand part of the diagram.
2. Humans are different in their perceptions of comfort and comfort-related factors. Accordingly, comfort surveys should not treat this complex phenomenon solely on the basis of a discrete or a bi-polar physiological scaling.
3. Some suggest that maximum comfort is virtually represented by the nude status of the human body, provided that the surrounding environment is supportive. This makes minimal awareness of fabric touching human body an automatic cause of comfort. In this regard, it should be realized that the early human was indeed naked and the reason for covering body with clothing was primarily for protection and not to add comfort.
4. At extreme discomfort (e.g. an under shirt made from very stiff fibers, or a totally closed hydrophobic fabric structure against the skin), the level of awareness is very high and the level of fuzziness is very low. On the other hand, an extremely comfortable status cannot be defined or identified with a high degree of objectivity; it is highly variable and immensely relative. Thus, in this state, which is almost imaginary, the

degree of fuzziness is at its highest level by virtue of subjectivity, variability and relativity.

5. The fuzziness scale should be based on human judgment of the comfort feel and not expert judgment.

6. The proposed neutral comfort status should be defined with respect to predetermined boundaries for physical activities and environmental conditions. This means that the awareness–fuzziness diagram should be developed with respect to the application in hand and the type of intended fibrous product.

Historically comfort research has been divided into two major approaches: (a) subjective research relying on psychological scaling with many descriptors determined by scientists and (b) objective research relying on measuring various parameters that are directly or indirectly related to comfort. The awareness–fuzziness diagram corresponds well to these two categories of research. When the focus is on minimizing discomfort by reducing awareness of objects contacting the body and providing a more comfortable environment (left-hand part of the diagram), objective research prevails and meaningful parameters can be derived, as discussed in Chapter 12. On the other hand, when the focus is on maximizing comfort by further reducing awareness of objects contacting the body and further providing a more comfortable environment (right-hand part of the diagram), subjective research prevails as human judgment is often associated with a high degree of fuzziness that requires extensive subject analysis.

Developing a design-oriented fabric comfort model

The ultimate complexity of the comfort phenomenon is experienced when the goal is to design a product for comfort. In Chapter 12, this phenomenon was discussed from a thermo-physiological viewpoint and a number of deterministic factors were discussed that can be useful in the design of fibrous products that meet the thermal criteria of comfort. These factors include: (1) the selection of an appropriate fiber type depending on whether the objective of the product is moisture absorption or moisture transfer, (2) the design of appropriate yarn and fabric structures that exhibit optimum pore size to allow the presence of static air for thermal insulation, (3) the use of special finishes that can result in hydrophobic or hydrophilic fluid reaction and (4) the use of multiple fabric layers.

The other key aspect of the comfort phenomenon is the tactile comfort reflected in the interaction between the fabric and the human body. This aspect is critical for the design of appropriate protective systems as failure to meet minimum acceptable levels of tactile comfort would mean a reluctance to use protective clothing in both mild and risky situations. In a previous study on fabric comfort,[41] the tactile aspect of comfort was extensively analyzed and a design-oriented tactile comfort model was developed. The general form of this model and the key design factors explored are discussed below.

The underlying basis for the design-oriented tactile comfort model was the conceptual awareness–fuzziness relationship discussed earlier. Based on this relationship, tactile comfort can be defined as 'the state at which the fabric/garment has a minimum mechanical interaction with the skin, and an optimum positive interaction with the environment; the environment here being the surrounding media and the localized against-skin media'.

In light of the above definition, the most critical factor determining tactile comfort is fabric/skin interaction. In the context of pure comfort, this interaction should be minimized to provide a great deal of unawareness of clothing contact with the skin. In the context of protection, this interaction should be optimized so that it is not too high to avoid irritation and loss of mobility and it is not too low to insure acceptable levels of protection. The question now is how to translate this factor into a characterization index of fabric comfort and how to establish design parameters for textile fabrics and garments that reflect this interaction. This question was addressed in the analysis using the so-called 'area ratio' defined by the ratio between the true area of fabric/skin contact and the corresponding apparent area:

$$\text{Area ratio } (AR) = \frac{\text{true area of contact}}{\text{apparent area of contact}} = \frac{A_t}{A_a} \tag{15.3}$$

One of the fundamental structural differences between fabrics and non-fibrous structures stems from the fact that what is perceived to be a flexible flat sheet (the fabric) actually never exhibits complete flatness when it comes into contact with other solid surfaces. In other words, the apparent area of contact between a fabric and another surface is typically much greater than the actual area of contact. This is a direct result of structural and deformation effects inherent in fibrous structures. As a result, garment conformity to body contours is achieved at the expense of a substantial gap between the skin and the fabric in contact with it. In the context of comfort, this can be considered to be a unique added-value phenomenon by virtue of the advantage that it provides with respect to the level of wearer awareness of fabric/skin contact. In the context of protection, the level of awareness is not a unidirectional effect; instead an optimum level of awareness (not too high and not too low) is required to achieve an optimum comfort/protection trade-off.

Figure 15.17 shows the different modes of contact between the human body and the worn fabric. On a macroscopic level, fabric natural irregulari-

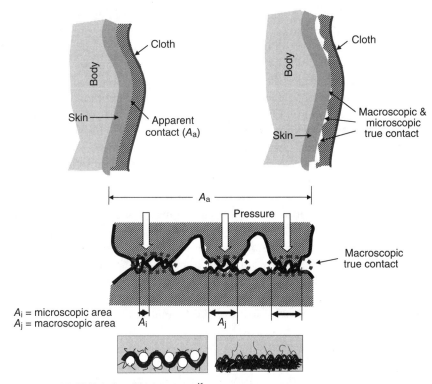

15.17 Fabric–skin contact.[41]

ties created by the fabric pattern (valleys and troughs) prevent the fabric from pure contoured contact. On a microscopic scale, surface disturbances such as projecting hairs or pills further decrease the true contact. When one considers lateral effects such as applied pressure, multiple fabric layers (weight and thickness) and human dynamics, one will see that the area ratio represents a key variable that can determine fabric performance in different applications.

The fact that the true area of fabric/skin contact is much smaller than the apparent area of contact represents one of the unique structural features of textile fabrics that have been largely overlooked in previous studies of fabric structure. In extreme cases, such as fabric puckering and dimensional instability, this feature is undesirable, as it results in great difficulty in handling fabrics. Under normal wearing conditions, the large difference between the apparent contact area and the true contact area can indeed provide many merits in several applications. For example, in relation to fabric comfort one can think of two extreme situations of fabric/skin contact: (a) an area ratio of approximately one and (b) an area ratio of zero. At an area ratio of one, the fabric is in total (complete) contact with the body skin. This virtual situation can only occur if the contact area is under extremely high pressure or if the fabric is virtually cemented to the body skin. This situation is totally undesirable as it completely hinders body mobility and closes the critical fabric/skin gaps required to demonstrate the critical effects of clothing such as thermal insulation, air permeability and breathability. On the other hand, if the area ratio is close to zero, a complete unawareness of the fabric will be felt. This situation can be positive if the body does not require any protection because the surrounding environment is 'cooperative'. In case of protective clothing, where the human body requires protection from the surrounding environment, an optimum fabric/skin contact will be required.

Detailed analysis leading to the design-oriented comfort model has been published[20,30,41] and it is outside the scope of this book. In principle, the analysis was initiated by considering the relationship between the contact area and the external pressure applied to the fabric at the microscopic level (or the asperities of contact). The shear behavior of the asperities of contact that are viscoelastic in nature was evaluated using Gupta and Elmogahzy friction model.[42] The general form of the true area of contact model was as follows:

$$A_t = C_M K^{-\gamma} m^{1-\gamma} N^{\gamma} \tag{15.4}$$

where $\gamma = 1/(1 + \alpha)$ is a deformation factor determining the pressure–area relationship, C_M is a constant dependent on the load distribution on the area of contact (approaching unity for most distributions), K is a surface rigidity index, m is the number of asperities in the area of contact and N is the normal load on the apparent area of contact.

The area ratio expression was expressed in a number of forms depending on the fabric factors considered in the design analysis of fabric for tactile comfort. The basic expression is as follows:

$$\frac{A_{true}}{A_{apparent}} = C_M K^{-\gamma} M_a^{1-\gamma} P^{\gamma} \tag{15.5}$$

where M_a is the number of asperities per unit area, expressing surface roughness and P is the applied pressure on the region of contact.

The above expression indicates that the main factors influencing the area ratio are:

1. the pressure (P) applied to the region of contact imposed by the garment weight, by some parts of the body contour or by multiple-layer garments
2. load distribution (C_M), as different parts of the body may exhibit different levels of pressure

3. fabric lateral rigidity (K)
4. fabric surface roughness (M_a)

From a design viewpoint, the above factors can be optimized through appropriate selection of fiber type, yarn type, yarn structure, fabric type, fabric structure and fabric finish.

15.6 References

1. BYRNE C, 'Technical textiles market–an overview', *Handbook of Technical Textiles*, Horrocks A R and Anand S C (editors), Woodhead Publishing Limited, Cambridge, UK, 2000, 1–23.
2. RIGBY A J and ANAND S C, 'Medical textiles', *Handbook of Technical Textiles*, Horrocks A R and Anand S C (editors), Woodhead Publishing Limited, Cambridge, UK, 2000, 407–23.
3. SHISHOO R, 'Safety and protective textiles: the opportunities and challenges', *Conference Proceedings, 4th International Conference on Safety and Protective Fabrics*, Industrial Fabric Association, IFI, Pittsburgh, USA, 2004, 3–17.
4. EL MOGAHZY Y, *Statistics and Quality Control for Engineers and Manufacturers: from Basic to Advanced Topics*, 2nd edition, Quality Press, Atlanta, GA, 2002.
5. ANANDJIWALA R D, 'Role of advanced textile materials in healthcare', *Medical Textiles and Biomaterials for Healthcare*, Anand S C, Kennedy J F, Miraftab M and Rajendran S (editors), Woodhead Publishing Limited, Cambridge, UK, 2000, 90–8.
6. KENNEDY J F and KNILL C J, 'Biomaterials utilized in medical textiles: an overview', *Medical Textiles and Biomaterials for Healthcare*, Anand S C, Kennedy J F, Miraftab M and Rajendran S (editors), Woodhead Publishing Limited, Cambridge, UK, 2000, 3–22.
7. AJMERI C J and AJMERI J R, 'Application of nonwovens in healthcare and hygiene sector', *Medical Textiles and Biomaterials for Healthcare*, Anand S C, Kennedy J F, Miraftab M and Rajendran S (editors), Woodhead Publishing Limited, Cambridge, UK, 2000, 80–9.
8. WALKER V, 'Nonwovens – The choice for the medical industry into the next millennium', *Medical Textiles*, Woodhead Publishing Limited, Cambridge, UK, 2001, 12–20.
9. LIN P H, HIRKO M K, VON FRAUNDHOFER J A and GREISLER H P, 'Wound healing and inflammatory response to biomaterials', *Wound Closure Biomaterials and Devices*, Chu C C, von Fraundhofer J A and Greisler H P (editors), CRC Press, NY, 1996, Chapter 2, 7–24.
10. COTRAN R S, KUMAR V and ROBBIN S L, 'Inflammation and repair', in *Robbin's Pathologic Basis of Disease*, W.B. Saunders, Philadelphia, 1989, 39.
11. CAHIN I I, GOLDSTEIN L M and SYDERMAN R, *Inflammation: Basic Principles and Clinical Correlates*, 2nd edition, Raven Press, New York, 1992.
12. COHEN I K, DIEGELMANN R E and LINDBLAD W J (editors), *Wound Healing: Biochemical and Clinical Aspects*, W.B. Saunders, Philadelphia, 1992.
13. AJMERI J R and AJMERI C J, 'Surgical sutures: The largest textile implant material', *Medical Textiles and Biomaterials for Healthcare*, Anand S G,

Kennedy J F, Miraftab M and Rajendran S (editors), Woodhead Publishing Limited, Cambridge, UK, 2000, 432–40.

14. KARACA E and HOCKENBERGER A S, 'Knot Performance of monofilament and braided polyamide sutures under different test conditions', *Medical Textiles and Biomaterials for Healthcare*, Anand S C, Kennedy J F, Miraftab M and Rajendran S (editors), Woodhead Publishing Limited, Cambridge, UK, 2000, 378–85.

15. THACKER J G, RODEHEAVER G, MOORE J W, KAUZLARICH J J, KURTZ L, EDGERTON M T and EDLICH R F, 'Mechanical performance of surgical sutures', *The American Journal of Surgery*, 1975, **130**(9), 374–80.

16. TOMITA N, TARNAI S, MORIHARA T, IKEUCHI K and IKADA Y, 'Handling characteristics of braided suture materials for tight tying', *Journal of Applied Biomaterials*, 1993, 461–5.

17. TRIMBOS J B, VAN RIJSSEL E J C and KLOPPER P J, 'Performance of sliding knots in monofilament and multifilament suture material', *Obstetrics & Gynecology*, 1986, **68**(3), 425–30.

18. GUPTA B S, MILAM B L and PATTY R R, 'Use of carbon dioxide lasers in improving knot security in polyester sutures', *Journal of Applied Biomaterials*, 1990, **1**, 121–5.

19. GUPTA B S and KASYANOV V A, 'Biomechanics of the human common carotid artery and design of novel hybrid textile compliant vascular grafts', *Journal of Biomedical Material Research*, 1997, **34**, 341–9.

20. ELMOGAHZY Y, KILINC F, HASSAN M and FARAG R, 'Protective clothing: the unresolved ultimate trade-off between protection and comfort – The concept of area ratio', *Conference Proceedings, 4th International Conference on Safety and Protective Fabrics*, Industrial Fabric Association, IFI, Pittsburgh, PA, 2004, 37–57.

21. HAVENITH G, 'Heat balance when wearing protective clothing', *Annals Occupational Hygiene*, 1999, **43**(5), 289–96.

22. EL MOGAHZY Y E, 'Understanding fabric comfort, human survey', *Textile Science 93 International Conference Proceedings*, Volume 1, Technical University of Liberec, Czech Republic, 1993.

23. BAJAJ P A and SENGGUPTA A K, 'Protective clothing', *Textile Progress*, 1992, **22**(2/3/4), 65.

24. GASPAR N A, 'Technical problems associated with protective clothing for military use', Paper presented at *36th International Man-made Fibers Conference*, Osterreichisches Chemiefaser Institut, Dombim, Austria, 17–19 September 1997.

25. RICHARD A SCOTT, 'Textiles in defense', *Handbook of Technical Textiles*, Horrocks A R and Anand S C (editors), Woodhead Publishing Limited, Cambridge, UK, 2000, 425–58.

26. HOLMES D A, 'Textiles for survival', *Handbook of Technical Textiles*, Horrocks A R and Anand S C (editors), Woodhead Publishing Limited, Cambridge, UK, 2000, 461–88.

27. HOLCOMBE B V and HOSCHKE B N, *Performance of Protective Clothing*, Barker R L and Coletta G C (editors), ASTM Special Technical Publication 900, Philadelphia, 1986, p. 327.

28. VAN KREVELEN D W, 'Flame resistance of chemical fibers', *Journal of Applied Polymer Science*, Applied Polymer Symposium, 1977, **31**, 269–92.

29. ZHANG S and HORROCKS A R, 'A review of flame retardant polypropylene fibres', *Progress in Polymer Science*, 2003, **28**(11), 1517–38.

30. EL MOGAHZY Y, 'Human survey of comfort and protection aspects under strenuous situations', *QBC Newsletter*, http://www.qualitybc.com/, No. 8, 2006.

31. UMBACH K H, 'Protective clothing against cold with a wide range of thermophysiological control, Part I', *Melliand Texilber*, 1981, **3**, 360–4 (English edition).

32. VOKAC Z, KOPKE V and KEUL P, 'Physiological responses and thermal, humidity, and comfort sensations in wear trials with cotton and polypropylene vests', *Textile Research Journal*, 1976, **46**, 30–8.

33. UMBACH K H, 'Protective clothing against cold with a wide range of thermophysiological control, Part II', *Melliand Texilber*, 1981, **4**, 456–62 (English edition).

34. HOLMER I, 'Heat exchange and thermal insulation compared in woolen and nylon garments during wear trials', *Textile Research Journal*, 1985, **55**, 511–18.

35. SPENCER-SMITH J L, 'Physical basis of clothing comfort, Part V: the behavior of clothing in transient conditions', *Clothing Research Journal*, 1978, 21–30.

36. WEHNER J A, MILLER B and REBENFELD L, 'Dynamics of water vapor transmission through fabric barriers', *Textile Research Journal*, 1988, 58.

37. BEHMANN F W, 'Influence of the sorption properties of clothing on sweating loss and the subjective feeling of sweating', *Journal of Applied Polymer Science*, 1971, **18**, 1477–82.

38. CASSIE A B D, ATKINS B E and KING G, 'Thermo-static action of textile fibers', *Nature*, 1939, **143**, 162.

39. CASSIE A B D, 'Fibers and fluids', *Journal of the Textile Institute*, 1962, **53**, P739–P745.

40. BACKER S, 'The relationship between the structural geometry of a textile fabric and its physical properties', *Textile Research Journal*, 1948, **18**(11), 650–8.

41. EL MOGAHZY Y, GUPTA B S, KRISHNA P, PASCOE D and KILINC F S, *Developing a Design-Oriented Fabric Comfort Model*, Project Final Report, National Textile Center, Project S01-AE32, Auburn University, Georgia Tech, NCSU, 2003.

42. GUPTA B S and EL MOGAHZY Y, 'Friction in fibrous materials. Part I: structural model', *Textile Research Journal*, 1991, **61**(9), 547–55.

Index